T0293193

Generalized Linear Models with Random Effects

Unified Analysis via H-likelihood

Second Edition

MONOGRAPHS ON STATISTICS AND APPLIED PROBABILITY

General Editors

F. Bunea, V. Isham, N. Keiding, T. Louis, R. L. Smith, and H. Tong

1.   Stochastic Population Models in Ecology and Epidemiology *M.S. Barlett* (1960)
2.   Queues *D.R. Cox and W.L. Smith* (1961)
3.   Monte Carlo Methods *J.M. Hammersley and D.C. Handscomb* (1964)
4.   The Statistical Analysis of Series of Events *D.R. Cox and P.A.W. Lewis* (1966)
5.   Population Genetics *W.J. Ewens* (1969)
6.   Probability, Statistics and Time *M.S. Barlett* (1975)
7.   Statistical Inference *S.D. Silvey* (1975)
8.   The Analysis of Contingency Tables *B.S. Everitt* (1977)
9.   Multivariate Analysis in Behavioural Research *A.E. Maxwell* (1977)
10. Stochastic Abundance Models *S. Engen* (1978)
11. Some Basic Theory for Statistical Inference *E.J.G. Pitman* (1979)
12. Point Processes *D.R. Cox and V. Isham* (1980)
13. Identification of Outliers *D.M. Hawkins* (1980)
14. Optimal Design *S.D. Silvey* (1980)
15. Finite Mixture Distributions *B.S. Everitt and D.J. Hand* (1981)
16. Classification *A.D. Gordon* (1981)
17. Distribution-Free Statistical Methods, 2nd edition *J.S. Maritz* (1995)
18. Residuals and Influence in Regression *R.D. Cook and S. Weisberg* (1982)
19. Applications of Queueing Theory, 2nd edition *G.F. Newell* (1982)
20. Risk Theory, 3rd edition *R.E. Beard, T. Pentikäinen and E. Pesonen* (1984)
21. Analysis of Survival Data *D.R. Cox and D. Oakes* (1984)
22. An Introduction to Latent Variable Models *B.S. Everitt* (1984)
23. Bandit Problems *D.A. Berry and B. Fristedt* (1985)
24. Stochastic Modelling and Control *M.H.A. Davis and R. Vinter* (1985)
25. The Statistical Analysis of Composition Data *J. Aitchison* (1986)
26. Density Estimation for Statistics and Data Analysis *B.W. Silverman* (1986)
27. Regression Analysis with Applications *G.B. Wetherill* (1986)
28. Sequential Methods in Statistics, 3rd edition *G.B. Wetherill and K.D. Glazebrook* (1986)
29. Tensor Methods in Statistics *P. McCullagh* (1987)
30. Transformation and Weighting in Regression *R.J. Carroll and D. Ruppert* (1988)
31. Asymptotic Techniques for Use in Statistics *O.E. Bandorff-Nielsen and D.R. Cox* (1989)
32. Analysis of Binary Data, 2nd edition *D.R. Cox and E.J. Snell* (1989)
33. Analysis of Infectious Disease Data *N.G. Becker* (1989)
34. Design and Analysis of Cross-Over Trials *B. Jones and M.G. Kenward* (1989)
35. Empirical Bayes Methods, 2nd edition *J.S. Maritz and T. Lwin* (1989)
36. Symmetric Multivariate and Related Distributions *K.T. Fang, S. Kotz and K.W. Ng* (1990)
37. Generalized Linear Models, 2nd edition *P. McCullagh and J.A. Nelder* (1989)
38. Cyclic and Computer Generated Designs, 2nd edition *J.A. John and E.R. Williams* (1995)
39. Analog Estimation Methods in Econometrics *C.F. Manski* (1988)
40. Subset Selection in Regression *A.J. Miller* (1990)
41. Analysis of Repeated Measures *M.J. Crowder and D.J. Hand* (1990)
42. Statistical Reasoning with Imprecise Probabilities *P. Walley* (1991)
43. Generalized Additive Models *T.J. Hastie and R.J. Tibshirani* (1990)
44. Inspection Errors for Attributes in Quality Control *N.L. Johnson, S. Kotz and X. Wu* (1991)
45. The Analysis of Contingency Tables, 2nd edition *B.S. Everitt* (1992)
46. The Analysis of Quantal Response Data *B.J.T. Morgan* (1992)
47. Longitudinal Data with Serial Correlation—A State-Space Approach *R.H. Jones* (1993)

Monographs on Statistics and Applied Probability 153

Generalized Linear Models with Random Effects

Unified Analysis via H-likelihood

Second Edition

Youngjo Lee
John A. Nelder
Yudi Pawitan

CRC Press
Taylor & Francis Group
Boca Raton London New York

CRC Press is an imprint of the
Taylor & Francis Group, an **informa** business

A CHAPMAN & HALL BOOK

CRC Press
Taylor & Francis Group
6000 Broken Sound Parkway NW, Suite 300
Boca Raton, FL 33487-2742

© 2017 by Taylor & Francis Group, LLC
CRC Press is an imprint of Taylor & Francis Group, an Informa business

No claim to original U.S. Government works

Printed on acid-free paper
Version Date: 20170704

International Standard Book Number-13: 978-1-4987-2061-8 (Hardback)

Library of Congress Cataloging-in-Publication Data

Names: Lee, Youngjo. | Nelder, John A. | Pawitan, Yudi.
Title: Generalized linear models with random effects : unified analysis via
h-likelihood / Youngjo Lee, John A. Nelder, and Yudi Pawitan.
Description: Second edition. | Boca Raton, Florida : CRC Press, [2017] |
Series: Chapman & hall/CRC monographs on statistics & applied probability
| Includes bibliographical references and index.
Identifiers: LCCN 2016033750| ISBN 9781498720618 (hardback : alk. paper) |
ISBN 9781498720625 (e-book)
Subjects: LCSH: Linear models (Statistics) | Generalized estimating equations.
Classification: LCC QA279 .L43 2017 | DDC 519.5--dc23
LC record available at https://lccn.loc.gov/2016033750

Visit the Taylor & Francis Web site at
http://www.taylorandfrancis.com

and the CRC Press Web site at
http://www.crcpress.com

Contents

List of notations

$f_\theta(y)$ probability density function including both discrete and continuous models. The argument determines the function, for example, $f_\theta(x)$ and $f_\theta(y)$ may refer to different densities.

$H(\beta, v; y, v)$ h-likelihood of (β, v) based on data (y, v).

$h(\beta, v; y, v)$ h-loglihood of (β, v) based on data (y, v).

$I(\theta)$ Fisher information.

$I(\widehat{\theta})$ observed Fisher information.

$\mathcal{I}(\theta)$ expected Fisher information.

$L(\theta; y)$ likelihood of θ based on data y.

$\ell(\theta; y)$ log-likelihood (loglihood) of θ based on data y.

$p_\eta(\ell)$ adjusted profile of a generic loglihood ℓ, after eliminating a generic nuisance parameter η.

$q(\mu; y)$ quasi–likelihood of model μ, based on data y.

$S(\theta)$ score statistic.

Preface to first edition

The class of generalized linear models has proved a useful generalization of classical normal models since its introduction in 1972. Three components characterize all members of the class: (1) the error distribution, which is assumed to come from a one-parameter exponential family; (2) the linear predictor, which describes the pattern of the systematic effects; and (3) the algorithm, iterative weighted least squares, which gives the maximum-likelihood estimates of those effects.

In this book the class is greatly extended, while at the same time retaining as much of the simplicity of the original as possible. First, to the fixed effects may be added one or more sets of random effects on the same linear scale; secondly GLMs may be fitted simultaneously to both mean and dispersion; thirdly the random effects may themselves be correlated, allowing the expression of models for both temporal and spatial correlation; lastly random effects may appear in the model for the dispersion as well as that for the mean.

To allow likelihood-based inferences for the new model class, the idea of h-likelihood is introduced as a criterion to be maximized. This allows a single algorithm, expressed as a set of interlinked GLMs, to be used for fitting all members of the class. The algorithm does not require the use of quadrature in the fitting, and neither are prior probabilities required. The result is that the algorithm is orders of magnitude faster than some existing alternatives.

The book will be useful to statisticians and researchers in a wide variety of fields. These include quality improvement experiments, combination of information from many trials (meta-analysis), frailty models in survival analysis, missing-data analysis, analysis of longitudinal data, analysis of spatial data on infection etc., and analysis of financial data using random effects in the dispersion. The theory, which requires competence in matrix theory and knowledge of elementary probability and likelihood theory, is illustrated by worked examples and many of these can be run by the reader using the code supplied on the accompanying CD. The flexibility of the code makes it easy for the user to try out alternative

analyses for him/herself. We hope that the development will be found
to be self-contained, within the constraint of monograph length.

Youngjo Lee, John Nelder and Yudi Pawitan
Seoul, London and Stockholm

Preface

Since the first paper on hierarchical generalized linear models appeared in 1996, interest in the topic grew to produce the first edition of this book in 2006. In last ten years, the topic has grown so much that it needs three separate books. The second edition of the first book adds some recent developments in HGLMs, while the other two books deal with survival analyses and applied data analyses.

The GLM class is greatly extended to HGLMs, allowing unobservable random variables such as random effects, missing data, and other factors while at the same time retaining as much of the simplicity of original GLMs. The classical likelihood is extended to the h-likelihood to allow statistical inferences for unobserved random variables and leads to a single algorithm, expressed as a set of interlinked and/or augmented GLMs, to be used for fitting broad class of new models with random effects. The second edition covers new developments in variable selection, multiple testing and structural equation models via random effect approach. We have also revised most of the previous figures and tables from the examples based on our new R package.

The book will be useful to statisticians and researchers in a wide variety of fields. The h-likelihood leads to a new paradigm in statistics, which requires new theory, algorithms and applications. An application book shows how to run R packages for the examples in this book and in the survival analysis book. The flexibility of the package makes it easy for the user to try alternative analyses.

In 2010, Professor John A. Nelder passed away; his wish was that the extended class of new models and associated likelihood theories would be well received. To achieve his wish, R packages have been developed for the practitioners to run the new models in their data analyses and to see why extended likelihood inferences are necessary for inferences about random unknowns.

Youngjo Lee and Yudi Pawitan
Seoul and Stockholm

xix

Introduction

We aim to build an extensive class of new statistical models by combining a small number of basic statistical ideas in diverse ways. Although we use (a very small part of) mathematics as our basic tool to describe the statistics, our aim is not primarily to develop theorems and proofs of theorems, but rather to provide the statistician with statistical tools for dealing with inferences from a wide range of data, which may be structured in many different ways. We develop and establish an extended likelihood framework, derived from classical likelihood, which is described in Chapter 1.

The starting point in our development of the model class is the idea of a generalized linear model (GLM). An account of GLMs in summary form appears in Chapter 2. The algorithm for fitting GLMs is iterative weighted least squares, and this forms the basis of fitting algorithms for our entire class, in that these can be reduced to fitting a set of interconnected GLMs. For inferences of GLMs, classical likelihood concepts can lead to all necessary inferential tools.

Two important extensions of GLMs, discussed in Chapter 3, involve the ideas of quasi–likelihood (QL) and extended quasi–likelihood (EQL). QL extends the scope of GLMs to errors defined by their mean and variance functions only, while EQL forms the pivotal quantity for the joint modelling of mean and dispersion; such models can be fitted with two interlinked GLMs. However, the QL approach may not give proper likelihood for dispersion parameters and analysis of correlated data, and could yield non-consistent estimation of parameters.

Chapter 4 presents an extended likelihood and discusses why it is necessary for inferences about unobserved random variables. Simple joint maximization of the extended likelihood can produce alleged counterexamples, which can be avoided by using the h-likelihood. It is an extension of Fisher likelihood to models of the GLM type with additional random effects in the linear predictor. H-likelihood leads to adequate estimation of both fixed and random effects with efficient fitting algorithms. An important feature of algorithms using h-likelihood for fitting is that quadrature is not required and again a reduction to interconnected GLMs suffices to fit the models. Methods requiring quadrature cannot be used for high-dimensional integration. In recent decades we have witnessed the emergence of several computational methods to overcome this difficulty, for example, Monte Carlo-type and/or expectation-maximization (EM)-type methods to compute the maximum likelihood estimates (MLE) for extended classes of models. It is now possible to compute them directly with h-likelihood without resorting to these computationally intensive methods. The method does not require the use of prior probabilities

Normal models with additional (normal) random effects are dealt with in Chapter 5. We compare marginal likelihood with h-likelihood for fitting the fixed effects, and show how restricted maximum likelihood (REML) can be described as maximizing an adjusted profile likelihood.

In Chapter 6 we bring together the GLM formulation with additional random effects in the linear predictor to form HGLMs. Special cases include GLMMs, where the random effects are assumed normal, and conjugate HGLMs, where the random effects are assumed to follow the conjugate distribution to that of the response variable. An adjusted profile h-likelihood gives a generalization of REML to non-normal GLMs.

HGLMs can be further extended by allowing the dispersion parameter of the response to have a structure defined by its own set of covariates. This brings together the HGLM class with the joint modelling of mean and dispersion described in Chapter 3, and this synthesis forms the basis of Chapter 7. HGLMs and conjugate distributions for arbitrary variance functions can be extended to quasi–likelihood HGLMs and quasi-conjugate distributions, respectively.

Many models for spatial and temporal data require the observations to be correlated. We show how these may be dealt with by transforming linearly the random terms in the linear predictor. Covariance structures may have no correlation parameters or those derived from covariance matrices or precision matrices; correlations derived from the various forms are illustrated in Chapter 8. Existing factor and structure equation models can be further extended via HGLMs.

Chapter 9 deals with smoothing, whereby a parametric term in the linear predictor may be replaced by a data-driven smooth curve called a spline. We show that splines are isomorphic to certain random effect models, so that they fit easily into the HGLM framework.

Chapter 10 deals with a further extension to HGLMs, whereby the dispersion model and the mean model, may have random effects in their linear predictors. These are shown to be relevant to, and indeed to extend, certain models proposed for the analysis of financial data. These double HGLMs (DHGLMs) represent the furthest extension of our model class, and the algorithm for fitting them still reduces to the fitting of interconnected GLMs . They provided robust analysis for various model misspecifications.

Chapter 11 provides a specific DHGLM to handle variable selections. This model can be viewed as an extension of ridge and LASSO methods, but it achieves selection consistency. This new random effect model approach makes the variable selection more transparent to explaining and extending various multivariate methods and structured regression with various hierarchy constraints.

In Chapter 12, further synthesis is achieved by multivariate HGLMs. HGLMs can be extended to analysis of multivariate responses. We show how missing mechanisms can be modelled as bivariate HGLMs. Furthermore, h-likelihood allows a fast imputation to provide a powerful algorithm for denoising signals.

In Chapter 13 we show that hypothesis testing can be viewed as a prediction of discrete random effects (e.g., 0 and 1 for the null and alternative hypotheses). We show that the ratio of h-likelihoods leads to the classical likelihood-ratio tests in single hypothesis testing and new efficient likelihood-ratio tests in multiple hypothesis testing.

In the last chapter we show how random effect models can be extended to survival data. We study two alternative models, namely frailty models and normal-normal HGLMs for censored data. We also show how to model interval-censored data. The h-likelihood provides useful inferences for the analysis of survival data.

The aim of this book is to establish and illustrate an extended likelihood framework for the analysis of various kinds of data. Many other existing statistical models fall into the HGLM class. We believe that many statistical areas covered by the classical likelihood framework fall into our extended framework.

We are grateful to Jeongseop Han, Daehan Kim, and Hyunsung Park for their proofreading, editorial assistance and comments.

Software

An R package has been developed for various classes of HGLMs, and detailed explanations and examples for the use of the R package are found in a separate applied data analysis book by Lee, Rönnegård and Noh (2017).

CHAPTER 1

Classical likelihood theory

1.1 Definition

'The problems of theoretical statistics,' wrote Fisher in 1921, 'fall into two main classes':

(a) To discover what quantities are required for the adequate description of a population, which involves the discussion of mathematical expressions by which frequency distributions may be represented.

(b) To determine how much information, and of what kind, respecting these population values is afforded by a random sample or a series of random samples.

It is clear that these two classes refer to statistical modelling and inference. In the same paper, for the first time, Fisher coined the term 'likelihood' explicitly and contrasted it with 'probability', two "radically distinct concepts [that] have been confused under the name of 'probability'...." Likelihood is a key concept in both modelling and inference, and throughout this book we shall rely greatly on this concept. This book aims to extend classical likelihood inferences to general classes of models, including unobservable random variables, because it is important to have a deep understanding of existing likelihood inferences. This chapter summarizes all the classical likelihood concepts from Pawitan (2001) that we need in this book for extensions; occasionally, for more details, we refer the reader to that book.

Definition 1.1 *Assuming a statistical model $f_\theta(y)$ parameterized by a fixed and unknown θ, the likelihood $L(\theta)$ is the probability of the observed data y considered as a function of θ.*

The generic data y include any set of observations we might get from an experiment of *any complexity*, and the model $f_\theta(y)$ should specify how the data could have been generated probabilistically. For discrete data, the definition is directly applicable since the probability is nonzero. For continuous data that are measured with good precision, the probability of observing data in a small neighbourhood of y is approximately equal to the density function times a small constant. We shall use the terms 'probability' or 'probability density' to cover both discrete and continuous outcomes. With many straightforward measurements, the likelihood is simply the probability density seen as a function of the parameter.

The θ in the definition is also a generic parameter that, in principle, can be of any dimension. However, in this chapter we shall restrict θ to consist of fixed parameters only. The purpose of the likelihood function is to convey information about unknown quantities. Its direct use for inference is controversial, but the most commonly used form of inference today is based on quantities derived from the likelihood function and justified using probabilistic properties of those quantities.

If y_1 and y_2 are independent data sets with probabilities $f_{1,\theta}(y_1)$ and $f_{2,\theta}(y_2)$ that share a common parameter θ, the likelihood from the combined data is

$$
\begin{aligned}
L(\theta) &= f_{1,\theta}(y_1) f_{2,\theta}(y_2) \\
&= L_1(\theta) L_2(\theta), \qquad\qquad (1.1)
\end{aligned}
$$

where $L_1(\theta)$ and $L_2(\theta)$ are the likelihoods from the individual data sets. On a log scale this property is a simple additive property

$$
\ell(\theta) \equiv \log L(\theta) = \log L_1(\theta) + \log L_2(\theta),
$$

giving a very convenient formula for combining information from independent experiments. Since the term 'log-likelihood' occurs often, we shorten it to 'loglihood.'

The simplest case occurs if y_1 and y_2 are an independent-and-identically-distributed (iid) sample from the same density $f_\theta(y)$, so

$$
L(\theta) = f_\theta(y_1) f_\theta(y_2),
$$

or $\ell(\theta) = \log f_\theta(y_1) + \log f_\theta(y_2)$. If $y_1, \ldots, y_n$ represent an iid sample from $f_\theta(y)$ we have

$$
L(\theta) = \prod_{i=1} f_\theta(y_i),
$$

or $\ell(\theta) = \sum_{i=1}^{n} \log f_\theta(y_i)$.

Example 1.1: Let $y_1, \ldots, y_n$ be an iid sample from $N(\theta, \sigma^2)$ with known σ^2. The contribution of y_i to the likelihood is

$$
L_i(\theta) = \frac{1}{\sqrt{2\pi\sigma^2}} \exp\left\{ -\frac{(y_i - \theta)^2}{2\sigma^2} \right\},
$$

and the total loglihood is

$$\ell(\theta) = \sum_{i=1}^{n} \log L_i(\theta)$$

$$= -\frac{n}{2} \log(2\pi\sigma^2) - \frac{1}{2\sigma^2} \sum_{i=1}^{n} (y_i - \theta)^2. \quad \square$$

Example 1.2: We observe the log ratios of a certain protein expression from $n = 10$ tissue samples compared to the reference sample:

-0.86 0.73 1.29 1.41 1.56 1.86 2.33 2.59 3.37 3.48.

The sample mean and standard deviation are 1.776 and 1.286, respectively. Assume that the measurements are a random sample from $N(\theta, \sigma^2 = s^2 = 1.286^2 = 1.65)$, where the variance is assumed known at the observed value. Assuming all the data are observed with good precision, the normal density is immediately applicable, and the likelihood of θ is given in the previous example. This is plotted as the solid curve in Figure 1.1. Typically, in plotting the function we set the maximum of the function to one.

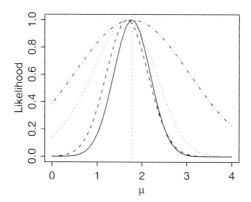

Figure 1.1 *Likelihood of the mean from the normal model, assuming all the data are observed (solid), the data were reported in grouped form (dashed), only the sample size and the maximum were reported (dotted), and a single observation equals to the sample mean (dashed-dotted).*

Now, suppose the data were reported in grouped form as the number of values that are ≤ 0, between 0 and 2, and > 2, thus

$$n_1 = 1, \; n_2 = 5, \; n_3 = 4.$$

Let us still assume that the original data came from $N(\theta, \sigma^2 = 1.65)$. The distribution of the counts is multinomial with probabilities

$$p_1 = \Phi\left(\frac{0-\theta}{\sigma}\right)$$

$$p_2 = \Phi\left(\frac{2-\theta}{\sigma}\right) - \Phi\left(\frac{0-\theta}{\sigma}\right)$$

$$p_3 = 1 - p_1 - p_2$$

where $\Phi(z)$ is the standard normal distribution function, and the likelihood is

$$L(\theta) = \frac{n!}{n_1!n_2!n_3!}p_1^{n_1}p_2^{n_2}p_3^{n_3},$$

shown as the dashed line in Figure 1.1. We can also consider this as the likelihood of interval data. It is interesting to note that it is very close to the likelihood of the original raw data, so there is only a minor loss of information in the grouped data about the mean parameter.

Now suppose that only the sample size $n = 10$ and the maximum $y_{(10)} = 3.48$ were reported. What is the likelihood of θ based on the same model above? If $y_1, \ldots, y_n$ is an identically and independently distributed (iid) sample from $N(\theta, \sigma^2)$, the distribution function of $y_{(n)}$ is

$$\begin{aligned} F(t) &= P(Y_{(n)} \leq t) \\ &= P(Y_i \leq t, \text{ for each } i) \\ &= \left\{\Phi\left(\frac{t-\theta}{\sigma}\right)\right\}^n. \end{aligned}$$

The likelihood based on observing $y_{(n)}$ is

$$L(\theta) = f_\theta(y_{(n)}) = n\left\{\Phi\left(\frac{t-\theta}{\sigma}\right)\right\}^{n-1}\phi\left(\frac{t-\theta}{\sigma}\right)\frac{1}{\sigma}.$$

Figure 1.1 shows this likelihood as a dotted line, indicating that more information is lost compared to the categorical data above. However, the maximum carries substantially more information than a *single* observation alone (assumed equal to the sample mean and having variance σ^2), as shown by the dashed-dotted line in the same figure.

1.1.1 Invariance principle and likelihood ratio

Suppose $y = g(x)$ is a one-to-one transformation of the observed data x; if x is continuous, the likelihood based on y is

$$L(\theta; y) = L(\theta; x)\left|\frac{\partial x}{\partial y}\right|.$$

Obviously x and y should carry the same information about θ, so to compare θ_1 and θ_2 only the likelihood ratio is relevant since it is invariant with respect to the transformation:

$$\frac{L(\theta_2; y)}{L(\theta_1; y)} = \frac{L(\theta_2; x)}{L(\theta_1; x)}.$$

More formally, we might add that the ratio is maximal invariant in the sense that any function of $L(\theta_1; y)$ and $L(\theta_2; y)$ that is invariant under different one-to-one transformations $y = g(x)$ must be a function of the likelihood ratio.

This means that proportional likelihoods are equivalent, i.e., carrying the same information about the parameter and to make it unique, especially for plotting, it is customary to normalize the likelihood function to have unit maximum.

That the likelihood ratio should be the same for different transformations of the data seems like a perfectly natural requirement. It seems reasonable also that the ratio should be the same for different transformations of the parameter itself. For example, it should not matter whether we analyse the dispersion parameter in terms of σ^2 or σ; the data should carry the same information, that is

$$\frac{L^*(\sigma_2^2; y)}{L^*(\sigma_1^2; y)} = \frac{L(\sigma_2; y)}{L(\sigma_1; y)}.$$

Since this requirement does not follow from any other principle, it should be regarded as an axiom, which is implicitly accepted by all statisticians except the so-called Bayesians. This axiom implies that in computing the likelihood of a transformed parameter, we cannot use a Jacobian term. Hence, fundamentally, the likelihood cannot be treated like a probability density function over the parameter space and it does not obey probability laws; for example, it does not have to integrate to one.

Any mention of Bayesianism touches on deeply philosophical issues that are beyond the scope of this book. Suffice it to say that the likelihood-based approaches we take in this book are fundamentally non-Bayesian. However, there are some computational similarities between likelihood and Bayesian methods, and these are discussed in Section 1.10.

1.1.2 Likelihood principle

Why should we start with the likelihood? One of the earliest theoretical results is that, given a statistical model $f_\theta(y)$ and observation y from it, the likelihood function is a minimal sufficient statistic (see, e.g., Pawitan, 2001, Chapter 3). Practically, this means that there is a corresponding minimal set of statistics that would be needed to draw the likelihood function, so this set would be sufficient and any further data reduction would involve some loss of information about the parameter. In the normal example above, for a given sample size and variance, knowing the sample mean is sufficient to draw the likelihood based on the whole sample, so the sample mean is minimally sufficient.

Birnbaum (1962) showed that a stronger statement is possible: any measure of evidence from an experiment depends on the data only through the likelihood function. Such a result has been elevated into a principle, the so-called 'likelihood principle,' intuitively stating that the likelihood contains all the evidence or information about θ in the data. Violation of the principle, i.e., by basing an analysis on something other than the likelihood, leads to a potential loss of information, or it can make the analysis open to contradictions (see Pawitan, Chapter 7). This means that in any statistical problem, whenever possible, it is always a good idea to start with the likelihood function.

Another strong justification comes from optimality considerations: under very general conditions, as the sample size increases, likelihood-based estimates are usually the best possible estimates of the unknown parameters. In finite or small samples, the performance of a likelihood-based method depends on the models, so that it is not possible to have a simple general statement about optimality.

We note, however, that these beneficial properties hold only when the presumed model is correct. In practice, we have no guarantee that the model is correct, so that any data analyst should fit a reasonable model and then perform model checking.

1.2 Quantities derived from the likelihood

In most regular problems, where the loglihood function is reasonably quadratic, its analysis can focus on the location of the maximum and the curvature around it. Such an analysis requires only the first two derivatives. Thus we define the *score function* $S(\theta)$ as the first derivative of the loglihood:

$$S(\theta) \equiv \frac{\partial}{\partial \theta} \ell(\theta),$$

and the maximum likelihood estimate (MLE) $\widehat{\theta}$ is the solution of the *score equation*

$$S(\theta) = 0.$$

At the maximum, the second derivative of the loglihood is negative, so we define the curvature at $\widehat{\theta}$ as $I(\widehat{\theta})$, where

$$I(\theta) \equiv -\frac{\partial^2}{\partial \theta^2} \ell(\theta).$$

A large curvature $I(\widehat{\theta})$ is associated with a tight or strong peak, intuitively indicating less uncertainty about θ. $I(\widehat{\theta})$ is called the *observed Fisher information*. For distinction we call $I(\theta)$ the Fisher information.

Under some regularity conditions — mostly ensuring valid interchange of derivative and integration operations — that hold for the models in this book, the score function has interesting properties:

$$E_\theta S(\theta) = 0$$
$$\text{var}_\theta S(\theta) = E_\theta I(\theta) \equiv \mathcal{I}(\theta).$$

The latter quantity $\mathcal{I}(\theta)$ is called the *expected Fisher information*. For simple proofs of these results and a discussion of the difference between observed and expected Fisher information, see Pawitan (2001, Chapter 8). To emphasize, we now have (at least) three distinct concepts: Fisher information $I(\theta)$, observed Fisher information $I(\widehat{\theta})$ and expected Fisher information $\mathcal{I}(\theta)$. We could also have 'estimated expected Fisher information' $\mathcal{I}(\widehat{\theta})$, which in general is different from the observed Fisher information, but such a term is never used in practice.

Example 1.3: Let $y_1, \ldots, y_n$ be an iid sample from $N(\theta, \sigma^2)$. For the moment assume that σ^2 is known. Ignoring irrelevant constant terms

$$\ell(\theta) = -\frac{1}{2\sigma^2} \sum_{i=1}^{n} (y_i - \theta)^2,$$

so we immediately get

$$S(\theta) = \frac{\partial}{\partial \theta} \log L(\theta) = \frac{1}{\sigma^2} \sum_{i=1}^{n} (y_i - \theta).$$

Solving $S(\theta) = 0$ produces $\widehat{\theta} = \overline{y}$ as the MLE of θ. The second derivative of the loglihood gives the observed Fisher information

$$I(\widehat{\theta}) = \frac{n}{\sigma^2}.$$

Here $\text{var}(\widehat{\theta}) = \sigma^2/n = I^{-1}(\widehat{\theta})$, so larger information implies a smaller variance. Furthermore, the standard error of $\widehat{\theta}$ is $\text{se}(\widehat{\theta}) = \sigma/\sqrt{n} = I^{-1/2}(\widehat{\theta})$. □

Example 1.4: Many commonly used distributions in statistical modelling such as the normal, Poisson, binomial, and gamma distributions belong to the so-called exponential family. It is an important family for its versatility, and

many theoretical results hold for this family. A p-parameter exponential family
has a log-density

$$\log f_\theta(y) \;=\; \sum_{i=1}^{p} \theta_i t_i(y) - b(\theta) + c(y),$$

where $\theta = (\theta_1, \ldots, \theta_p)$ and $t_1(y), \ldots, t_p(y)$ are known functions of y. The θ_i
values are called the *canonical parameters*; if these parameters comprise p free
parameters, the family is called a full exponential family. For most models, the
commonly used parameterization will not be in canonical form. For example,
for the Poisson model with mean μ

$$\log f_\theta(y) = y \log \mu - \mu - \log y!$$

so the canonical parameter is $\theta = \log \mu$, and $b(\theta) = \mu = \exp(\theta)$. The canonical
form often leads to simpler formulae. Since the moment–generating function
is

$$m(\eta) = E \exp(\eta y) = \exp\{b(\theta + \eta) - b(\theta)\},$$

an exponential family is characterized by the $b()$ function.

From the moment generating function, we can immediately show an important
result about the relationship between the mean and variance of an exponential
family

$$\begin{aligned}
\mu &\equiv& Ey = b'(\theta) \\
V(\mu) &\equiv& \mathrm{var}(y) = b''(\theta).
\end{aligned}$$

This means that the mean-variance relationship determines the $b()$ function
by a differential equation

$$V(b'(\theta)) = b''(\theta).$$

For example, for Poisson model $V(\mu) = \mu$, so $b()$ must satisfy

$$b'(\theta) = b''(\theta),$$

giving $b(\theta) = \exp(\theta)$.

Suppose $y_1, \ldots, y_n$ comprise an independent sample, where y_i comes from a
one-parameter exponential family with canonical parameter θ_i, $t(y_i) = y_i$ and
having a common function $b()$. The joint density is also of exponential family
form

$$\log f_\theta(y) \;=\; \sum_{i=1}^{n} \{\theta_i y_i - b(\theta_i) + c(y_i)\}.$$

In regression problems the θ_is are a function of common structural parameter
β, i.e., $\theta_i \equiv \theta_i(\beta)$, where the common parameter β is p-dimensional. Then the
score function for β is given by

$$\begin{aligned}
S(\beta) &=& \sum_{i=1}^{n} \frac{\partial \theta_i}{\partial \beta} \{y_i - b'(\theta_i)\} \\
&=& \sum_{i=1}^{n} \frac{\partial \theta_i}{\partial \beta} (y_i - \mu_i)
\end{aligned}$$

using the above result that $\mu_i = Ey_i = b'(\theta_i)$ and the Fisher information matrix by

$$I(\beta) = \sum_{i=1}^{n}\left[-\frac{\partial^2\theta_i}{\partial\beta\partial\beta'}(y_i - \mu_i) + b''(\theta_i)\frac{\partial\theta_i}{\partial\beta}\frac{\partial\theta_i}{\partial\beta'}\right]. \qquad (1.2)$$

The expected Fisher information is

$$\mathcal{I}(\beta) = \sum_{i=1}^{n}b''(\theta_i)\frac{\partial\theta_i}{\partial\beta}\frac{\partial\theta_i}{\partial\beta'}.$$

In general $I(\beta) \neq \mathcal{I}(\beta)$, but equality occurs if the canonical parameter θ_i is a linear function of β, since in this case the first term of $I(\beta)$ in (1.2) becomes zero. However, in general, this first term has zero mean, and as n gets large, it tends to be much smaller than the second term. $\square$

1.2.1 Regularity and quadratic approximation of the loglihood

Using a second-order Taylor's expansion around $\widehat{\theta}$

$$\ell(\theta) \approx \log L(\widehat{\theta}) + S(\widehat{\theta})(\theta - \widehat{\theta}) - \frac{1}{2}(\theta - \widehat{\theta})^t I(\widehat{\theta})(\theta - \widehat{\theta})$$

we get

$$\log\frac{L(\theta)}{L(\widehat{\theta})} \approx -\frac{1}{2}(\theta - \widehat{\theta})^t I(\widehat{\theta})(\theta - \widehat{\theta}),$$

providing a quadratic approximation of the normalized loglihood around $\widehat{\theta}$. We can judge the accuracy of the quadratic approximation by plotting the true loglihood and the approximation together. In a loglihood plot, we set the maximum of the loglihood to zero and check a range of θ such that the loglihood is approximately between -4 and 0. In the normal example above (Example 1.3) the quadratic approximation is exact:

$$\log\frac{L(\theta)}{L(\widehat{\theta})} = -\frac{1}{2}(\theta - \widehat{\theta})^t I(\widehat{\theta})(\theta - \widehat{\theta}),$$

so a quadratic approximation of the loglihood corresponds to a normal approximation for $\widehat{\theta}$. *We shall say the loglihood is regular if it is well approximated by a quadratic.*

The standard error of an estimate is defined as its estimated standard deviation and it is used as a measure of precision. For scalar parameters, the most common formula is

$$se = I^{-1/2}(\widehat{\theta}).$$

In the vector case, the standard error for each parameter is given by the square root of diagonal terms of the inverse Fisher information. If the likelihood function is not very quadratic, the standard error is not meaningful. In this case, a set of likelihood or confidence intervals is a better supplement to the MLE.

1.3 Profile likelihood

While the definition of likelihood covers multiparameter models, the resulting multidimensional likelihood function can be difficult to describe or to communicate. Even when we are interested in several parameters, it is always easier to describe one parameter at a time. The problem also arises in cases where we may be interested in only a subset of the parameters. In the normal model, for example, we might be interested only in the mean μ, while σ^2 is a nuisance used only to make the model able to adapt to data variability. A method is needed to concentrate the likelihood on the parameter of interest by eliminating the nuisance parameter.

Accounting for the extra uncertainty due to unknown nuisance parameters is an essential consideration, especially with small samples and a naive plug-in method for the unknown nuisance parameter is often unsatisfactory. A likelihood approach to eliminating a nuisance parameter is to replace it by its MLE at each fixed value of the parameter of interest. The result is called the *profile likelihood*.

Let (θ, η) be the full parameter and θ the parameter of interest.

Definition 1.2 *Given the joint likelihood $L(\theta, \eta)$ the* profile likelihood *of θ is given by*

$$L(\theta) = \max_{\eta} L(\theta, \eta),$$

where the maximization is performed at a fixed value of θ.

It should be emphasized that at fixed θ the MLE of η is generally a function of θ, so we can also write

$$L(\theta) = L(\theta, \widehat{\eta}_\theta).$$

The profile likelihood is then treated like the standard likelihood function.

The profile likelihood usually gives a reasonable likelihood for each component parameter, especially if the number of nuisance parameters is small relative to sample size. The main problem comes from the fact that it is not a proper likelihood function in the sense that it is not based on a probability of some observed quantity. For example, the score statistic derived from the profile likelihood does not have the usual properties of a score statistic in that it often has nonzero mean and its variance does not match the expected Fisher information. These problems typically lead to biased estimation and over-precision of the standard errors. Adjustments to the profile likelihood needed to overcome these problems are discussed in Section 1.9.

Example 1.5: Suppose $y_1, \ldots, y_n$ are iid samples from $N(\mu, \sigma^2)$ with both parameters unknown. The likelihood function of (μ, σ^2) is given by

$$L(\mu, \sigma^2) = \left(\frac{1}{\sqrt{2\pi\sigma^2}} \right)^n \exp \left\{ -\frac{1}{2\sigma^2} \sum_i (y_i - \mu)^2 \right\}.$$

A likelihood of μ without reference to σ^2 is not an immediately meaningful quantity, since it is very different at various values of σ^2. As an example, suppose we observe

$$0.88 \ 1.07 \ 1.27 \ 1.54 \ 1.91 \ 2.27 \ 3.84 \ 4.50 \ 4.64 \ 9.41.$$

The MLEs are $\hat{\mu} = 3.13$ and $\hat{\sigma}^2 = 6.16$. Figure 1.2(a) plots the contours of the likelihood function at 90%, 70%, 50%, 30% and 10% cutoffs. There is a need to plot the likelihood of each parameter individually: it is more difficult to describe or report a multiparameter likelihood function, and usually we are not interested in a simultaneous inference of μ and σ^2.

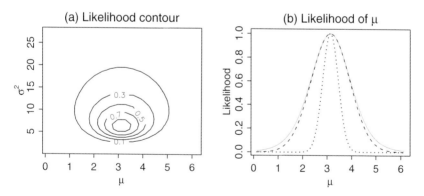

Figure 1.2 *(a) Likelihood function of (μ, σ^2). The contour lines are plotted at 90%, 70%, 50%, 30% and 10% cutoffs. (b) Profile likelihood of the mean μ (solid), $L(\mu, \sigma^2 = \hat{\sigma}^2)$ (dashed), and $L(\mu, \sigma^2 = 1)$ (dotted).*

The profile likelihood function of μ is computed as follows. For fixed μ, the maximum likelihood estimate of σ^2 is

$$\hat{\sigma}_\mu^2 = \frac{1}{n} \sum_i (y_i - \mu)^2,$$

so the profile likelihood of μ is

$$L(\mu) \quad = \quad \text{constant} \times (\widehat{\sigma}_\mu^2)^{-n/2}.$$

This is not the same as

$$L(\mu, \sigma^2 = \widehat{\sigma}^2) = \text{constant} \times \exp\left\{-\frac{1}{2\widehat{\sigma}^2}\sum_i (y_i - \mu)^2\right\},$$

the slice of $L(\mu, \sigma^2)$ at $\sigma^2 = \widehat{\sigma}^2$; this is known as an estimated likelihood. These two likelihoods will be close if σ^2 is well estimated, otherwise the profile likelihood is to be preferred.

For the observed data, $L(\mu)$ and $L(\mu, \sigma^2 = \widehat{\sigma}^2)$ are plotted in Figure 1.2(b). It is obvious that ignoring the unknown variability, e.g., by assuming $\sigma^2 = 1$, would give a wrong inference. So, in general, a nuisance parameter is needed to allow for a better model, but it has to be eliminated properly in order to concentrate the inference on the parameter of interest.

The profile likelihood of σ^2 is given by

$$L(\sigma^2) \quad = \quad \text{constant} \times (\sigma^2)^{-n/2} \exp\left\{-\frac{1}{2\sigma^2}\sum_i (y_i - \bar{y})^2\right\}$$

$$= \quad \text{constant} \times (\sigma^2)^{-n/2} \exp\{-n\widehat{\sigma}^2/(2\sigma^2)\}. \quad \square$$

1.4 Distribution of the likelihood ratio statistic

Traditional inference on an unknown parameter θ relies on the distribution theory of its estimate $\widehat{\theta}$. A large-sample theory is needed in the general case, but in the normal mean model, from Example 1.3 we have

$$\log \frac{L(\theta)}{L(\widehat{\theta})} = -\frac{n}{2\sigma^2}(\bar{y} - \theta)^2.$$

Now, we know $\bar{y}$ is $N(\theta, \sigma^2/n)$, so

$$\frac{n}{\sigma^2}(\bar{y} - \theta)^2 \sim \chi_1^2,$$

or

$$W \equiv 2\log\frac{L(\widehat{\theta})}{L(\theta)} \sim \chi_1^2. \tag{1.3}$$

W is called Wilks's likelihood ratio statistic. Its χ^2 distribution is exact in the normal mean model and approximate in general cases; see below. This is the key distribution theory needed to calibrate the likelihood.

One of the most useful inferential quantities we can derive from the likelihood function is an interval containing parameters with the largest likelihoods:

$$\left\{\theta;\ \frac{L(\theta)}{L(\widehat{\theta})} > c\right\},$$

which is the basis of likelihood-based confidence intervals. In view of
(1.3), for an unknown but fixed θ, the probability that the likelihood
interval covers θ is

$$P\left(\frac{L(\theta)}{L(\widehat{\theta})} > c\right) = P\left(2\log\frac{L(\widehat{\theta})}{L(\theta)} < -2\log c\right)$$

$$= P(\chi_1^2 < -2\log c).$$

If for some $0 < \alpha < 1$ we choose a cutoff

$$c = e^{-\frac{1}{2}\chi_{1,(1-\alpha)}^2},\tag{1.4}$$

where $\chi_{1,(1-\alpha)}^2$ is the $100(1-\alpha)$ percentile of χ_1^2, we have

$$P\left(\frac{L(\theta)}{L(\widehat{\theta})} > c\right) = P(\chi_1^2 < \chi_{1,(1-\alpha)}^2) = 1 - \alpha.$$

This means that by choosing c in (1.4) the likelihood interval

$$\left\{\theta, \frac{L(\theta)}{L(\widehat{\theta})} > c\right\}$$

is a $100(1-\alpha)\%$ confidence interval for θ.

In particular, for $\alpha = 0.05$ and 0.01, formula (1.4) gives $c = 0.15$ and
0.04. We arrive at the important conclusion that, in the normal mean
case, we get an exact 95% or 99% confidence interval for the mean by
choosing a cutoff of 15% or 4%, respectively. This same confidence interval interpretation is approximately true for reasonably regular problems.

1.4.1 Large-sample approximation theory

Using a second-order expansion around $\widehat{\theta}$ as before

$$L(\theta) \approx \text{constant} \times \exp\left\{-\frac{1}{2}(\theta - \widehat{\theta})^t I(\widehat{\theta})(\theta - \widehat{\theta})\right\},$$

which can be seen immediately as the likelihood based on a single observation $\widehat{\theta}$ taken from $N(\theta, I^{-1}(\widehat{\theta}))$, so intuitively

$$W \equiv 2\log\frac{L(\widehat{\theta})}{L(\theta)}$$

$$\approx (\widehat{\theta} - \theta)^t I(\widehat{\theta})(\widehat{\theta} - \theta) \sim \chi_p^2.$$

A practical guide to judge the accuracy of the approximation is that the likelihood is reasonably regular.

The distribution theory may be used to get an approximate P-value for testing $H_0 : \theta = \theta_0$ versus $H_1 : \theta \neq \theta_0$. Specifically, on observing a normalized likelihood

$$\frac{L(\theta_0)}{L(\widehat{\theta})} = r,$$

we compute $w = -2 \log r$, and

$$\text{P-value} = P(W \geq w),$$

where W has a χ_p^2 distribution.

From the distribution theory we can also set an approximate $100(1-\alpha)\%$ confidence region for θ as

$$\text{CR} = \left\{ \theta; \ 2 \log \frac{L(\widehat{\theta})}{L(\theta)} < \chi^2_{p,(1-\alpha)} \right\}.$$

For example, an approximate 95% CI is

$$\begin{aligned}
\text{CI} &= \left\{ \theta; \ 2 \log \frac{L(\widehat{\theta})}{L(\theta)} < 3.84 \right\} \\
&= \{ \theta; L(\theta) > 0.15 \times L(\widehat{\theta}) \}.
\end{aligned}$$

Such a confidence region is unlikely to be useful for $p > 2$ because of the display problem. *The case of $p = 2$ is particularly simple*, since the $100(1-\alpha)\%$ likelihood cutoff has an approximate $100(1-\alpha)\%$ confidence level. This is true since

$$\exp\left\{ -\frac{1}{2}\chi^2_{p,(1-\alpha)} \right\} = \alpha,$$

so the contour $\{\theta; L(\theta) = \alpha L(\widehat{\theta})\}$ defines an approximate $100(1-\alpha)\%$ confidence region.

If there are nuisance parameters, we use the profile likelihood method to remove them. Let $\theta = (\theta_1, \theta_2) \in R^p$, where $\theta_1 \in R^q$ is the parameter of interest and $\theta_2 \in R^r$ is the nuisance parameter, so $p = q + r$. Given the likelihood $L(\theta_1, \theta_2)$, we compute the profile likelihood as

$$\begin{aligned}
L(\theta_1) &\equiv \max_{\theta_2} L(\theta_1, \theta_2) \\
&\equiv L(\theta_1, \widehat{\theta}_2(\theta_1)),
\end{aligned}$$

where $\widehat{\theta}_2(\theta_1)$ is the MLE of θ_2 at a fixed value of θ_1.

The theory (Pawitan, 2001, Chapter 9) indicates that we can treat $L(\theta_1)$

as if it were a true likelihood; in particular, the profile likelihood ratio follows the usual asymptotic theory:

$$W = 2\log\frac{L(\widehat{\theta}_1)}{L(\theta_1)} \xrightarrow{d} \chi_q^2 = \chi_{p-r}^2. \tag{1.5}$$

Here is another way of looking at the profile likelihood ratio from the point of view of testing $H_0: \theta_1 = \theta_{10}$. This is useful for dealing with hypotheses that are not easily parameterized, for example, testing for independence in a two-way table. By definition,

$$
\begin{aligned}
L(\theta_{10}) &= \max_{\theta_2, \theta_1 = \theta_{10}} L(\theta_1, \theta_2) \\
&= \max_{H_0} L(\theta) \\
L(\widehat{\theta}_1) &= \max_{\theta_1}\{\max_{\theta_2} L(\theta_1, \theta_2)\} \\
&= \max_{\theta} L(\theta).
\end{aligned}
$$

Therefore,

$$W = 2\log\left\{\frac{\max L(\theta), \text{ no restriction on } \theta}{\max L(\theta), \theta \in H_0}\right\}.$$

A large value of W means H_0 has a small likelihood or there are other values with higher support, so we should reject H_0.

How large is large will be determined by the sampling distribution of W. We can interpret p and r as

$$
\begin{aligned}
p &= \text{dimension of the whole parameter space } \theta \\
&= \text{total number of free parameters} \\
&= \text{total degrees of freedom of the parameter space} \\
r &= \text{dimension of the parameter space under } H_0 \\
&= \text{number of free parameters under } H_0 \\
&= \text{degrees of freedom of the model under } H_0
\end{aligned}
$$

Hence the number of degrees of freedom in (1.5) is the change in the dimension of the parameter space from the whole space to the one under H_0. For easy reference, the main result is stated as

Theorem 1.1 *Assuming regularity conditions, under $H_0: \theta_1 = \theta_{10}$*

$$W = 2\log\left\{\frac{\max L(\theta)}{\max_{H_0} L(\theta)}\right\} \to \chi_{p-r}^2.$$

In some applications it is possible to get an exact distribution for W.

For example, many normal-based classical tests, such as the t-test or F-test, are exact likelihood ratio tests.

Example 1.6: Let $y_1, \ldots, y_n$ be an iid sample from $N(\mu, \sigma^2)$ with σ^2 unknown and we are interested in testing H_0: $\mu = \mu_0$ versus H_1: $\mu \neq \mu_0$. Under H_0 the MLE of σ^2 is

$$\widehat{\sigma}^2 = \frac{1}{n}\sum_i (y_i - \mu_0)^2.$$

Up to a constant term,

$$\max_{H_0} L(\theta) \quad \propto \quad \left\{ \frac{1}{n}\sum_i (y_i - \mu_0)^2 \right\}^{-n/2}$$

$$\max L(\theta) \quad \propto \quad \left\{ \frac{1}{n}\sum_i (y_i - \bar{y})^2 \right\}^{-n/2}$$

and

$$\begin{aligned}
W &= n \log \frac{\sum_i (y_i - \mu_0)^2}{\sum_i (y_i - \bar{y})^2} \\
&= n \log \frac{\sum_i (y_i - \bar{y})^2 + n(\bar{y} - \mu_0)^2}{\sum_i (y_i - \bar{y})^2} \\
&= n \log \left(1 + \frac{t^2}{n-1} \right),
\end{aligned}$$

where $t = \sqrt{n}(\bar{y} - \mu_0)/s$ and s^2 is the sample variance. W is monotone increasing in t^2, so we reject H_0 for large values of t^2 or $|t|$. This is the usual t-test. A critical value or P-value can be determined from the t_{n-1}-distribution. □

1.5 Distribution of the MLE and the Wald statistic

As defined earlier, let the expected Fisher information be

$$\mathcal{I}(\theta) = E_\theta I(\theta)$$

where the expected value assumes θ is the true parameter. For independent observations, we can show that information is additive, so for n iid observations we have

$$\mathcal{I}(\theta) = n\mathcal{I}_1(\theta),$$

where $\mathcal{I}_1(\theta_0)$ is the expected Fisher information from a single observation.

We first state the scalar case: Let $y_1, \ldots, y_n$ be an iid sample from $f_{\theta_0}(y)$, and assume that the MLE $\widehat{\theta}$ is consistent in the sense that

$$P(\theta_0 - \epsilon < \widehat{\theta} < \theta_0 + \epsilon) \to 1$$

for all $\epsilon > 0$ as n gets large. Then, under some regularity conditions,

$$\sqrt{n}(\widehat{\theta} - \theta_0) \to N(0, 1/\mathcal{I}_1(\theta_0)). \tag{1.6}$$

For a complete list of standard regularity conditions see Lehmann (1983, Chapter 6). One important condition is that θ must not be a boundary parameter; for example, the parameter θ in Uniform$(0, \theta)$ is a boundary parameter for which the likelihood cannot be regular.

We give only an outline of the proof: a linear approximation of the score function $S(\theta)$ around θ_0 gives

$$S(\theta) \approx S(\theta_0) - I(\theta_0)(\theta - \theta_0)$$

and since $S(\widehat{\theta}) = 0$, we have

$$\sqrt{n}(\widehat{\theta} - \theta_0) \approx \{I(\theta_0)/n\}^{-1} S(\theta_0)/\sqrt{n}.$$

The result follows using the law of large numbers:

$$I(\theta_0)/n \to \mathcal{I}_1(\theta_0)$$

and the central limit theorem (CLT):

$$S(\theta_0)/\sqrt{n} \to N\{0, \mathcal{I}_1(\theta_0)\}.$$

We can then show that all the following are true:

$$\begin{aligned}
\sqrt{\mathcal{I}(\theta_0)}(\widehat{\theta} - \theta_0) &\to N(0, 1) \\
\sqrt{I(\theta_0)}(\widehat{\theta} - \theta_0) &\to N(0, 1) \\
\sqrt{\mathcal{I}(\widehat{\theta})}(\widehat{\theta} - \theta_0) &\to N(0, 1) \\
\sqrt{I(\widehat{\theta})}(\widehat{\theta} - \theta_0) &\to N(0, 1).
\end{aligned}$$

These statements in fact hold more generally than (1.6) since, under similar general conditions, they also apply for independent but non–identical outcomes, as long as the CLT holds for the score statistic. The last two forms are the most practical, and informally we say

$$\begin{aligned}
\widehat{\theta} &\sim N(\theta_0, 1/\mathcal{I}(\widehat{\theta})) \\
\widehat{\theta} &\sim N(\theta_0, 1/I(\widehat{\theta})).
\end{aligned}$$

In the full exponential family (Example 1.4) these two versions are identical. However, in more complex cases where $\mathcal{I}(\widehat{\theta}) \neq I(\widehat{\theta})$, the use of $I(\widehat{\theta})$ is preferable (Pawitan, 2001, Chapter 9).

In the multiparameter case, the asymptotic distribution of the MLE $\widehat{\theta}$

is given by the following equivalent results:

$$\sqrt{n}(\widehat{\theta} - \theta) \quad \to \quad N(0, \mathcal{I}_1(\theta)^{-1})$$
$$\mathcal{I}(\theta)^{1/2}(\widehat{\theta} - \theta) \quad \to \quad N(0, I_p)$$
$$I(\widehat{\theta})^{1/2}(\widehat{\theta} - \theta) \quad \to \quad N(0, I_p)$$
$$(\widehat{\theta} - \theta)^t I(\widehat{\theta})(\widehat{\theta} - \theta) \quad \to \quad \chi_p^2,$$

where I_p is an identity matrix of order p. In practice, we would use

$$\widehat{\theta} \sim N(\theta, I(\widehat{\theta})^{-1}).$$

The standard error of $\widehat{\theta}_i$ is given by the estimated standard deviation

$$\mathrm{se}(\widehat{\theta}_i) = \sqrt{I^{ii}},$$

where I^{ii} is the ith diagonal term of $I(\widehat{\theta})^{-1}$. A test of an individual parameter $H_0 : \theta_i = \theta_{i0}$ is given by the Wald statistic

$$z_i = \frac{\widehat{\theta}_i - \theta_{i0}}{\mathrm{se}(\widehat{\theta}_i)},$$

whose null distribution is approximately standard normal.

Wald confidence interval

As stated previously, the normal approximation is closely associated with the quadratic approximation of the loglihood. In these regular cases where the quadratic approximation works well and $I(\widehat{\theta})$ is meaningful, we have

$$\log \frac{L(\theta)}{L(\widehat{\theta})} \approx -\frac{1}{2} I(\widehat{\theta})(\theta - \widehat{\theta})^2$$

so the likelihood interval $\{\theta, \ L(\theta)/L(\widehat{\theta}) > c\}$ is approximately

$$\widehat{\theta} \pm \sqrt{-2 \log c} \times I(\widehat{\theta})^{-1/2}.$$

In the normal mean model in Example 1.3 this is an exact confidence interval (CI) with confidence level

$$P(\chi_1^2 < -2 \log c).$$

For example,

$$\widehat{\theta} \pm 1.96 \ I(\widehat{\theta})^{-1/2}$$

is an exact 95% CI. The asymptotic theory justifies its use in non–normal cases as an approximate 95% CI. Comparing with likelihood-based intervals, note the two levels of approximation to set up this interval: the loglihood is approximated by a quadratic and the confidence level is approximate.

What if the loglihood function is far from quadratic? See Figure 1.3. From the likelihood view, the Wald interval is deficient, since it includes values with lower likelihood compared to values outside the interval.

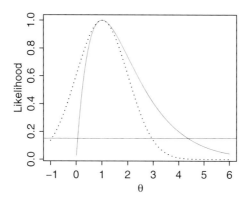

Figure 1.3 *Poor quadratic approximation (dotted) of a likelihood function (solid).*

Wald intervals might be called MLE-based intervals. To be clear, confidence intervals based on $\{\theta, \ L(\theta)/L(\hat{\theta}) > c\}$ will be called *likelihood-based* confidence intervals. Wald intervals are always symmetric, but likelihood-based intervals can be asymmetric. Computationally the Wald interval is much easier to compute than the likelihood-based interval. If the likelihood is regular, the two intervals will be similar. However, if they are not similar, the likelihood-based CI is preferable.

One problem with the Wald interval is that it is not invariant with respect to parameter transformation: if (θ_L, θ_U) is the Wald 95% CI for θ, $(g(\theta_L), g(\theta_U))$ is not the Wald 95% CI for $g(\theta)$, unless $g()$ is linear. This means Wald intervals work well only on one particular scale where the estimate is most normally distributed. In contrast, the likelihood-based CI is transformation invariant, so it works similarly in any scale.

Example 1.7: Suppose $y = 8$ is a binomial sample with $n = 10$ and probability θ. We can show graphically that the quadratic approximation is poor. The standard error of $\hat{\theta}$ is $I(\hat{\theta})^{-1/2} = 1/\sqrt{62.5} = 0.13$, so the Wald 95% CI is

$$0.8 \pm 1.96/\sqrt{62.5},$$

giving $0.55 < \theta < 1.05$, clearly inappropriate for a probability. For $n = 100$

and $y = 80$, the standard error for $\widehat{\theta}$ is $I(\widehat{\theta})^{-1/2} = 1/\sqrt{625} = 0.04$. Here we have a much better quadratic approximation, with the Wald 95% CI

$$0.8 \pm 1.96/\sqrt{625}$$

or $0.72 < \theta < 0.88$, compared with $0.72 < \theta < 0.87$ from the exact likelihood. □

1.6 Model selection

The simplest model selection problem occurs when we are comparing nested models. For example, suppose we want to compare two models A: $\mu = \beta_0 + \beta_1 x$ versus B: $\mu = \beta_0$. The models are nested in the sense that B is a submodel of A. The problem is immediately recognizable as a hypothesis testing of H_0: $\beta_1 = 0$, so many standard methodologies apply. This is the case with all nested comparisons.

A non-nested comparison is not so easy, since we cannot reduce it to a standard hypothesis testing problem. For example, to model positive outcome data we might consider two competing models:

(a):   Generalized linear model (GLM) with normal family and identity link function with possible heteroscedastic errors

(b):   GLM with gamma family and log-link function

(These models are discussed in the next chapter and are stated here only for illustration of a model selection problem, so no specific understanding is expected.) In this case the usual parameters in the models take the role of nuisance parameters in the model comparison.

We can imagine maximizing the likelihood over all potential models. Suppose we want to compare K models $f_k(\cdot, \theta_k)$ for $k = 1, \ldots, K$, given data $y_1, \ldots, y_n$, where θ_k is the parameter associated with model k. The parameter dimension is allowed to vary between models, and the interpretation of the parameter can be model dependent. We then

(a) Find the best θ_k in each model via the standard maximum likelihood, i.e. by choosing

$$\widehat{\theta}_k = \operatorname{argmax}_\theta \sum_i \log f_k(y_i, \theta_k).$$

This is the likelihood profiling step.

(b) Choose the model k that maximizes the log-likelihood

$$\log L(\widehat{\theta}_k) \equiv \sum_i \log f_k(y_i, \widehat{\theta}_k).$$

As a crucial difference with the comparison of nested models, here *all the constants in the density function must be kept*. This can be problematic when we use likelihood values reported by standard statistical software, since it is not always clear whether they are based on the full definition of the density, including all the constant terms.

We cannot naively compare the maximized log-likelihood $\log L(\widehat{\theta}_k)$ since it is a biased quantity: the same data are used to compute $\widehat{\theta}_k$ by maximizing the likelihood. For example, we can always increase the maximized likelihood by enlarging the parameter set, even though a model with more parameters is not necessarily better. The Akaike information criterion (AIC) formula tries to correct this bias by a simple adjustment determined by the number of free parameters only. Thus

$$\text{AIC}(k) = -2 \sum_i \log f_k(y_i, \widehat{\theta}_k) + 2p,$$

where p is the number of parameters, so the model with minimum AIC is considered the best one. We can interpret the first term in $\text{AIC}(k)$ as a measure of data fit and the second as a penalty. If we are comparing models with the same number of parameters, we need only compare the maximized likelihood.

In summary, if we are comparing nested models we can always use the standard likelihood ratio test and its associated probability-based inference. In non-nested comparisons, we use the AIC and will not attach any probability-based assessment such as the P-value.

1.7 Marginal and conditional likelihoods

As a general method, consider a transformation of the data y to (v, w) such that either the marginal distribution of v or the conditional distribution of v given w depends only on the parameter of interest θ. Let the total parameter be (θ, η). In the first case

$$
\begin{aligned}
L(\theta, \eta) &= f_{\theta,\eta}(v, w) \\
&= f_\theta(v) f_{\theta,\eta}(w|v) \\
&\equiv L_1(\theta) L_2(\theta, \eta),
\end{aligned}
$$

so the *marginal likelihood* of θ is defined as

$$L_1(\theta) = f_\theta(v).$$

In the second case

$$
\begin{aligned}
L(\theta, \eta) &= f_\theta(v|w) f_{\theta,\eta}(w) \\
&\equiv L_1(\theta) L_2(\theta, \eta),
\end{aligned}
$$

where the *conditional likelihood* is defined as

$$L_1(\theta) = f_\theta(v|w).$$

The question of which one is applicable has to be decided on a case-by-case basis. If v and w are independent the two likelihood functions coincide. Marginal or conditional likelihoods are useful if

- $f_\theta(v)$ or $f_\theta(v|w)$ is simpler than the original model $f_{\theta,\eta}(y)$.
- Not much information is lost by ignoring $L_2(\theta, \eta)$.
- The use of full likelihood is inconsistent.

Proving the second condition is often difficult, so it is usually argued informally on an intuitive basis. If the last condition applies, the use of marginal or conditional likelihood is essential. When available, these likelihoods are true likelihoods in the sense that they correspond to a probability of the observed data; this is their main advantage over profile likelihood. However, it is not always obvious how to transform the data to arrive at a model that is free of the nuisance parameter.

Example 1.8: Suppose y_{i1} and y_{i2} are an iid sample from $N(\mu_i, \sigma^2)$, for $i = 1, \ldots, N$, and they are all independent over index i; the parameter of interest is σ^2. This matched-pair data set is one example of highly stratified or clustered data. The key feature here is that the number of nuisance parameters N is of the same order as the number of observations $2N$. To compute the profile likelihood of σ^2, we first show that $\hat{\mu}_i = \bar{y}_i$. Using the residual sum of squares $SSE = \sum_i \sum_j (y_{ij} - \bar{y}_i)^2$, the profile likelihood of σ^2 is given by

$$
\begin{aligned}
\log L(\sigma^2) &= \max_{\mu_1,\ldots,\mu_N} \log L(\mu_1, \ldots, \mu_N, \sigma^2) \\
&= -N \log \sigma^2 - \frac{SSE}{2\sigma^2}
\end{aligned}
$$

and the MLE is

$$\hat{\sigma}^2 = \frac{SSE}{2N}.$$

It is clear that the SSE is $\sigma^2 \chi^2_N$, so $E\hat{\sigma}^2 = \sigma^2/2$ for any N.

To get an unbiased inference for σ^2, consider the following transformations:

$$
\begin{aligned}
v_i &= (y_{i1} - y_{i2})/\sqrt{2} \\
w_i &= (y_{i1} + y_{i2})/\sqrt{2}.
\end{aligned}
$$

Clearly v_is are iid $N(0, \sigma^2)$, and w_is are iid $N(\sqrt{2}\mu_i, \sigma^2)$. The likelihood of σ^2 based on v_is is a marginal likelihood given by

$$L_v(\sigma^2) = \left(\frac{1}{\sqrt{2\pi\sigma^2}}\right)^N \exp\left(-\frac{1}{2\sigma^2} \sum_{i=1}^N v_i^2\right).$$

Since v_i and w_i are independent, in this case it is also a conditional likelihood. The MLE from the marginal likelihood is given by

$$\widehat{\sigma}^2 = \frac{1}{N} \sum_{i=1}^{N} v_i^2 = \frac{SSE}{N},$$

which is now an unbiased and consistent estimator.

How much information is lost by ignoring the w_i? The answer depends on the structure of μ_i. It is clear that the maximum loss occurs if $\mu_i \equiv \mu$ for all i, since in this case we should have used $2N$ data points to estimate σ^2, so there is 50% loss of information by conditioning. Hence we should expect that if the variation between the μ_is is small relative to within-pair variation, an unconditional analysis that assumes some structure on them should improve on the conditional analysis.

Now suppose, for $i = 1, \ldots, n$, y_{i1} is $N(\mu_{i1}, \sigma^2)$ and y_{i2} is $N(\mu_{i2}, \sigma^2)$, with σ^2 *known* and these are all independent. In practice, these might be observations from a study where two treatments are randomized within each pair. Assume a common mean difference across the pairs, so that

$$\mu_{i1} = \mu_{i2} + \theta.$$

Again, since v_i is iid $N(\theta/\sqrt{2}, \sigma^2)$, the conditional or marginal inference based on v_i is free of the nuisance pair effects, and the conditional MLE of θ is

$$\widehat{\theta} = \frac{1}{N} \sum_{i=1}^{N} (y_{i1} - y_{i2}).$$

The main implication of conditional analysis is that we are using only information from within the pair. However, in this case, regardless of what values μ_{1i}s take, $\widehat{\theta}$ is the same as the unconditional MLE of θ from the full data. This means that, in contrast to the estimation of σ^2, in the estimation of the mean difference θ, we incur no loss of information by conditioning. This normal theory result occurs in balanced cases; otherwise an unconditional analysis may carry more information than the conditional analysis, a situation called recovery of inter-block information (Yates, 1939); see also Section 5.1.

Note, however, that inference of θ requires knowledge of σ^2, and from the previous argument, when σ^2 is unknown, conditioning may lose information. In a practical problem where σ^2 is unknown, even in balanced cases, it is possible to beat the conditional analysis by an unconditional analysis. $\square$

Example 1.9: Suppose that y_1 and y_2 are independent Poisson with means μ_1 and μ_2, and we are interested in the ratio $\theta = \mu_2/\mu_1$. The conditional distribution of y_2 given the sum $y_1 + y_2$ is binomial with parameters $n = y_1 + y_2$ and probability

$$\pi = \frac{\mu_2}{\mu_1 + \mu_2} = \frac{\theta}{1 + \theta},$$

which is free of nuisance parameters. Alternatively, we may write

$$\frac{\pi}{1 - \pi} = \theta,$$

so that the MLE of θ from the conditional model is clearly y_2/y_1. This result shows a connection between odd-ratios for the binomial and intensity ratios for the Poisson that is useful for modelling of paired Poisson data.

Intuitively it seems that there will be little information in the sum n about the ratio parameter; in fact, in this case there is no information loss. This can be seen as follows. If we have a collection of paired Poisson data (y_{i1}, y_{i2}) with mean $(\mu_i, \theta\mu_i)$, y_{i2} conditionally given the sum $n_i = y_{i1} + y_{i2}$ is independent binomial with parameter (n_i, π). Conditioning has removed the pair effects, and from the conditional model we get the MLEs

$$\widehat{\pi} = \frac{\sum_i y_{i2}}{\sum_i n_i}$$

$$\widehat{\theta} = \frac{\sum_i y_{i2}}{\sum_i y_{i1}}.$$

Now, assuming there are no pair effects so that $\mu_i \equiv \mu$, based on the same data set, *unconditionally* the MLE of θ is also $\widehat{\theta} = \sum_i y_{i2} / \sum_i y_{i1}$. In contrast with the estimation of σ^2 but similar to the estimation of mean difference θ in Example 1.8 above, there is no loss of information by conditioning. This can be extended to more than two Poisson means, where conditioning gives the multinomial distribution. □

Example 1.10: Suppose y_1 is binomial $B(n_1, \pi_1)$ and independently y_2 is $B(n_2, \pi_2)$, say measuring the number of people with certain conditions. The data are tabulated into a 2-by-2 table

	Group 1	Group 2	Total
Present	y_1	y_2	t
Absent	$n_1 - y_1$	$n_2 - y_2$	u
Total	n_1	n_2	n

As the parameter of interest, we consider the log odds ratio θ defined by

$$\theta = \log \frac{\pi_1/(1 - \pi_1)}{\pi_2/(1 - \pi_2)}.$$

Now we make the following transformation: (y_1, y_2) to $(y_1, y_1 + y_2)$. The conditional probability of $Y_1 = y_1$ given $Y_1 + Y_2 = t$ is

$$P(Y_1 = y_1 | Y_1 + Y_2 = t) = \frac{P(Y_1 = y_1, Y_1 + Y_2 = t)}{P(Y_1 + Y_2 = t)}.$$

The numerator is equal to

$$\binom{n_1}{y_1} \binom{n_2}{t - y_1} e^{\theta y_1} \left(\frac{\pi_2}{1 - \pi_2}\right)^t (1 - \pi_1)^{n_1} (1 - \pi_2)^{n_2},$$

so that the conditional probability is

$$P(Y_1 = y_1 | Y_1 + Y_2 = t) = \frac{\binom{n_1}{y_1} \binom{n_2}{t - y_1} e^{\theta y_1}}{\sum_{s=0}^{t} \binom{n_1}{s} \binom{n_2}{t - s} e^{\theta s}},$$

which is free of any nuisance parameters. The common hypothesis of interest H_0: $\pi_1 = \pi_2$ is equivalent to H_0: $\theta = 0$, and it leads to the so-called Fisher's exact test with the hypergeometric null distribution.

If we have a collection of paired binomials (y_{i1}, y_{i2}) with parameters (n_{i1}, π_{i1}) and (n_{i2}, π_{i2}) with a common odds ratio θ, a reasonable inference on θ can be based on the conditional argument above. However, when there is no pair effect, so that $\pi_{i1} = \pi_1$, the situation is more complicated than in the Poisson case of Example 1.9. We have no closed-form solutions here, and the conditional and unconditional estimates of θ are no longer the same, indicating a potential loss of information due to conditioning.

In the important special case of binary matched pairs, where $n_{i1} = n_{i2} = 1$, it is possible to write more explicitly. The sum $y_{1i} + y_{2i} = t_i$ is either 0, 1 or 2, corresponding to paired outcomes $(0,0)$, $\{(0,1)$ or $(1,0)\}$ and $(1,1)$. If $t_i = 0$, both y_{i1} and y_{i2} are 0, and if $t_i = 2$, both y_{i1} and y_{i2} are 1. In the conditional analysis these concordant pairs do not contribute any information. If $t_i = 1$, the likelihood contribution is determined by

$$p = P(y_{i1} = 1 | y_{i1} + y_{i2} = 1) = \frac{e^{\theta}}{1 + e^{\theta}}$$

or

$$\log \frac{p}{1 - p} = \theta, \tag{1.7}$$

so that θ can be easily estimated from the discordant pairs only as the log ratio of the number of $(1, 0)$ over $(0, 1)$ pairs.

The matched-pair design allows general predictors x_{ij}, so that starting with the model

$$\log \frac{P(y_{ij} = 1)}{1 - P(y_{ij} = 1)} = x_{ij}^t \beta + v_i$$

we can follow the previous derivation and get

$$p_i = P(y_{i1} = 1 | y_{i1} + y_{i2} = 1) = \frac{e^{(x_{i1} - x_{i2})^t \beta}}{1 + e^{(x_{i1} - x_{i2})^t \beta}}$$

$$= \frac{e^{x_{i1}^t \beta}}{e^{x_{i1}^t \beta} + e^{x_{i2}^t \beta}}$$

or

$$\log \frac{p_i}{1 - p_i} = (x_{i1} - x_{i2})^t \beta, \tag{1.8}$$

so the conditioning gets rid of the pair effects v_is. When θ is the only parameter as in model (1.7), Lindsay (1983) shows the conditional analysis is asymptotically close to an unconditional analysis, meaning that there is no loss of information due to conditioning. However, this may not be true if there are other covariates; see Kang *et al.* (2005). In particular, in model (1.8), the conditional approach allows only covariates that vary within the pair (e.g., two treatments assigned within the pair), but there is a complete loss of information on covariates that vary only between the pairs (e.g., age of a twin pair). $\square$

Example 1.11: In general, an exact conditional likelihood is available if both the parameter of interest and the nuisance parameter are the canonical parameters of an exponential family model; see Example 1.4. Suppose y is in the $(q + r)$-parameter exponential family with log-density

$$\log f_{\theta,\eta}(y) = \theta^t t_1(y) + \eta^t t_2(y) - b(\theta, \eta) + c(y),$$

where θ is a q-vector of parameters of interest, and η is an r-vector of nuisance parameters. The marginal log density of $t_1(y)$ is of the form

$$\log f_{\theta,\eta}(t_1) = \theta^t t_1(y) - A(\theta, \eta) + c_1(t_1, \eta),$$

which involves both θ and η, but the conditional density of t_1 given t_2 depends only on θ, according to

$$\log f_{\theta}(t_1|t_2) = \theta^t t_1(y) - A_1(\theta, t_2) + h_1(t_1, t_2),$$

for some (potentially complicated) functions $A_1(\cdot)$ and $h_1(\cdot)$. An approximation to the conditional likelihood can be made using the adjusted profile likelihood given in Section 1.9.

However, even in the exponential family, parameters of interest can appear in a form that cannot be isolated using conditioning or marginalizing. Let y_1 and y_2 be independent exponential variates with means η and $\theta\eta$, respectively, and suppose that the parameter of interest θ is the mean ratio. Here

$$\log f(y_1, y_2) = -\log \theta - 2\log \eta - y_1/\eta - y_2/(\theta\eta).$$

The parameter of interest is not canonical and the conditional distribution of y_2 given y_1 is not free of η. An approximate conditional inference using an adjusted profile likelihood is given in Section 1.9. $\square$

1.8 Higher-order approximations

Likelihood theory is an important route to many higher-order approximations in statistics. From the standard theory we have approximately

$$\widehat{\theta} \sim N\{\theta, I(\widehat{\theta})^{-1}\}$$

so that the approximate density of $\widehat{\theta}$ is

$$f_{\theta}(\widehat{\theta}) \approx |I(\widehat{\theta})/(2\pi)|^{1/2} \exp\left\{-(\widehat{\theta} - \theta)^t I(\widehat{\theta})(\widehat{\theta} - \theta)/2\right\}. \tag{1.9}$$

We have also shown the quadratic approximation

$$\log \frac{L(\theta)}{L(\widehat{\theta})} \approx -(\widehat{\theta} - \theta)^t I(\widehat{\theta})(\widehat{\theta} - \theta)/2,$$

so immediately we have another approximate density

$$f_\theta(\widehat{\theta}) \approx |I(\widehat{\theta})/(2\pi)|^{1/2} \frac{L(\theta)}{L(\widehat{\theta})}. \qquad (1.10)$$

We shall refer to this as the likelihood-based p-formula, which turns out to be much more accurate than the normal-based formula (1.9). Even though we are using a likelihood ratio, it should be understood that (1.10) is a formula for a sampling density: θ is fixed and $\widehat{\theta}$ varies.

Example 1.12: Let $y_1, \ldots, y_n$ be an iid sample from $N(\theta, \sigma^2)$ with σ^2 known. Here we know that $\widehat{\theta} = \bar{y}$ is $N(\theta, \sigma^2/n)$. To use formula (1.10) we need

$$\begin{aligned}
\log L(\theta) &= -\frac{1}{2\sigma^2}\left\{\sum_i (y_i - \bar{y})^2 + n(\bar{y} - \theta)^2\right\} \\
\log L(\widehat{\theta}) &= -\frac{1}{2\sigma^2}\sum_i (y_i - \bar{y})^2 \\
I(\widehat{\theta}) &= n/\sigma^2,
\end{aligned}$$

so

$$f_\theta(\bar{y}) \approx |2\pi(\sigma^2/n)|^{-1/2} \exp\left\{-\frac{n}{2\sigma^2}(\bar{y} - \theta)^2\right\},$$

exactly the density of the normal distribution $N(\theta, \sigma^2/n)$. □

Example 1.13: Let y be Poisson with mean θ. The MLE of θ is $\widehat{\theta} = y$, and the Fisher information is $I(\widehat{\theta}) = 1/\widehat{\theta} = 1/y$. The p-formula (1.10) is

$$\begin{aligned}
f_\theta(y) &\approx (2\pi)^{-1/2}(1/y)^{1/2}\frac{e^{-\theta}\theta^y/y!}{e^{-y}y^y/y!} \\
&= \frac{e^{-\theta}\theta^y}{(2\pi y)^{1/2}e^{-y}y^y},
\end{aligned}$$

so in effect we have approximated the Poisson probability by replacing $y!$ with its Stirling approximation. The approximation is excellent for $y > 3$, but not so good for $y \le 3$. Nelder and Pregibon (1987) suggested a simple modification of the denominator to

$$(2\pi(y + 1/6))^{1/2}e^{-y}y^y,$$

which works remarkably well for all $y \ge 0$. □

The p-formula can be further improved by a generic normalizing constant to make the density integrate to one. The formula

$$p_\theta^*(\widehat{\theta}) = c(\theta)|I(\widehat{\theta})/(2\pi)|^{1/2}\frac{L(\theta)}{L(\widehat{\theta})} \qquad (1.11)$$

is called Barndorff-Nielsen's (1983) p^*-formula. As we should expect, in many cases $c(\theta)$ is very nearly one; in fact, $c(\theta) \approx 1 + B(\theta)/n$, where $B(\theta)$ is bounded over n. If difficult to derive analytically, $c(\theta)$ can be computed numerically. For likelihood approximations the p-formula is more convenient.

1.9 Adjusted profile likelihood

In general problems, exact marginal or conditional likelihoods are often unavailable. Even when theoretically available, the exact form may be difficult to derive (see Example 1.11). It turns out that an approximate marginal or conditional likelihood can be found by adjusting the ordinary profile likelihood (Barndorff-Nielsen, 1983). We shall provide here a heuristic derivation only.

First recall the likelihood-based p-formula from Section 1.8 that provides an approximate density for $\widehat{\theta}$:

$$f_\theta(\widehat{\theta}) \approx |I(\widehat{\theta})/(2\pi)|^{1/2} \frac{L(\theta)}{L(\widehat{\theta})}.$$

In the multiparameter case, let $(\widehat{\theta}, \widehat{\eta})$ be the MLE of (θ, η); then we have the approximate density

$$f(\widehat{\theta}, \widehat{\eta}) \approx |I(\widehat{\theta}, \widehat{\eta})/(2\pi)|^{1/2} \frac{L(\theta, \eta)}{L(\widehat{\theta}, \widehat{\eta})}.$$

Let $\widehat{\eta}_\theta$ be the MLE of η at a fixed value of θ, and $I(\widehat{\eta}_\theta)$ the corresponding observed Fisher information. Given θ, the approximate density of $\widehat{\eta}_\theta$ is

$$f(\widehat{\eta}_\theta) \approx |I(\widehat{\eta}_\theta)/(2\pi)|^{1/2} \frac{L(\theta, \eta)}{L(\theta, \widehat{\eta}_\theta)},$$

where $L(\theta, \widehat{\eta}_\theta)$ is the profile likelihood of θ. So, the marginal density of $\widehat{\eta}$ is

$$
\begin{aligned}
f(\widehat{\eta}) &= f(\widehat{\eta}_\theta) \left| \frac{\partial \widehat{\eta}_\theta}{\partial \widehat{\eta}} \right| \\
&\approx |I(\widehat{\eta}_\theta)/(2\pi)|^{1/2} \frac{L(\theta, \eta)}{L(\theta, \widehat{\eta}_\theta)} \left| \frac{\partial \widehat{\eta}_\theta}{\partial \widehat{\eta}} \right|.
\end{aligned}
\tag{1.12}
$$

The conditional distribution of $\widehat{\theta}$ given $\widehat{\eta}$ is

$$
\begin{aligned}
f(\widehat{\theta}|\widehat{\eta}) &= \frac{f(\widehat{\theta}, \widehat{\eta})}{f(\widehat{\eta})} \\
&\approx |I(\widehat{\eta}_\theta)/(2\pi)|^{-1/2} \frac{L(\theta, \widehat{\eta}_\theta)}{L(\widehat{\theta}, \widehat{\eta})} \left| \frac{\partial \widehat{\eta}}{\partial \widehat{\eta}_\theta} \right|,
\end{aligned}
$$

where we have used the p-formula on both the numerator and the denominator. Hence, up to the constant, the approximate conditional loglihood of θ is

$$\ell_m(\theta) = \ell(\theta, \widehat{\eta}_\theta) - \frac{1}{2} \log |I(\widehat{\eta}_\theta)/(2\pi)| + \log \left| \frac{\partial \widehat{\eta}}{\partial \widehat{\eta}_\theta} \right|. \qquad (1.13)$$

We can arrive at the same formula using a marginal distribution of $\widehat{\theta}$. Note here that the constant 2π is kept so that the formula is as close as possible to log of a proper density function. It is especially important when comparing non-nested models using the AIC, where all the constants in the density must be kept in the likelihood computation (Section 1.6). In certain models, such as the variance component models studied in later chapters, the constant 2π is also necessary even when we want to compare likelihoods from nested models where the formula does not allow simply setting certain parameter values to zero.

The quantity $\frac{1}{2} \log |I(\widehat{\eta}_\theta)/(2\pi)|$ can be interpreted as the information concerning θ carried by $\widehat{\eta}_\theta$ in the ordinary profile likelihood. The Jacobian term $|\partial \widehat{\eta}/\partial \widehat{\eta}_\theta|$ keeps the modified profile likelihood invariant with respect to transformations of the nuisance parameter. In lucky situations, we might have orthogonal parameters in the sense $\widehat{\eta}_\theta = \widehat{\eta}$, implying $|\partial \widehat{\eta}/\partial \widehat{\eta}_\theta| = 1$, so that the last term of (1.13) vanishes. Cox and Reid (1987) showed that if θ is scalar it is possible to set the nuisance parameter η such that $|\partial \widehat{\eta}/\partial \widehat{\eta}_\theta| \approx 1$. We shall use their adjusted profile likelihood formula heavily with the notation

$$p_\eta(\ell|\theta) \equiv \ell(\theta, \widehat{\eta}_\theta) - \frac{1}{2} \log |I(\widehat{\eta}_\theta)/(2\pi)|, \qquad (1.14)$$

to emphasize that we are profiling the loglihood ℓ over the nuisance parameter η. When obvious from the context, the parameter θ is dropped, so the adjusted profile likelihood becomes $p_\eta(\ell)$. In some models, for computational convenience, we may use the expected Fisher information for the adjustment term.

Example 1.14: Suppose the outcome vector y is normal with mean μ and variance V, where

$$\mu = X\beta$$

for a known design matrix X, and $V \equiv V(\theta)$. In practice θ will contain the variance component parameters. The overall likelihood is

$$\ell(\beta, \theta) = -\frac{1}{2} \log |2\pi V| - \frac{1}{2}(y - X\beta)^t V^{-1}(y - X\beta).$$

Given θ, the MLE of β is the generalized least squares estimate

$$\widehat{\beta}_\theta = (X^t V^{-1} X)^{-1} X^t V^{-1} y,$$

and the profile likelihood of θ is

$$\ell_p(\theta) = -\frac{1}{2}\log|2\pi V| - \frac{1}{2}(y - X\widehat{\beta}_\theta)^t V^{-1}(y - X\widehat{\beta}_\theta).$$

Here the observed and expected Fisher information are the same and given by

$$I(\widehat{\beta}_\theta) = X^t V^{-1} X.$$

We can check that

$$E\left\{\frac{\partial^2}{\partial\beta\partial\theta_i}\log L(\beta,\theta)\right\} = E\left\{-X^t V^{-1}\frac{\partial V}{\partial\theta_i}V^{-1}(Y - X\beta)\right\} = 0$$

for any θ_i, so that β and θ are information orthogonal. Hence the adjusted profile likelihood is

$$p_\beta(\ell|\theta) = \ell_p(\theta) - \frac{1}{2}\log|X^t V^{-1} X/(2\pi)|. \qquad (1.15)$$

This matches exactly the so-called restricted maximum likelihood (REML), derived by Patterson and Thompson (1971) and Harville (1974), using the marginal distribution of the error term $y - X\widehat{\beta}_\theta$.

Since the adjustment term does not involve $\widehat{\beta}_\theta$, *computationally* we have an interesting coincidence that the two-step estimation procedure of β and θ is equivalent to a single-step joint optimization of an objective function

$$Q(\beta,\theta) = -\frac{1}{2}\log|2\pi V| - \frac{1}{2}(y - X\beta)^t V^{-1}(y - X\beta) - \frac{1}{2}\log|X^t V^{-1} X/(2\pi)|. \qquad (1.16)$$

However, strictly speaking this function $Q(\beta,\theta)$ is no longer a loglihood function as it does not correspond to an exact log density or an approximation to one. Furthermore, if the adjustment term for the profile likelihood (1.15) were a function of $\widehat{\beta}_\theta$, then the joint optimization of $Q(\beta,\theta)$ is no longer equivalent to the original two-step optimization involving the full likelihood and the adjusted profile likelihood. Thus, when we refer to REML adjustment we refer to adjusting the likelihood of the variance components given in equation (1.15). □

1.10 Bayesian and likelihood methods

We shall now describe very briefly the similarities and differences between the likelihood and Bayesian approaches. In Bayesian computations we begin with a prior $f(\theta)$ and compute the posterior

$$\begin{aligned} f(\theta|y) &= \text{constant} \times f(\theta)f(y|\theta) \\ &= \text{constant} \times f(\theta)L(\theta), \qquad (1.17) \end{aligned}$$

where, to follow Bayesian thinking we use $f(y|\theta) \equiv f_\theta(y)$. Comparing (1.17) with (1.1) we see that the Bayesian method achieves the same effect as the likelihood method: it combines the information from the prior and the current likelihoods by a simple multiplication.

If we treat the prior $f(\theta)$ as a prior likelihood, the posterior is a combined likelihood. However, without putting much of the Bayesian philosophical and intellectual investment in the prior, if *we know absolutely nothing about* θ prior to observing $X = x$, the *prior likelihood* is always $f(\theta) \equiv 1$, and the likelihood function then expresses the current information on θ after observing y. As a corollary, if we knew anything about θ prior to observing y, we should feel encouraged to include it in the consideration. Using a uniform prior, the Bayesian posterior density and the likelihood functions would be the same up to a constant term. Note, however, that the likelihood is not a probability density function, so it does not have to integrate to one, and there is no such thing as an improper prior likelihood.

Bayesians eliminate all unwanted parameters by integrating them out; that is consistent with their view that parameters have regular density functions. However, the likelihood function is not a probability density function, and it does not obey probability laws (see Section 1.1), so integrating out a parameter in a likelihood function is not a meaningful operation. It turns out, however, there is a close connection between the Bayesian integrated likelihood and an adjusted profile likelihood (Section 1.9).

For scalar parameters, the quadratic approximation

$$\log \frac{L(\theta)}{L(\widehat{\theta})} \approx -\frac{1}{2} I(\widehat{\theta})(\theta - \widehat{\theta})^2$$

implies

$$\int L(\theta)d\theta \approx L(\widehat{\theta}) \int e^{-\frac{1}{2}I(\widehat{\theta})(\theta-\widehat{\theta})^2} d\theta \qquad (1.18)$$

$$= L(\widehat{\theta})|I(\widehat{\theta})/(2\pi)|^{-1/2}. \qquad (1.19)$$

This is known as Laplace's integral approximation; it is highly accurate if $\ell(\theta)$ is well approximated by a quadratic. For a two-parameter model with joint likelihood $L(\theta, \eta)$, we immediately have

$$L_{int}(\theta) \equiv \int L(\theta, \eta)d\eta \approx L(\theta, \widehat{\eta}_\theta)|I(\widehat{\eta}_\theta)/(2\pi)|^{-1/2},$$

and

$$\ell_{int}(\theta) \approx \ell(\theta, \widehat{\eta}_\theta) - \frac{1}{2} \log |I(\widehat{\eta}_\theta)/(2\pi)|, \qquad (1.20)$$

so the integrated likelihood is approximately the adjusted profile likelihood in the case of orthogonal parameters (Cox and Reid, 1987).

Recent advances in Bayesian computation have generated many powerful methodologies such as Gibbs sampling or Markov chain Monte Carlo

(MCMC)] methods for computing posterior distributions. Similarities between likelihood and posterior densities mean that the same methodologies can be used to compute likelihood. However, from the classical likelihood perspectives, there are other computational methods based on optimization, often more efficient than MCMC, that we can use to get the relevant inferential quantities.

1.11 Confidence distribution

Mainstream statistical inference relies heavily on the confidence concept, e.g., as applied to the confidence interval. While it is possible to construct confidence intervals purely from the sampling distribution theory, the idea of confidence distribution turns out to be fruitful. We start with a sensible estimate T (assumed continuous for now) of a scalar parameter θ. For a fixed $T = t$ and *seen as a function of* θ, the one-sided P-value

$$C(\theta) \equiv P_\theta(T \geq t)$$

is monotone increasing function from zero to one, so it satisfies requirements of a distribution function, which we shall call the *confidence distribution* of θ. The reason is obvious: let θ_α be the 100α percentile of the confidence distribution, i.e.

$$C(\theta_\alpha) = P_{\theta_\alpha}(T \geq t) = \alpha. \qquad (1.21)$$

Then θ_α is in fact the $100(1-\alpha)\%$ *lower confidence bound* for the unknown fixed θ with coverage probability

$$P_\theta(\theta_\alpha < \theta) = 1 - \alpha. \qquad (1.22)$$

The potential for confusion is enormous here, primarily about what is fixed and what varies. In the confidence distribution, θ varies and θ_α is fixed, but in the probability (1.22), θ is fixed while θ_α varies as the random variable over different data sets. To show (1.22), think of a fixed θ and random θ_α. The event $[\theta_\alpha < \theta]$ is equivalent to $[T < t_{1-\alpha}(\theta)]$, where the quantile $t_{1-\alpha}(\theta)$ is defined such that

$$P_\theta(T < t_{1-\alpha}(\theta)) = 1 - \alpha.$$

Equivalently, we have $P_\theta(\theta \leq \theta_\alpha) = \alpha$, so if we read the function $C(\cdot)$ as 'the confidence of' and define

$$C(\theta \leq \theta_\alpha) \equiv C(\theta_\alpha) = \alpha$$

we arrive at a simple but powerful conclusion that confidence statements from $C(\theta)$ have a justified coverage probability. Thus, while the confidence distribution is a distribution of our (subjective) confidence about where θ is, it is stronger than the likelihood since it corresponds to an objective coverage probability over repeated sampling. The *confidence density* of θ is defined as the derivative of $C(\theta)$ is

$$c(\theta) = \frac{\partial C(\theta)}{\partial \theta}.$$

Example 1.15: All these concepts are clear in the normal case. Suppose we observe $T = t$ from $N(\theta, \sigma^2)$ with known σ^2. The confidence distribution is

$$P_\theta(T \geq t) = P(Z \geq (t - \theta)/\sigma)$$
$$= 1 - \Phi\left(\frac{t - \theta}{\sigma}\right),$$

and the confidence density is

$$c(\theta) = \frac{1}{\sigma}\phi\left(\frac{t - \theta}{\sigma}\right).$$

This is exactly the normal density centred at the observed $\widehat{\theta} = t$, which is also the likelihood function (if normalized to integrate to one). It is easy to see here how to derive all types of confidence intervals for θ (e.g., one-sided or two-sided) from the confidence density. If T is approximately normal, then the confidence density is also approximately normal. $\square$

As a note for the future, the correspondence between confidence and coverage probability justifies the use the confidence density as weight in integration, which we will do in Chapter 4. Another useful property is that if there exists a normalizing transform for T, the confidence distribution can be approximated by the bootstrap distribution (Pawitan, 2001, Chapter 5). Getting CIs from this bootstrap distribution is called the percentile method (Efron, 1979).

Finally, if T is discrete, the confidence distribution is computed the same way once the one-sided P-value is defined, but generally we cannot get an exact coverage probability. The typical solution is to guarantee a minimum coverage; see Pawitan (2001, Chapter 5) on how this is achieved for example for the Poisson and binomial cases.

CHAPTER 2

Generalized linear models

2.1 Linear models

We begin with what is probably the main workhorse of statisticians: the linear model, also known as the regression model. It has been the subject of many books, so rather than going into the mathematics of linear models, we shall discuss mainly the statistical aspects. We begin with a formulation adapted to the generalizations that follow later. The components are

(a) A response variate y, a vector of length n, whose elements are assumed to follow identical and independent normal distribution with mean vector μ and constant variance σ^2:

(b) A set of explanatory variates $x_1, x_2, ..., x_p$, all of length n, which may be regarded as forming the columns of a $n \times p$ model matrix X; it is assumed that the elements of μ are expressible as a linear combination of effects of the explanatory variables, so that we may write $\mu_i = \sum_j \beta_j x_{ij}$ or in matrix form $\mu = X\beta$. Although it is common to write this model in the form

$$y = X\beta + e$$

where e is a vector of errors having normal distributions with mean zero, this formulation does not lead naturally to the generalizations we shall be making in the next section.

This definition of a linear model is based on several important assumptions:

- For the *systematic* part of the model, a first assumption is additivity of effects; the individual effects of the explanatory variables are assumed to combine additively to form the joint effect. The second assumption is linearity: the effect of each explanatory variable is assumed to be linear in the sense that doubling the value of x will double the contribution of that x to the mean μ.

- For the *random* part of the model, a first assumption is that the errors associated with the response variable are independent and secondly that the variance of the response is constant, and, in particular, does not depend upon the mean. The assumption of normality, although important as the basis for an exact finite-sample theory, becomes less relevant in large samples. The theory of least squares can be developed using assumptions about the first two moments only, without requiring a normality assumption. The first-moment assumption is the key to the unbiasedness of estimates of β, and the second moment to their optimality. The comparison between second-moment assumptions and fully specified distributional assumptions is discussed in Chapter 3.

2.1.1 Types of terms in the linear predictor

The vector quantity $X\beta$ is known in GLM parlance as the linear predictor and its structure may involve the combination of different types of terms, some of which interact with each other. There are two basic types of terms in the linear predictor: continuous covariates and categorical covariates; the latter are also known as factors. From these can be constructed various forms of compound terms. The table below shows some examples with two representations, one algebraic and one as a model formula, using the notation of Wilkinson and Rogers (1973). In the latter, X is a single vector, while A represents a factor with a set of dummy variates, one for each level. Terms in a model formula define vector spaces without explicit definition of the corresponding parameters. For a full discussion see Chapter 3 of McCullagh and Nelder (1989).

Type of term	Algebraic	Model formula
Continuous Covariate	λx	X
Factor	α_i	A
Mixed	$\lambda_i x$	$A \cdot X$
Compound	$(\alpha\beta)_{ij}$	$A \cdot B$
Compound mixed	$\lambda_{ij} x$	$A \cdot B \cdot X$

2.1.2 Aliasing

The vector spaces defined by two terms, say P and Q, in a model formula are often linearly independent. If p and q are the respective dimensions, the dimension of $P + Q$ is $p + q$. Such terms are described as *unaliased* with one another. If P and Q span the same space, they are said to be *completely aliased*. If they share a common subspace, they are *partly aliased*. Note that if aliasing occurs, the corresponding parameter estimates will be formed from the same combinations of the response, so that there will be no separate information about them in the data. It is important to be aware of the aliasing structures of the columns of a data matrix when fitting any model. There are two forms of aliasing: *extrinsic* and *intrinsic*. Extrinsic aliasing depends on the particular form of the data matrix, and intrinsic aliasing depends on the relationship between the terms in a model formula.

Extrinsic aliasing

This form of aliasing arises from the particular form of the data matrix for a data set. Complete aliasing will occur, for example, if we attempt to fit x_1, x_2 and $x_3 = x_1 + x_2$. Similarly, if two three-level factors A and B index a model matrix with five units according to the pattern below with the additive model $\alpha_i + \beta_j$, the dummy variate vectors for α_3 and β_3 are identical and so produce extrinsic aliasing. Thus only four parameters are estimable. If we change the fourth unit (2,2) to (3,2) the aliasing disappears, and the five distinct parameters of the additive model are all estimable.

A	B
1	1
1	2
2	1
2	2
3	3

Intrinsic aliasing

A distinct type of aliasing can occur with models containing factors, a simple case shown by the model formula

$$1 + A,$$

where A is a factor and 1 stands for the dummy variate for the intercept with all elements of ones. The algebraic form is

$$\mu + \alpha_i,$$

where i indexes the groups defined by A. Here the term α_i can be written in vector notation as

$$\alpha_1 \mathbf{u}_1 + \alpha_2 \mathbf{u}_2 + \dots + \alpha_k \mathbf{u}_k,$$

where the dummy vector $\mathbf{u}_i$ has one wherever the level of the factor A is i, else zero. The dummy vectors for A add up to that for one, so that μ is aliased with $\sum \alpha_i$ whatever the allocation of units to the groups; such aliasing is called intrinsic, because it occurs whatever the pattern of the model matrix.

Note that the relationship between μ and α_i is asymmetric because while μ lies in the space of the dummy variates α_i, the reverse is not true. Such a relationship is described by saying that μ is *marginal* to α_i, and in consequence the terms μ and α_i are ordered as the result of the marginality relationship. An important consequence of this ordering is that it makes no sense to consider the hypothesis that $\mu = 0$ when the α_i are not assumed known.

The linear predictor $\eta_i = \mu + \alpha_i$ is unchanged if we subtract a constant c to μ and add it to each of the αs. Any contrast $\sum b_i \alpha_i$ with $\sum b_i = 0$, for example $\alpha_1 - \alpha_2$, is unaffected by this transformation and is said to be *estimable*, whereas μ, α_i and $\alpha_1 + \alpha_2$ are not estimable. Only estimable contrasts are relevant in any analysis and to obtain values for $\hat{\mu}$ and $\hat{\alpha}_1$ it is necessary to impose constraints on these estimates. Two common forms of constraint are

- To set $\hat{\alpha}_1$ to zero
- To set $\sum \hat{\alpha}_i$ to zero

The values of estimable contrasts are independent of the constraints imposed. It is important to stress that imposition of constraints to obtain unique estimates of the parameters does *not* imply that constraints should be imposed on the parameters themselves, as distinct from their estimates. We now consider the marginality relations in the important case of a two-way classification.

Intrinsic aliasing in the two-way classification

We deal with the linear predictor

$$1 + A + B + A \cdot B,$$

expressed algebraically as

$$\eta_{ij} = \mu + \alpha_i + \beta_j + \gamma_{ij}.$$

From the dummy vectors for the four terms we find

$$\sum \alpha_i \equiv \mu, \quad \sum \beta_j \equiv \mu, \quad \sum_j \gamma_{ij} \equiv \alpha_i, \quad \sum_i \gamma_{ij} \equiv \beta_j,$$

from which

$$\sum_{ij} \gamma_{ij} \equiv \mu.$$

The corresponding marginality relations are

$$\mu \text{ is marginal to } \alpha_i, \beta_j \text{ and } \gamma_{ij},$$

$$\alpha_i \text{ and } \beta_j \text{ are marginal to } \gamma_{ij}$$

These give a partial ordering: first μ, then α_i and β_j together, then γ_{ij}.

It is important to note that as a result of this ordering the interpretation of $A \cdot B$ depends on the marginal terms that precede it. Table 2.1 shows four possibilities.

Table 2.1 *The interpretation of $A \cdot B$ term in various model formula.*

Model formula	Interpretation
$A \cdot B$	$(\alpha\beta)_{ij}$, i.e., a separate effect for each combination of i and j
$A + A \cdot B$	Effects of B within each level of A
$B + A \cdot B$	Effects of A within each level of B
$A + B + A \cdot B$	Interaction between A and B after eliminating main effects of A and B

The marginality relations mean that it makes no sense to postulate that a main effect of A or B is null when $A \cdot B$ is not assumed zero. Attempts to fit models which ignore marginality relations result from imposing constraints on the parameters because they have been imposed on the parameter estimates. See Nelder (1994) for a full discussion.

2.1.3 Functional marginality

Functional marginality is primarily concerned with relations between continuous covariates, particularly as they occur in polynomials. The rules are similar, but not identical, to those for models based on factors.

Consider the quadratic polynomial

$$y = \beta_0 + \beta_1 x + \beta_2 x^2$$

where x and x^2 are linearly independent (true if more than two distinct values occur). There is thus no aliasing, but there is still an implied ordering governing the fitting of the terms. Thus the model $\beta_1 x$ makes sense only if it is known *a priori* that y is zero when x is zero, i.e., that $x = 0$ is a special point on the scale. Similarly, fitting the model

$$\beta_0 + \beta_2 x^2$$

makes sense only if the maximum or minimum of the response is known *a priori* to occur at $x = 0$. With product terms like $x_1 x_2$, the linear predictor must include terms in x_1 and x_2 unless the point $(0,0)$ is known *a priori* to be a saddlepoint on the surface. Where no special points exist on any of the x scales, models must be well formed polynomials, i.e., any compound term must have all its marginal terms included. Unless this is done, the model will have the undesirable property that a linear transformation of an x will change its goodness of fit, so that, for example, the fit will change if a temperature is changed from degrees F to degrees C. An attempt to relax this rule by allowing just one of the linear terms of, say, $x_1 x_2$, to appear in the model (the so-called weak heredity principle, Brinkley, Meyer and Lu, 1996) was shown by Nelder (1998) to be unsustainable. For a general discussion of marginality and functional marginality see McCullagh and Nelder (1989).

2.2 Generalized linear models

Generalized linear models (GLMs) can be derived from classical normal models by two extensions, one to the random part and one to the systematic part. Random elements may now come from a one-parameter exponential family, of which the normal distribution is a special case. Distributions in this class include Poisson, binomial, gamma and inverse Gaussian and normal; see also Example 1.4. They have a form for the log-likelihood function (abbreviated from now on to *loglihood*) given by

$$\sum [\{y\theta - b(\theta)\}/\phi + c(y, \phi)], \qquad (2.1)$$

where y is the response, θ is the canonical parameter and ϕ is the dispersion parameter. For these distributions the mean and variance are given by

$$\mu = b'(\theta)$$

and

$$\mathrm{var}(y) = \phi b''(\theta).$$

Since the mean depends only on θ, in a standard GLM the term $c(y, \phi)$ can be left unspecified without affecting the likelihood-based estimation of the regression parameters.

The form of the variance is important: it consists of two terms, the first of which depends only on ϕ, the dispersion parameter, and the second is a function of θ, the canonical parameter, and hence of μ. The second term, expressed as a function of μ, is the variance function and is written $V(\mu)$. The variance function defines a distribution in the GLM class of families if one exists, and the function $b(\theta)$ is the cumulant generating function. The forms of these two functions for the five main GLM distributions are as follows:

	$V(\mu)$	θ	$b(\theta)$	Canonical link
Normal	1	μ	$\theta^2/2$	Identity
Poisson	μ	$\log \mu$	$\exp(\theta)$	Log
Binomial*	$\mu(m-\mu)/m$	$\log\{\mu/(m-\mu)\}$	$m\log(1+exp(\theta))$	Logit
Gamma	μ^2	$-1/\mu$	$-\log(-\theta)$	reciprocal
Inverse Gaussian	μ^3	$-1/(2\mu^2)$	$-(-2\theta)^{1/2}$	$1/\mu^2$

Note: m is the binomial denominator.

The generalization of the systematic part consists in allowing the linear predictor to be a monotone function of the mean. We write

$$\eta = g(\mu)$$

where $g()$ is called the *link function*. If $\eta = \theta$, the canonical parameter, we have the *canonical link*. Canonical links give rise to simple sufficient statistics, but there is often no reason why they should be specially relevant in forming models.

Some prefer to use the linear model with normal errors after first transforming the response variable. However, a single data transformation may fail to satisfy all the properties necessary for an analysis. With GLMs, the identification of the mean-variance relationship and the choice of the scale on which the effects are to be measured can be done separately, thus overcoming the shortcomings of the data transformation approach.

GLMs transform the parameters to achieve the linear additivity. In Poisson GLMs for count data, we cannot transform the data to $\log y$, which is not defined for $y = 0$. In GLMs, $\log \mu = X\beta$ is used, which causes no problem when $y = 0$.

2.2.1 Iterative weighted least squares

The underlying procedure for fitting GLMs by maximum likelihood takes the form of iterative weighted least squares (IWLS) involving an adjusted dependent variable z, and an iterative weight W. Given a starting value of the mean $\hat{\mu}_0$ and linear predictor $\hat{\eta}_0 = g(\hat{\mu}_0)$, we compute z and W as

$$z = \hat{\eta}_0 + (y - \hat{\mu}_0)(d\eta/d\mu)_0$$

where the derivative is evaluated at $\hat{\mu}_0$, and

$$W^{-1} = (d\eta/d\mu)_0^2 V_0,$$

where V_0 is the variance function evaluated at $\hat{\mu}_0$. We now regress z on the covariates $x_1, x_2, ..., x_p$ with weight W to produce revised estimates $\hat{\beta}_1$ of the parameters, from which we get a new estimate $\hat{\eta}_1$ of the linear predictor. We then iterate until the changes are sufficiently small. Although non-linear, the algorithm has a simple starting procedure by which the data themselves are used as a first estimate of $\hat{\mu}_0$. Simple adjustments to the starting values are needed for extreme values such as zeros in count data.

Given the dispersion parameter ϕ, the ML estimators for β are obtained by solving the IWLS equation

$$X^t \Sigma^{-1} X \hat{\beta} = X^t \Sigma^{-1} z, \qquad (2.2)$$

where $\Sigma = \phi W^{-1}$, and the variance-covariance estimators are obtained from

$$\operatorname{var}(\hat{\beta}) = (X^t \Sigma^{-1} X)^{-1} = \phi(X^t W X)^{-1}. \qquad (2.3)$$

In IWLS equations $1/\phi$ plays the part of a prior weight. A more detailed derivation of IWLS is given in Section 3.2.

We may view the IWLS equations (2.2) as WLS equations from the linear model

$$z = X\beta + e,$$

where $e = (y - \mu)(\partial\eta/\partial\mu) \sim N(0, \Sigma)$. Note here that $\mathcal{I} = X^t \Sigma^{-1} X$ is the expected Fisher information and the IWLS equations (2.2) are obtained by the Fisher scoring method, which uses the expected Fisher information matrix in the Newton-Raphson method. The Fisher scoring and Newton-Raphson methods reduce to the same algorithm for the canonical link, because the expected and observed informations coincide. Computationally the IWLS procedure provides a numerically stable algorithm. For a detailed derivation of this algorithm, see McCullagh and Nelder (1989, Section 2.5).

In linear models we have

$$z = y, \quad \phi = \sigma^2 \quad \text{and} \quad W = I,$$

so that the ordinary least squares estimators (OLS) for β can be obtained via the normal equations

$$X^t X \hat{\beta} = X^t y,$$

and their variance-covariance estimators are given by

$$\text{var}(\hat{\beta}) = \sigma^2 (X^t X)^{-1}.$$

These can be solved without iteration. The OLS estimators are the best linear unbiased estimators (BLUEs). Under normality they become best unbiased estimators. These properties hold for all sample sizes. This theory covers only linear parametrizations of β. Under the normality assumptions $\hat{\beta}$ is the ML estimator and asymptotically best among consistent estimators for all parameterizations. The IWLS procedure is an extension of the OLS procedure to non-normal models, and now requires iteration.

2.2.2 Deviance for the goodness of fit

For a measure of goodness of fit analogous to the residual sum of squares for normal models, two such measures are in common use: the first is the generalized Pearson X^2 statistic, and the second the log likelihood ratio statistic, called the deviance in GLMs. These take the form

$$X^2 = \sum (y - \hat{\mu})^2 / V(\hat{\mu})$$

and

$$D = 2\phi \{ \ell(y; y) - \ell(\hat{\mu}; y) \}$$

where ℓ is the loglihood of the distribution. For normal models, the scaled deviances X^2/ϕ and D/ϕ are identical and become the scaled residual sum of squares, having an exact χ^2 distribution with $n - p$ degrees of freedom. In general they are different and we rely on asymptotic results for other distributions. When the asymptotic approximation is doubtful, for example for binary data with $\phi = 1$, the deviance cannot be used to give an absolute goodness-of-fit test. For grouped data, e.g., binomial with large enough n, we can often justify assuming that X^2 and D are approximately χ^2.

The deviance has a general advantage as a measure of discrepancy in that it is additive for nested sets of models, leading to likelihood ratio tests. Furthermore, the χ^2 approximation is usually quite accurate for the differences of deviances even though it could be inaccurate for the deviances themselves. Another advantage of the deviance over the X^2 is that it leads to the best normalizing residuals (Pierce and Schafer, 1986).

2.2.3 Estimation of the dispersion parameter

It remains to estimate the dispersion parameter ϕ for those distributions where it is not fixed (ϕ is fixed at one for the Poisson and binomial distributions). If the term $c(y, \phi)$ in the model (2.1) is available explicitly, we can use the full likelihood to estimate β and ϕ jointly. But often $c(y, \phi)$ is not available, so estimation of ϕ needs special consideration. We discuss this fully in Section 3.5. For the moment, we simply state that ϕ may be estimated using either X^2 or D, divided by the appropriate degrees of freedom. While X^2 is asymptotically unbiased (given the correct model) D is not. However, D often has smaller sampling variance, so that, in terms of MSE, neither is uniformly better (Lee and Nelder, 1992).

If ϕ is estimated by the REML method (Chapter 3) based upon X^2 and D, the scaled deviances $X^2/\hat{\phi}$ and $D/\hat{\phi}$ become the degrees of freedom $n - p$, so that the scaled deviance test for lack of fit is not useful when ϕ is estimated, but it can indicate that a proper convergence has been reached in estimating ϕ.

2.2.4 Residuals

In GLMs the deviance is represented by sum of deviance components

$$D = \sum d_i,$$

where the deviance component $d_i = 2 \int_{\hat{\mu}_i}^{y_i} (y_i - s) / V(s)\, ds$. The forms of deviance components for the GLM distributions are as follows:

Normal	$(y_i - \hat{\mu}_i)^2$
Poisson	$2\{y_i \log(y_i/\hat{\mu}_i) - (y_i - \hat{\mu}_i)\}$
Binomial	$2[y_i \log(y_i/\hat{\mu}_i) + (m_i - y_i) \log\{(m_i - y_i)/(m_i - \hat{\mu}_i)\}]$
Gamma	$2\{-\log(y_i/\hat{\mu}_i) + (y_i - \hat{\mu}_i)/\hat{\mu}_i\}$
Inverse Gaussian	$(y_i - \hat{\mu}_i)^2/(\hat{\mu}_i^2 y_i)$

Two forms of residuals are based on the signed square root of the components of X^2 or D. One is the Pearson residual

$$r_P = \frac{y - \mu}{\sqrt{V(\mu)}}$$

and the other is the deviance residual

$$r_D = \text{sign}(y - \mu)\sqrt{d}.$$

For example, for the Poisson distribution we have

$$r_P = \frac{y - \mu}{\sqrt{\mu}}$$

and

$$r_D = \text{sign}(y - \mu)\sqrt{2\{y\log(y/\mu) - (y - \mu)\}}.$$

The residual sum of squares for Pearson residuals is the famous Pearson χ^2 statistic for the goodness of fit. Since

$$r_D = r_P + O_p(\mu^{-1/2}),$$

both residuals are similar for large μ, in which the distribution tends to normality. For small μ they can be somewhat different. Deviance residuals as a set are usually more nearly normal with non-normal GLM distributions than Pearson residuals (Pierce and Schafer, 1986) and are therefore to be preferred for normal plots. Other definitions of residuals have been given.

2.2.5 Special families of GLMs

We give brief summaries of the principal GLM families.

Poisson   Log-linear models, which use the Poisson distribution for errors and the log link, are useful for data that take the form of counts. The choice of link means that we are modelling frequency ratios rather than, say, frequency differences. To make sense, the fitted values must be non-negative and the log link deals with this. The canonical link is the log and the variance function $V(\mu) = \mu$. The adjusted dependent variable is given by $z = \eta + (y - \mu)/\mu$ and the iterative weight by μ.

Binomial   Data in the form of proportions, where each observation takes the form of r cases responding out of m, are naturally modelled by the binomial distribution for errors. The canonical link is $\text{logit}(\pi) = \log(\pi/(1 - \pi))$, in which effects are assumed additive on the log-odds scale. Two other links are in common use: the probit link based on the cumulative normal distribution, and the complementary log-log (CLL) link based on the extreme value distribution. For many data sets the fit from using the probit scale will be close to that from the logit. Both these links are symmetrical, whereas the complementary log-log link is not; as π approaches one, the CLL link approaches infinity much more slowly than either the probit or the logit. It has proved useful in modelling plant infection rates involving a large population of spores, each with a very small chance of infecting a plant.

Multinomial   Polytomous data, where each individual is classified in one of k categories of some response factor R, and classified by explanatory factors A, B, C, say, can be analysed by using a special log-linear model. We define F as a factor with levels for each combination of A, B and C. Any analysis must be conditional on the F margin, and we fix the F margin by fitting it as a minimal model. This constrains the fitted F totals to equal the corresponding observed totals. If we compare F with $F \cdot R$, we are looking for overall constant relative frequencies of the response in respect of variation in A, B and C. To look at the effect of A on the relative frequencies of R, add the term $R \cdot A$ to the linear predictor. To test for additive effects (on the log scale) of A, B and C, fit $F + R \cdot (A + B + C)$. If the response factor has only two levels, fitting a log-linear model of the above type is entirely equivalent to fitting a binomial model to the first level of R with the marginal total as the binomial denominator. For more than two levels of response, we must use the log-linear form, and there is an assumption that the levels of R are unordered. For response factors with ordered levels, it is better to use models based on cumulative frequencies (McCullagh, 1980; McCullagh and Nelder, 1989).

Gamma   The gamma distribution is continuous and likely to be useful for modelling non-negative continuous responses. These are indeed more common than responses that cover the whole real line. It has been common to deal with the former by transforming the data by taking logs. In GLMs, we transform the mean μ and the gamma distribution is a natural choice for the error distribution. Its variance function is $V(\mu) = \mu^2$. The canonical link is given by $\eta = 1/\mu$, but the log link is frequently used. Analyses with a log transformation of the data and normal errors usually give parameter estimates very close to those derived from a GLM with gamma errors and a log link. Because η must be positive, the values of each term in the linear predictor should also be positive, so that if a covariate x is positive, the corresponding β must also be positive. In gamma models, the proportional standard deviation is constant, so that ideally any rounding errors in the response should also be proportional. This is very seldom the case in which case weights calculated for small values of the response will be relatively too large. There is a useful set of response surfaces for use with the canonical link, namely inverse polynomials. These take the form $x/\mu = P(x)$, where P denotes a polynomial in x. The inverse linear curve goes through the origin and rises to an asymptote, whereas the inverse quadratic rises to a maximum and then falls to zero. Generalizations to polynomials in more than one variable are straightforward; see McCullagh and Nelder (1989) for more details and examples.

Inverse Gaussian This distribution is rarely used as a basis for GLM errors, mainly because data showing a variance function of $V(\mu) = \mu^3$ seem to be rare. We shall not consider such models further here.

2.3 Model checking

In simple (or simplistic) data analyses, the sequence of operations is a one-way process and takes the form

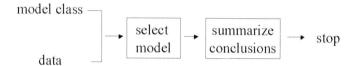

The underlying assumption here is that the model class chosen is correct for the data, so that all that is required is to fit the model by maximizing a criterion such as maximum likelihood, and present estimates with their standard errors. In more complex (or more responsible) analyses, we need to modify the diagram to the following:

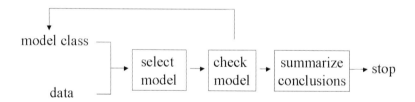

The introduction of the feedback loop for model selection changes the process of analysis profoundly.

The analysis process above consists of two main activities: the first is model selection, which aims to find parsimonious well-fitting models for the basic responses measured, and the second is model prediction, where the output from the primary analysis is used to derive summarizing quantities of interest together with their uncertainties (Lane and Nelder, 1982). In this formulation, it is clear that summarizing statistics are quantities of interest belonging to the prediction stage, and thus cannot serve as a response in model selection.

Discrepancies between the data and the fitted values produced by the model fall into two main classes: isolated and systematic.

2.3.1 Isolated discrepancy

Isolated discrepancies appear when only a few observations have large
residuals. Such residuals can occur if the observations are simply wrong,
for instance where 129 has been recorded as 192. Such errors are under-
standable if data are hand recorded, but even automatically recorded
data are not immune. Robust methods were introduced partly to cope
with the possibility of such errors; for a description of robust regression
in a likelihood context see Pawitan (2001, Chapters 6 and 14). Observations with large residuals are systematically downweighted so that the
more extreme the value, the smaller the weight it gets. Total rejection of
extreme observations (*outliers*) can be regarded as a special case of robust methods. Robust methods are data driven, and to that extent they
may not indicate any causes of the discrepancies. A useful alternative is
to model isolated discrepancies as caused by variation in the dispersion
and seek covariates that may account for them. The techniques of joint
modelling of mean and dispersion developed in this book (see Chapter 3)
make such exploration straightforward. Furthermore, if a covariate can
be found to account for the discrepancies, a model-based solution which
can be checked in the future.

Outliers are observations which have large discrepancies on the y-axis.
For the x-axis there is a commonly used measure, the so-called *leverage*.
In linear models, it can be measured by the diagonal elements of the
projection or *hat matrix*,

$$P_X = X(X^t X)^{-1} X^t,$$

an element of which is

$$q_i = x_i^t (X^t X)^{-1} x_i,$$

where x_i^t is the ith row of the model matrix X. The leverage can be
extended to GLMs by using the diagonal elements of

$$X(X^t W X)^{-1} X^t W,$$

the element of which is

$$q_i = x_i^t (X^t W X)^{-1} x_i w_i,$$

where w_i is the ith diagonal element of the GLM weight W.

If estimates from a regression analysis, for example the slope estimate,
change greatly by a deletion of data points we say that these are *influential*. Outliers or data points with large leverages tend to be potentially
influential.

2.3.2 Systematic discrepancy

Systematic discrepancies in the fit of a model imply that the model is deficient rather than the data. There are several types of systematic discrepancy, some of which may mimic the effects of others. For this reason, it is hard, perhaps impossible, to give a foolproof set of rules for identifying the different types. Consider, for example, a simple regression model with a response y and a single covariate x. We fit a linear relation with constant variance normal errors: discrepancies in the fit might require any of the following:

(a) Replacing x by $f(x)$ to produce linearity

(b) The link for y should not be the identity.

(c) Both (a) and (b) should be transformed to give linearity.

(d) The errors are non–normal and require a different distribution.

(e) The errors are not independent and require specification of some kind of correlation between them.

(f) An extra term should be added in the mode.

GLMs allow for a series of checks on different aspects of the model. Thus we can check the assumed form of the variance function, the link or the scale of the covariates in the linear predictor. A general technique is to embed the assumed value of, say, the variance function in a family indexed by a parameter, fit the extended model and compare the best fit with respect to the original fit for a fixed value of the parameter.

2.3.3 Model checking plots

Residuals based on
$$r = y - \hat{\mu}$$
play a major role in model checking for normal models. Different types of residuals have been extended to cover GLMs. These include standardized, Studentized and deletion residuals. We propose to use Studentized residuals from component GLMs for checking assumptions about components. Note that
$$\mathrm{var}(r) = \phi(1 - q),$$
so that a residual with a high leverage tends to have small variance. The Studentized residuals for normal models are
$$r = \frac{y - \hat{\mu}}{\sqrt{\phi(1 - q)}}.$$

The Studentized Pearson residual for GLMs is given by

$$r_p^s = \frac{r_p}{\sqrt{\phi(1-q)}} = \frac{y - \hat{\mu}}{\sqrt{\phi V(\hat{\mu})(1-q)}}.$$

Similarly, the Studentized deviance residual for GLMs is given by

$$r_d^s = \frac{r_D}{\sqrt{\phi(1-q)}}.$$

In this book we use deviance residuals since they give a good approximation to normality for all GLM distributions (Pierce and Schafer, 1986), excluding extreme cases such as binary data. With the use of deviance residuals, the normal probability plot can be used for model checking.

We apply the model checking plots of Lee and Nelder (1998) to GLMs. In a normal probability plot, ordered values of Studentized residuals are plotted against the expected order statistics of the standard normal sample. In the absence of outliers, this plot is approximately linear. Besides the normal probability plot for detecting outliers, two other plots are used: (a) the plot of residuals against fitted values on the constant information scale (Nelder, 1990), and (b) the plot of absolute residuals similarly.

For a satisfactory model, these two plots should show running means that are approximately straight and flat. Marked curvature in the first plot indicates either an unsatisfactory link function or missing terms in the linear predictor or both. If the first plot is satisfactory, the second plot may be used to check the choice of variance function for the distributional assumption. If, for example, the second plot shows a marked downward trend, this implies that the residuals are falling in absolute value as the mean increases, i.e., the assumed variance function is increasing too rapidly with the mean.

We also use the histogram of residuals. If the distributional assumption is correct, it shows symmetry provided the deviance residual is the best normalizing transformation. GLM responses are independent, so that these model checking plots assume that residuals are almost independent. Care will be necessary when we extend these residuals to correlated errors in later chapters.

2.3.4 Specification of GLMs

Throughout the remaining chapters, we build extended model classes for which GLMs provide the building blocks. Each component GLM has as attributes a response variable y, a variance function $V()$, a link function

$g()$, a linear predictor $X\beta$ and a prior weight $1/\phi$ and its attributes are shown in Table 2.2. This means that all the GLM procedures in this section for model fitting and checking can be applied to the extended model classes.

Table 2.2 *Attributes for GLMs.*

Components	β (fixed)
Response	y
Mean	μ
Variance	$\phi V(\mu)$
Link	$\eta = g(\mu)$
Linear predictor	$X\beta$
Deviance	d
Prior weight	$1/\phi$

$$d = 2 \int_{\hat{\mu}}^{y} (y - s)/V(s)\ ds$$

2.4 Examples

There are many available books on GLMs and their application to data analysis, including Agresti (2002), Aitkin *et al.* (1989), McCullagh and Nelder (1989), Myers *et al.* (2002). However, in this section we briefly illustrate likelihood inferences from GLMs because these tools will be needed for the extended model classes.

2.4.1 The stack loss data

This data set, first reported by Brownlee (1960), is the most analysed in the regression literature and is famous for producing outliers; there is even a canonical set, namely observations 1, 3, 4, ... Dodge (1997) found that 90 distinct papers have been published on the data set, and that the union of all the sets of apparent outliers contains all but 5 of the 21 observations!

The data consist of 21 units with 3 explanatory variables x_1, x_2, x_3 and a response. They come from a plant for the oxidation of ammonia to nitric acid, the response being the percentage of the ingoing ammonia lost escaping with the unabsorbed nitric oxides, called the stack loss.

The response is multiplied by 10 for analysis. The three explanatory variables are x_1 = airflow, x_2 = inlet temperature of the cooling water in degrees C and x_3 = acid concentration, coded by subtracting 50 and multiplying by 10.

Brownlee, in his original analysis, fitted a linear regression with linear predictor $x_1+x_2+x_3$, dropping x_3 after t-tests on individual parameters. Figure 2.1 shows the model checking plots for this model; they have several unsatisfactory features. The running mean in the plot of residuals against fitted values shows marked curvature, and the plot of absolute residuals has a marked positive slope, indicating that the variance is not constant but is increasing with the mean. The normal plot shows a discrepant unit, no. 21, whose residual is -2.64 as against the expected -1.89 from expected ordered statistics. In addition, the histogram of residuals is markedly skewed to the left. These defects indicate something wrong with the model, rather than the presence of a few outliers.

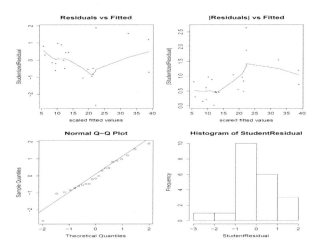

Figure 2.1 *The model checking plots for the linear regression model of the stack loss data.*

We seek to remove these defects by moving to a GLM with gamma errors and a log link. The additive model is still unsatisfactory, and we find that the cross term x_1x_3 is also required. The model checking plots are appreciably better than for the normal model and can be further improved by expressing x_1 and x_3 on the log scale to match the response (x_2, being a temperature, has an arbitrary origin). The resulting plots are shown in Figure 2.2. The approach in this analysis has been to include as much variation in the model as possible, as distinct from downweighting individual yields to produce an unsuitable model fit.

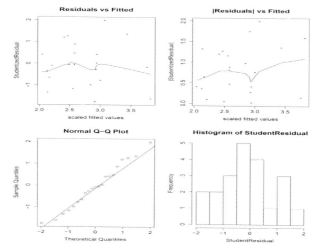

Figure 2.2 *The model checking plots for the gamma GLM of the stack loss data.*

Statistics from the gamma GLM are given in Table 2.3.

Table 2.3 *Results from gamma fit.*

	Estimate	Standard error	t test result
Constant	-244.53	113.94	-2.15
$\log(x_1)$	61.69	28.61	2.16
x_2	0.07	0.02	4.04
$\log(x_3)$	52.91	25.50	2.08
$\log(x_1) \cdot \log(x_3)$	-13.29	6.41	-2.08

2.4.2 Job satisfaction

The data are taken from Stokes *et al.* (Section 14.3.5, 1995) and related to job satisfaction. The response is the job satisfaction of workers (W) and possible explanatory factors are quality of the management (M) and the job satisfaction of the supervisor (S). The data are in Table 2.4 with the sequential analysis of deviance in Table 2.5.

Both the explanatory factors have marked effects, and act additively on the log scale, as is shown by the small size of the last deviance. In Table 2.4 by fixing margins of M·S we can fit the logistic model

Table 2.4 *Job satisfaction data.*

Management quality	Supervisor job satisfaction	Worker job satisfaction	
		Low	High
Poor	Low	103	87
Poor	High	32	42
Good	Low	59	109
Good	High	78	205

Table 2.5 *Analysis of deviance.*

Log-linear terms	Deviance	Degrees of freedom	Binomial terms
M · S (minimal)	76.89	4	
M · S + W	35.60	3	Intercept
M · S + W · M	5.39	2	M
M · S + W · S	19.71	2	S
M · S + W · (M + S)	0.07	1	M + S

M = Management quality, S = Supervisor job satisfaction,
W = Worker job satisfaction.

with the term M+S, given in Table 2.6. This model can be fitted by the Poisson model with the term M·S+W·(M+S) as in Table 2.7; the last three rows of the table correspond to the logistic regression fit. Here we treat the low level of W as the baseline, so that the estimates have opposite signs. There are two important things to note. The first is that by conditioning the number of terms is reduced in the binomial regression. This means that the nuisance parameters can be effectively eliminated by conditioning. The second is that the conditioning does not result in a loss of information in the log-linear models; the analysis from the Poisson and binomial GLMs is identical, having the same deviance etc. This is not so in general beyond GLMs. For further analysis of this example, see Nelder (2000).

2.4.3 General health questionnaire score

Silvapulle (1981) analysed data from a psychiatric study to investigate the relationship between psychiatric diagnosis (as *case*, requiring psychi-

Table 2.6 *Results from logistic fit.*

	Estimate	Standard error	t test result
Constant	0.15	0.13	1.16
M	−0.75	0.17	−4.42
S	−0.39	0.17	−2.31

M = Management quality, S = Supervisor job satisfaction.

Table 2.7 *Results from Poisson fit*

	Estimate	Standard error	t test error
Constant	4.63	0.09	48.83
M	−0.54	0.14	−3.74
S	−1.14	0.16	−6.97
M · S	1.40	0.17	8.19
W	−0.15	0.13	−1.16
W · M	0.75	0.17	4.42
W · S	0.39	0.17	2.31

M = Management quality, S = Supervisor job satisfaction,
W = Worker job satisfaction.

atric treatment or as a *non-case*) and the value of the score on a 12-item general health questionnaire (GHQ), for 120 patients attending a clinic. Each patient was administered the GHQ, resulting in a score between 0 and 12, and was subsequently given a full psychiatric examination by a psychiatrist who did not know the patient's GHQ score. The patient was classified by the psychiatrist as a case or a non-case. The GHQ score could be obtained from the patient without need for trained psychiatric staff. The question of interest was whether the GHQ score could indicate the need of psychiatric treatment. Specifically, given the value of GHQ score for a patient, what can be said about the probability that the patient is a psychiatric case? Patients were heavily concentrated at the low end of the GHQ scale, where the majority were classified as non-cases in Table 2.8 below. A small number of cases spread over medium and high values. Sex of the patient was an additional variable.

Though the true cell probabilities would be positive, some cells are empty due to the small sample size. Suppose that we fit a logistic regression. Cells with zero total for both C (case) and NC (non-case) are uninfor-

Table 2.8 *Number of patients classified by GHQ score and outcome of standardized psychiatric interview.*

GHQ score		0	1	2	3	4	5	6	7	8	9	10	11	12
M	C	0	0	1	0	1	3	0	2	0	0	1	0	0
	NC	18	8	1	0	0	0	0	0	0	0	0	0	0
	Total	18	8	2	0	1	3	0	2	0	0	1	0	0
F	C	2	2	4	3	2	3	1	1	3	1	0	0	0
	NC	42	14	5	1	1	0	0	0	0	0	0	0	0
	Total	44	16	9	4	3	3	1	1	3	1	0	0	0

M = Male, F = Female, C = Case, NC = Non-case.

mative in likelihood inferences and are not counted as observations in logistic regression. Suppose that we fit the model

$$\text{SEX+GHQX},$$

where GHQX is the GHQ score as a continuous covariate and SEX is the factor for sex. This gives a deviance 4.942 with 14 degrees of freedom. This suggests that the fit is adequate, although large-sample approximations of the deviance are of doubtful value here. The results of the fitting are given in Table 2.9. For sex effects, males are set as the reference group, so the positive parameter estimate means that females have a higher probability of being cases but the result is not significant. The GHQ score covariate is significant; a one-point increase in the GHQ score results in an increase of the odds $p/(1-p)$ for 95 % confidence of

$$\exp(1.433 \pm 1.96 * 0.2898) = (2.375, 7.397).$$

We also fit the model

$$\text{SEX+GHQI},$$

where GHQI is the GHQ score as a factor (categorical covariate). This gives a deviance 3.162 with five degrees of freedom, so that a linear trend model is adequate.

We can fit the logistic regression above by using a Poisson model with a model formula

$$\text{SEX} \cdot \text{GHQI} + \text{CASE} \cdot (\text{SEX+GHQX}),$$

where the factor CASE is equal to 1 for case and 2 for non-case. The term SEX·GHQI is the minimal model needed to fix the model-based

Table 2.9 *Results from logistic fit.*

	Estimate	Standard error	t test result
Constant	−4.07	0.98	−4.18
SEX	0.79	0.93	0.86
GHQX	1.43	0.29	4.94

Table 2.10 *Results from Poisson fit.*

	estimate	s.e.	t
Constant	−1.20	0.99	−1.21
SEX	1.67	0.95	1.75
GHQI	0.57	0.52	1.10
GHQI	0.42	0.95	0.45
⋮			
CASE	4.07	0.98	4.17
CASE · SEX	−0.79	0.93	−0.85
CASE · GHQX	−1.43	0.29	−4.93

estimated totals to be the same as the observed totals in Table 2.8, while the second term gives the equivalent model for SEX + GHQX in the logistic regression. The result is given in Table 2.10, where the last three lines correspond to Table 2.9. with opposite signs because cases are the baseline group; to get the same signs we can set case = 2 and noncase = 1. We can see slight differences in t-values between two fits.

Note that when a marginal total is zero, the cells contributing to that total have linear predictors tending to $-\infty$ in the GLM algorithm and hence zero contributions to the deviance. However, the software reports the deviance as 4.942 with the 23 degrees of freedom and non-zero deviances for cells with zero marginal totals because estimates cannot reach $-\infty$ in a finite number of iterations and the software reports wrong degrees of freedom. Thus the apparent 23 degrees of freedom for the deviance should be reduced by 9 to give 14, the same as in the logistic regression. To avoid such an embarrassment, cells with zero marginal totals should not be included in the data. We can see that results equivalent to the logistic regression can be obtained by using the Poisson fit.

2.4.4 Ozone relation to meteorological variables

The data give ozone concentrations in relation to nine meteorological variables for 330 units. These data have been used by Breiman (1995), among others. Breiman used them to illustrate a model selection technique which he called the nn-garotte, and he derived the following models from subset selection and his new method, respectively:

$$x_6 + x_2 x_4 + x_2 x_5 + x_4^2 + x_6^2 \tag{2.4}$$

and

$$x_1 + x_5 + x_2^2 + x_4^2 + x_6^2 + x_2 x_4 + x_5 x_7. \tag{2.5}$$

Neither of these polynomials is well formed; for example they contain product and squared terms without the corresponding main effects. If we now add in the missing marginal terms we get:

$$x_2 + x_4 + x_5 + x_6 + x_2 x_4 + x_2 x_5 + x_4^2 + x_6^2 \tag{2.6}$$

and

$$x_1 + x_2 + x_4 + x_5 + x_6 + x_7 + x_2^2 + x_4^2 + x_6^2 + x_2 x_4 + x_5 x_7. \tag{2.7}$$

The table below shows the residual sum of squares (RSS), the degrees of freedom and the residual mean square (RMS) for the four models.

Model	RSS	d.f.	RMS	C_p
(2.4)	6505	324	20.08	6727.40
(2.5)	6485	322	20.14	6781.26
(2.6)	5978	321	18.62	6311.37
(2.7)	5894	318	18.53	6338.76

To compare the fits, we must make allowance for the fact that the well formed models have more parameters than the original ones. In this standard regression setting, we can use the C_p criterion

$$C_p = \text{RSS} + 2p\sigma^2,$$

where p is the number of parameters and σ^2 is the error variance. The most unbiased estimate for σ^2 is given by the smallest RMS in the table, namely 18.53. The C_p criterion indicates that the well formed models give substantially better fits. Two of the xs, namely x_4 and x_7, are temperatures in degrees Fahrenheit. To illustrate the effect of rescaling the x variables on non–well formed models, we change the scales of x_4 and x_7 to degrees Celsius. The RSS for the model (2.5) is now 6404, whereas that for the model (2.7) remains at 5894.

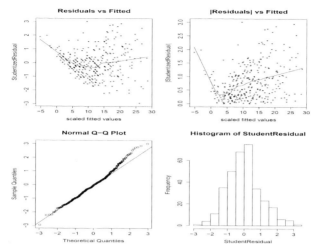

Figure 2.3 *The model checking plots for model (2.4).*

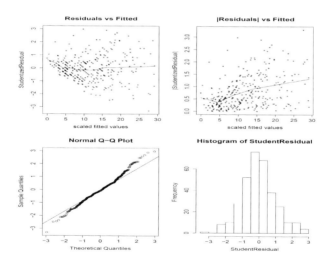

Figure 2.4 *The model checking plots for model (2.5).*

This example shows that no advantage is obtained by not using well formed polynomials; however, there is a more serious complaint to be made about the models fitted so far. Model checking plots show clearly that the assumption of normality of the errors with constant variance is incorrect. For example, all the four models in the table above produce some negative fitted values for the response, which by definition must be non-negative. We try using a GLM with a gamma error and a log link, and we find that a simple additive model

$$x_2 + x_4 + x_7 + x_8 + x_9 + x_8^2 + x_9^2 \qquad (2.8)$$

with five linear terms and two quadratics and no cross terms gives good model selection plots and, of course, no negative fitted values for the response.

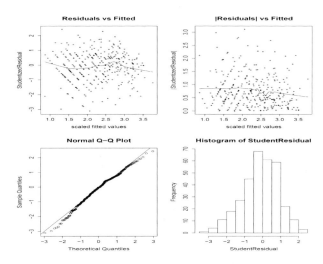

Figure 2.5 *The model checking plots for model (2.8).*

Because the gamma is not a standard linear model, we use the AIC for model comparison, and Table 2.11 strongly indicates the improvement in fit from the gamma GLM over the normal models.

Table 2.11 *Akaike information criterion (AIC).*

Model	(2.4)	(2.5)	(2.6)	(2.7)	(2.8)
AIC	1932.4	1935.4	1910.5	1912.0	1741.4

Table 2.12 *Results from model (2.8).*

	Estimate	Standard error	t test result
Constant	1.211	0.144	8.398
x_2	-0.0301	0.0101	-2.986
x_4	0.01003	0.00352	2.853
x_7	0.003348	0.000583	5.742
x_8	-0.005245	0.000934	-5.612
x_9	0.009610	0.00113	8.473
x_8^2	0.00001125	0.00000280	4.023
x_9^2	-0.00002902	0.00000297	-9.759

CHAPTER 3

Quasi–likelihood

One of the few points on which theoretical statisticians of all persuasions are agreed is the role played by the likelihood function in statistical modelling and inference. We have devoted the first two chapters to illustrating this point. Given a statistical model, we prefer to use likelihood inferences. However, there are many practical problems for which a complete probability mechanism (statistical model) is too complicated to specify fully or is not available, except perhaps for assumptions about the first two moments, hence precluding a classical likelihood approach.

Typical examples are structured dispersions of non-Gaussian data for modelling the mean and dispersion. These are actually quite common in practice: for example, in the analysis of count data the standard Poisson regression assumes the variance is equal to the mean: $V(\mu) = \mu$. However, often there is extra-Poisson variation so we would like to fit $V(\mu) = \phi\mu$ with $\phi > 1$, but it is now no longer obvious what distribution to use. In fact, Jørgensen (1986) showed no GLM family on the integers that satisfies this mean-variance relationship, so there is no simple adjustment to the standard Poisson density to allow overdispersion. Wedderburn's (1974) quasi–likelihood approach deals with this problem, and the analyst needs to specify only the mean-variance relationship rather than a full distribution for the data.

Suppose we have independent responses $y_1, \ldots, y_n$ with means $E(y_i) = \mu_i$ and variance $\mathrm{var}(y_i) = \phi V(\mu_i)$, where μ_i is a function of unknown regression parameters $\beta = (\beta_1, \ldots, \beta_p)$ and $V()$ is a known function. In this chapter, given the variance function $V()$, we study the use of the quasi–likelihood function analogous to a likelihood from the GLM family for inferences from models that do not belong to the GLM family.

Wedderburn defined the quasi–likelihood (QL, strictly a quasi–loglihood) as a function $q(\mu_i; y_i)$ satisfying

$$\frac{\partial q(\mu_i; y_i)}{\partial \mu_i} = \frac{y_i - \mu_i}{\phi V(\mu_i)}, \tag{3.1}$$

and, for independent data, the total quasi–likelihood is $\sum_i q(\mu_i; y_i)$. The regression estimate $\hat{\beta}$ satisfies the GLM-type score equations

$$\sum_i \frac{\partial q(\mu_i; y_i)}{\partial \beta} = \sum_i \frac{\partial \mu_i}{\partial \beta} \frac{(y_i - \mu_i)}{\phi V(\mu_i)} = 0. \qquad (3.2)$$

It is possible to treat the quasi–likelihood approach simply as an estimating equation approach, i.e., not considering the estimating function as a score function. We still derive the estimate using the same estimating equation, but for inference we do not use likelihood-based quantities such as the likelihood ratio statistic, and instead rely on the distributional properties of the estimate directly.

We shall investigate the use of the quasi–likelihood approach in several contexts:

- There exists an implied probability structure, a quasi–distribution from a GLM family of distributions that may not match the underlying distribution. For example, the true distribution may be the negativebinomial, while the quasi–distribution is Poisson. Also, a quasi–distribution might exist on a continuous scale when the true distribution is supported on a discrete scale or vice versa.

- There does not exist an implied probability structure, but a quasi–likelihood is available, i.e., there exists a real valued function $q(\mu_i; y_i)$, whose derivatives are as in (3.2).

- The estimating equations (3.2) can be further extended to correlated responses; a real valued function $q(\mu_i; y_i)$ may not even exist.

The original quasi–likelihood approach was developed to cover the first two contexts and has two notable features:

- In contrast to the full likelihood approach, we are not specifying any probability structure, but only assumptions about the first two moments. This relaxed requirement increases the flexibility of the QL approach substantially.

- The estimation is for the regression parameters for the mean only. For a likelihood-based approach to the estimation of the dispersion parameter ϕ, some extra principles are needed.

Wedderburn (1974) derived some properties of QL, but note that his theory assumes ϕ is known; in the following it is set to unity. With this assumption we see that the quasi–likelihood is a true loglihood if and only if the response y_i comes from a one-parameter exponential family model (GLM family with $\phi = 1$) with log-density

$$q(\mu; y) = \theta y - b(\theta) + c(y), \qquad (3.3)$$

where $\mu = b'(\theta)$ and $V(\mu) = b''(\theta)$. Choice of a particular mean-variance relationship is equivalent to choosing a function $b(\theta)$. In other words, the quasi–distribution associated with the quasi–likelihood is in the exponential family, as long as one exists. There is no guarantee that the implied function $b()$ leads to a proper log-density in (3.3). As far as first-order inferences are concerned (e.g., up to the asymptotic normality of the regression estimates), the quasi–likelihood implied by a mean-variance relationship behaves largely like a true likelihood. For example, we have

$$E\left(\frac{\partial q}{\partial \mu}\right) = 0$$

and

$$E\left(\frac{\partial q}{\partial \mu}\right)^2 = -E\left(\frac{\partial^2 q}{\partial \mu^2}\right) = \frac{1}{V(\mu)}.$$

If the true loglihood is $\ell(\mu)$, by the Cramér-Rao lower-bound theorem,

$$-E\left(\frac{\partial^2 q}{\partial \mu^2}\right) = \frac{1}{V(\mu)} \leq -E\left(\frac{\partial^2 \ell}{\partial \mu^2}\right),$$

with equality, if the true likelihood has the exponential family form. If ϕ is not known, the quasi–distribution is in general not in the exponential family. We discuss GLM families with unknown ϕ in Section 3.4.

Although in the QL approach we specify only the mean-variance relationship, the adopted estimating equation implies a quasi-distribution that has specific higher-order moments. Lee and Nelder (1999) pointed out that since the QL estimating equations are the score equations derived from a QL, the shape of the distribution follows approximately a pattern of higher-order cumulants that would arise from a GLM family if one existed. Among distributions having a given mean-variance relationship, the GLM family has a special position as follows:

- We can think of $-E(\partial^2 q/\partial \mu^2)$ as the information available in y concerning μ when only the mean-variance relationship is known. In this informational sense, the GLM family is the weakest assumption of distribution we can make in that it uses no information beyond the mean-variance relationship.

- The QL equations for the mean parameters involve only $(y_i - \mu_i)$ terms, not higher order terms $(y_i - \mu_i)^d$ for $d = 2, 3, \dots$. Among estimating equations involving only $(y_i - \mu_i)$, the QL equations are optimal in providing asymptotically minimum variance estimators (McCullagh, 1984). If there exists a GLM family with a specified mean-variance relationship, QL estimators are fully efficient under this GLM family. However, when the true distribution does not belong to the GLM family, the QL estimator may lose some efficiency.

Recovery of information is possible from the higher-order terms (Godambe and Thompson, 1989). Lee and Nelder (1999) showed under what circumstances the QL estimators have low efficiency.

- Finally, the ML estimator can be said to use all the information available if the true model were known, while given the mean-variance relationship only, the QL estimators are most robust against misspecification of skewness (Lee and Nelder, 1999).

Some of the most commonly used variance functions and associated exponential family models are given in Table 3.1.

3.1 Examples

One-sample problems

Estimating a population mean is the simplest non–trivial statistical problem. Given an iid sample $y_1, \ldots, y_n$, assume that

$$
\begin{aligned}
E(y_i) &= \mu \\
\mathrm{var}(y_i) &= \sigma^2.
\end{aligned}
$$

From (3.2), $\widehat{\mu}$ is the solution of

$$
\sum_{i=1}^{n}(y_i - \mu)/\sigma^2 = 0,
$$

which yields $\widehat{\mu} = \overline{y}$.

The quasi–distribution in this case is the normal distribution, so for the QL approach to work well, we should check whether it is a reasonable assumption. If the data are skewed, for example, it might be better to use a distribution with a different variance function. Alternatively, we might view the approach simply as an estimating equation approach. This example shows clearly the advantages and disadvantages of estimating equations (EE) and QL, compared with the full likelihood approach:

- The estimate is consistent for a very wide class of underlying distributions, namely any distribution with mean μ. In fact, the sample responses do not even have to be independent.
- We have to base inferences on asymptotic considerations, since there is no small-sample inference. With the QL approach we might use the REML method described in Section 3.6.1 to improve inference.

Table 3.1 *Commonly used variance functions and associated exponential family models.*

Function	$V(\mu)$	$q(\mu;y)$	Restriction
Normal	1	$-(y-\mu)^2/2$	–
Overdispersed Poisson	μ	$y\log\mu - \mu$	$\mu \geq 0,\, y \geq 0$
Overdispersed binomial	$\mu(1-\mu)$	$y\log\left(\frac{\mu}{1-\mu}\right) + \log(1-\mu)$	$0 \leq \mu \leq 1,\, 0 \leq y \leq 1$
Gamma	μ^2	$-y/\mu - \log\mu$	$\mu \geq 0,\, y \geq 0$
Power variance	μ^p	$\mu^{-p}\left(\frac{y\mu}{1-p} - \frac{\mu^2}{2-p}\right)$	$\mu \geq 0,\, y \geq 0,\, p \neq 0, 1, 2$

- There is a potential loss of efficiency compared with full likelihood inference when the true distribution does not belong to the GLM family.

- Even if the true distribution is symmetric, the sample mean is not robust against outliers. However, it is robust against skewed alternatives, but robust estimators proposed for symmetric but heavy-tailed distribution, may not be robust against skewness. Outliers can be caused by a skewness of distribution.

- There is no standard prescription for estimating the variance parameter σ^2. Other principles may be needed, for example, use of the method-of-moments estimate

$$\widehat{\sigma}^2 = \frac{1}{n}\sum_i (y_i - \bar{y})^2,$$

still without making any distributional assumption. Estimating equations can be extended to include the dispersion parameter; see extended quasi–likelihood below. The moment method can encounter a difficulty in semi–parametric models (having many nuisance parameters), while the likelihood approach does not (see Chapter 11).

Linear models

Given an independent sample (y_i, x_i) for $i = 1, \ldots, n$, let

$$\begin{aligned} E(y_i) &= x_i^t\beta \equiv \mu_i(\beta) \\ \mathrm{var}(y_i) &= \sigma_i^2 \equiv V_i(\beta) = V_i. \end{aligned}$$

The estimating equation for β is

$$\sum_i x_i \sigma_i^{-2}(y_i - x_i^t\beta) = 0,$$

giving us the weighted least squares estimate

$$\begin{aligned} \widehat{\beta} &= (\sum_i x_i x_i^t/\sigma_i^2)^{-1}\sum_i x_i y_i/\sigma_i^2 \\ &= (X^t V^{-1} X)^{-1} X^t V^{-1} y, \end{aligned}$$

where X is the $n \times p$ model matrix $[x_1 \ldots x_n]^t$, V the variance matrix $\mathrm{diag}[\sigma_i^2]$, and y the response vector $(y_1, \ldots, y_n)^t$.

Poisson regression

For independent count data y_i with predictor vector x_i, suppose we assume that

$$
\begin{aligned}
E(y_i) &= \mu_i = e^{x_i^t \beta} \\
\mathrm{var}(y_i) &= \phi \mu_i \equiv V_i(\beta, \phi).
\end{aligned}
$$

The estimating equation (3.2) for β is

$$
\sum_i e^{x_i^t \beta} x_i e^{-x_i^t \beta} (y_i - \mu_i)/\phi = 0
$$

or

$$
\sum_i x_i (y_i - e^{x_i^t \beta}) = 0,
$$

requiring a numerical solution (Section 3.2). The estimating equation here is exactly the score equation under the Poisson model. An interesting aspect of the QL approach is that we can use this model even for continuous responses as long as the variance can be reasonably modelled as proportional to the mean.

There are two ways of interpreting this. First, among the family of distributions satisfying the Poisson mean-variance relationship, the estimate based on Poisson quasi–likelihood is robust with respect to the distributional assumption. Second, the estimating-equation method is efficient (i.e., producing a result equal to the best estimate, which is the MLE), if the true distribution is Poisson. This is a typical instance of the robustness and efficiency of the quasi–likelihood method.

Models with constant coefficient of variation

Suppose $y_1, \ldots, y_n$ are independent responses with means

$$
\begin{aligned}
E(y_i) &= \mu_i = e^{x_i^t \beta} \\
\mathrm{var}(y_i) &= \phi \mu_i^2 \equiv V_i(\beta, \phi).
\end{aligned}
$$

The estimating equation (3.2) for β is

$$
\sum_i x_i (y_i - \mu_i)/(\phi e^{x_i^t \beta}) = 0.
$$

This model can be motivated by assuming the responses have a gamma distribution, but is applicable to any outcome where we believe the coefficient of variation is approximately constant. This method is fully efficient if the true model is the gamma among the family of distributions having constant coefficients of variation.

General QL models

With the general quasi–likelihood approach for a response y_i and predictor x_i, we specify known functions $f(\cdot)$ and $V(\cdot)$

$$E(y_i) = \mu_i = f(x_i^t \beta)$$

or

$$g(\mu_i) = x_i^t \beta,$$

where $g(\mu_i)$ is the link function, and

$$\text{var}(y_i) = \phi V(\mu_i) \equiv V_i(\beta, \phi).$$

We can generate a GLM using either the quasi– or full–likelihood approach. The QL extends the standard GLM by (a) allowing a dispersion parameter ϕ to common models such as Poisson and binomial and (b) allowing a more flexible and direct modelling of the variance function.

3.2 Iterative weighted least squares

The main computational algorithm for QL estimates of the regression parameters can be expressed as iterative weighted least squares (IWLS). It can be derived as a Gauss-Newton algorithm to solve the estimating equations. This is a general algorithm for solving nonlinear equations. We solve

$$\sum_i \frac{\partial \mu_i}{\partial \beta} V_i^{-1}(y_i - \mu_i) = 0$$

by first linearizing μ_i around an initial estimate β^0 and evaluating V_i at the initial estimate. Let $\eta_i = g(\mu_i) = x_i^t \beta$ be the linear predictor scale. Then

$$\frac{\partial \mu_i}{\partial \beta} = \frac{\partial \mu_i}{\partial \eta_i} \frac{\partial \eta_i}{\partial \beta} = \frac{\partial \mu_i}{\partial \eta_i} x_i$$

so

$$\begin{aligned}
\mu_i &\approx \mu_i^0 + \frac{\partial \mu_i}{\partial \beta}(\beta - \beta^0) \\
&= \mu_i^0 + \frac{\partial \mu_i}{\partial \eta_i} x_i^t (\beta - \beta^0)
\end{aligned}$$

and

$$y_i - \mu_i = y_i - \mu_i^0 - \frac{\partial \mu_i}{\partial \eta_i} x_i^t (\beta - \beta^0).$$

Putting these into the estimating equation, we obtain

$$\sum_i \frac{\partial \mu_i}{\partial \eta_i} V_i^{-1} x_i \{y_i - \mu_i^0 - \frac{\partial \mu_i}{\partial \eta_i} x_i^t (\beta - \beta^0)\} = 0$$

which we solve for β as the next iterate, giving an updating formula

$$\beta^1 = (X^t \Sigma^{-1} X)^{-1} X^t \Sigma^{-1} z,$$

where X is the model matrix of the predictor variables, Σ is a diagonal matrix with elements

$$\Sigma_{ii} = \left(\frac{\partial \eta_i}{\partial \mu_i}\right)^2 V_i,$$

where $V_i = \phi V(\mu_i^0)$, and z is the adjusted dependent variable

$$z_i = x_i^t \beta^0 + \frac{\partial \eta_i}{\partial \mu_i}(y_i - \mu_i^0).$$

Note that the constant dispersion parameter ϕ is not used in the IWLS algorithm.

3.3 Asymptotic inference

Quasi–likelihood leads to two different variance estimators:

- The natural formula using the Hessian matrix yields efficient estimates when the specification of mean-variance relationship is correct.
- The so-called 'sandwich formula' using the method of moments yields a robust estimate without assuming the correctness of the mean-variance specification.

With QL, as long as the mean function is correctly specified, the quasi–score statistic has zero mean and the estimation of β is consistent. The choice of the mean-variance relationship will affect the efficiency of the estimate.

Assuming correct variance

Assuming independent responses $y_1, \ldots, y_n$ with means $\mu_1, \ldots, \mu_n$ and variance $\text{var}(y_i) = \phi V(\mu_i) \equiv V_i(\beta, \phi)$, the quasi–score statistic $S(\beta) = \partial q / \partial \beta$ has mean zero and variance

$$
\begin{aligned}
\text{var}(S) &= \sum_i \frac{\partial \mu_i}{\partial \beta} V_i^{-1} \text{var}(y_i) V_i^{-1} \frac{\partial \mu_i}{\partial \beta^t} \\
&= \sum_i \frac{\partial \mu_i}{\partial \beta} V_i^{-1} V_i V_i^{-1} \frac{\partial \mu_i}{\partial \beta^t} \\
&= \sum_i \frac{\partial \mu_i}{\partial \beta} V_i^{-1} \frac{\partial \mu_i}{\partial \beta^t} \\
&= X^t \Sigma^{-1} X,
\end{aligned}
$$

using X and Σ as defined previously. The natural likelihood estimator is based upon the expected Hessian matrix

$$-E(\frac{\partial^2 q}{\partial \beta \partial \beta^t}) \equiv D = X^t \Sigma^{-1} X,$$

so the usual likelihood result holds that the variance of the score is equal to the expected second derivative. Using the first-order approximation

$$\begin{aligned} S(\beta) &\approx S(\widehat{\beta}) - D(\beta - \widehat{\beta}) \\ &= -D(\beta - \widehat{\beta}), \end{aligned}$$

since $\widehat{\beta}$ solves $S(\widehat{\beta}) = 0$. So

$$\widehat{\beta} \approx \beta + D^{-1} S(\beta), \tag{3.4}$$

and

$$\text{var}(\widehat{\beta}) \approx (X^t \Sigma^{-1} X)^{-1}.$$

Since $S(\beta)$ is a sum of independent variates, we expect the central limit theorem (CLT) to hold, so that approximately

$$\widehat{\beta} \sim N(\beta, (X^t \Sigma^{-1} X)^{-1}).$$

For the CLT to hold, it is usually sufficient that the matrix $X^t \Sigma^{-1} X$ grows large in some sense, e.g., its minimum eigenvalue grows large as the sample increases.

Example 3.1: (Poisson regression) Suppose we specify the standard Poisson model (dispersion parameter $\phi = 1$) with a log-link function for our outcome y_i:

$$\begin{aligned} E y_i &= \mu_i = e^{x_i^t \beta} \\ \text{var}(y_i) &= V_i = \mu_i. \end{aligned}$$

Then the working variance is

$$\begin{aligned} \Sigma_{ii} &= \left(\frac{\partial \eta_i}{\partial \mu_i}\right)^2 V_i \\ &= (1/\mu_i)^2 \mu_i = 1/\mu_i, \end{aligned}$$

so approximately

$$\widehat{\beta} \sim N(\beta, (\sum_i \mu_i x_i x_i^t)^{-1}),$$

where observations with large means get large weights. □

Not assuming the variance specification is correct

If the variance specification is not assumed to be correct, we get a slightly more complicated formula. The variance of the quasi-score is

$$
\begin{aligned}
\text{var}(S) &= \sum_i \frac{\partial \mu_i}{\partial \beta} V_i^{-1} \text{var}(y_i) V_i^{-1} \frac{\partial \mu_i}{\partial \beta^t} \\
&= X^t \Sigma^{-1} \Sigma_z \Sigma^{-1} X,
\end{aligned}
$$

where Σ_z is the true variance of the adjusted dependent variable, with elements

$$
\Sigma_{zii} = \left(\frac{\partial \eta_i}{\partial \mu_i} \right)^2 \text{var}(y_i).
$$

The complication occurs because $\Sigma_z \neq \Sigma$.

Assuming regularity conditions from (3.4), we expect $\widehat{\beta}$ to be approximately normal with mean β and variance

$$
\text{var}(\widehat{\beta}) = (X^t \Sigma^{-1} X)^{-1} (X^t \Sigma^{-1} \Sigma_z \Sigma^{-1} X)(X^t \Sigma^{-1} X)^{-1}.
$$

This formula does not simplify any further; it is called the sandwich variance formula, which is asymptotically correct even if the assumed variance of y_i is not correct. In practice we can estimate Σ_z by a diagonal matrix with elements

$$
(e_i^*)^2 = \left(\frac{\partial \eta_i}{\partial \mu_i} \right)^2 (y_i - \mu_i)^2
$$

so that the middle term of the variance formula can be estimated by

$$
X^t \widehat{\Sigma^{-1} \Sigma_z \Sigma^{-1}} X = \sum_i \frac{(e_i^*)^2}{\Sigma_{ii}^2} x_i x_i^t.
$$

The sandwich estimator and the natural estimator will be close when the variance specification is near correct and the sample size n is large.

Approach via generalized estimating equations

For independent scalar responses, we developed the quasi–likelihood estimating equations as the score equation from the QL. By replacing the variance with a general covariance matrix of responses in the QL estimating equations (3.2), this approach can be extended to correlated responses (Liang and Zeger, 1986 and Zeger and Liang, 1986). Suppose that y_i is a vector of repeated measurements from the same subject, as we often encounter in longitudinal studies. Let

$$
\begin{aligned}
E(y_i) &= x_i^t \beta \equiv \mu_i(\beta) \\
\text{var}(y_i) &\equiv V_i(\beta, \phi).
\end{aligned}
$$

For simplicity of argument let ϕ be known. Then, the estimating equations (3.2) for the rth element of β can be generalised to produce the generalised estimating equations (GEEs)

$$S(\beta_r) = \sum_i \frac{\partial \mu_i}{\partial \beta_r} V_i^{-1} (y_i - \mu_i) = 0.$$

However, as discussed in McCullagh and Nelder (1989, pp. 333–336), for correlated responses the estimating equation may not be a score equation since the mixed derivatives are not generally equal:

$$\frac{\partial S(\beta_r)}{\partial \beta_s} \neq \frac{\partial S(\beta_s)}{\partial \beta_r},$$

or in terms of the quasi–likelihood q:

$$\frac{\partial^2 q}{\partial \beta_r \partial \beta_s} \neq \frac{\partial^2 q}{\partial \beta_s \partial \beta_r},$$

which is not possible if there exists a real-valued function q. For the quasi–likelihood q to exist, special conditions are required for the function V_i.

In general, only the sandwich estimator is available for the variance estimator of the GEE estimates. To have a sandwich estimator it is not necessary for the variance V_i to be correctly specified. Thus, in the GEE approach, V_i is often called the working correlation matrix. The sandwich estimator may work well in longitudinal studies where the number of subjects is large and subjects are uncorrelated. However, the sandwich estimator is not applicable when the whole response vector y is correlated. For example, it cannot be applied to data from the row-column designs in agriculture experiments: see Lee (2002a). We study the QL models for such cases in Chapter 7.

The GEE approach has been widely used in longitudinal studies because of its robustness; it is typically claimed to provide a consistent estimator for β as long as the mean function is correctly specified (Zeger *et al.*, 1988). However, this has been shown by Crowder (1995) to be incompletely established. Chaganty and Joe (2004) show that the GEE estimator can be inefficient if Σ is very different from Σ_z.

A major concern about the GEE approach is its lack of any likelihood basis. Likelihood plays an important role in model selection and checking. For example, in modelling count data, the deviance test gives a goodness-of-fit test to assess whether the assumed model fits the data. In GLMs, besides the Pearson residuals, the deviance residuals are available and give the best normalized transformation. However, with GEE only the Pearson-type residuals would be available and their distributions are hardly known because they are highly correlated. The likelihood approach also allows non-nested model comparisons using AIC, while with the GEE approach only nested comparisons are possible. Without proper model checking, there is no simple empirical means of discovering whether the regression for the mean has been correctly, or more exactly, adequately specified. Estimates can of course be biased if important covariates are omitted. Lindsey and Lambert (1998) discuss advantages of the likelihood approach over the purely estimating approach of GEE. If there is no true model, it is awkward to say a consistency of GEE estimators: see Lee and Nelder (2004). Consistency presumes an existence of the true statistical model.

It is possible to extend Wedderburn's (1974) QL approach to models with correlated errors while retaining the likelihood basis and yielding orthogonal error components as shown in Chapter 7. All the likelihood methods are available for inference, for example, both natural and sandwich estimators are possible, together with goodness-of-fit tests, REML adjustments and deviance residuals.

3.4 Dispersion models

Wedderburn's original theory of quasi–likelihood assumes the dispersion parameter ϕ to be known, so his quasi–distribution belongs to the one-parameter exponential family. For unknown ϕ, the statement that 'QL is a true loglihood if and only if the distribution is in the exponential family' is not generally correct. In practice, of course, the dispersion parameter is rarely known, except for standard models such as the binomial or Poisson, and even in these cases the assumption that $\phi = 1$ is often questionable. However, the classical QL approach does not tell us how to estimate ϕ from the QL. This is because, in general, the quasi–distribution implied by the QL, having log-density

$$\log f(y_i; \mu_i, \phi) = \frac{y_i \theta_i - b(\theta_i)}{\phi} + c(y_i, \phi), \qquad (3.5)$$

contains a term $c(y, \phi)$ which may not be available explicitly. Jørgensen (1987) called this GLM family the exponential dispersion family, originally investigated by Tweedie (1947). In fact, there is no guarantee that (3.5) is a proper density function. Note that $b(\theta)$ must satisfy analytical conditions to get such a guarantee, but these conditions are too technical to describe here. For example, for the power variance function $V(\mu) = \mu^p$, for $p \neq 1, 2$, from $\mu = b'(\theta)$, we find that the function $b(\theta)$

satisfies the differential equation

$$b''(\theta) = V(b'(\theta)) = \{b'(\theta)\}^p,$$

giving the solution

$$b(\theta) = \alpha^{-1}(\alpha - 1)^{1-\alpha}\theta^\alpha$$

for $\alpha = (p - 2)/(p - 1)$. For $0 < p < 1$, the formula (3.5) is not a proper density; see Jørgensen (1987) for detailed discussion of the power variance model and other related issues.

For the Poisson variance function $V(\mu) = \mu$, we have $b''(\theta) = b'(\theta)$, with the solution $b(\theta) = \exp(\theta)$. For general $\phi > 0$, there exists a proper distribution with this variance function. Let $y = \phi z$, with

$$z \sim \text{Poisson}(\mu/\phi),$$

then y satisfies the variance formula $\phi\mu$ and has a log density of the form (3.5). However, this distribution is supported not on the integer sample space $\{0, 1, ...\}$, but on $\{0, \phi, 2\phi, ...\}$. In general, Jørgensen (1986) showed no discrete exponential dispersion model supported on the integers. This means that in fitting an overdispersed Poisson model using QL, in the sense of using $\text{var}(y_i) = \phi\mu_i$ with $\phi > 1$, there is a mismatch between the quasi–distribution and the underlying distribution.

The function $c(y_i, \phi)$ is available explicitly only in few special cases such as the normal, inverse normal and gamma distributions. In these cases, $c(y_i, \phi)$ is of the form $c_1(y_i) + c_2(\phi)$, so these distributions are in fact in the exponential family. If $c(y, \phi)$ is not available explicitly, even when (3.5) is a proper density, a likelihood-based estimation of ϕ is not immediate. Approximations are needed, and are provided by the extended QL given in the next section.

Double exponential family

Model (3.5) gives us little insight into overdispersed models defined on the natural sample space, for example, the overdispersed Poisson on the integers $\{0, 1, 2, ...\}$. Efron (1986) suggested the double exponential family, which has more intuitive formulae and has similar properties to (3.5). Suppose $f(y; \mu)$ is the density associated with the loglihood (3.3), i.e., the dispersion parameter is set to one. The double exponential family is defined as a mixture model

$$f(y; \mu, \phi) = c(\mu, \phi)\phi^{-1/2}f(y; \mu)^{1/\phi}f(y; y)^{1-1/\phi}. \qquad (3.6)$$

For a wide range of models and parameter values, the normalizing constant $c(\mu, \phi)$ is close to one, so Efron suggested the unnormalized form for practical inference.

For example, a double Poisson model $P(\mu, \phi)$ has an approximate density

$$f(y) = \phi^{-1/2}e^{-\mu/\phi}\left(\frac{e^{-y}y^y}{y!}\right)\left(\frac{\mu e}{y}\right)^{y/\phi}, \quad y = 0, 1, 2, \ldots \quad (3.7)$$

$$\approx \phi^{-1}e^{-\mu/\phi}\frac{(\mu/\phi)^{y/\phi}}{(y/\phi)!}, \quad y = 0, 1, 2, \ldots \quad (3.8)$$

with the factorial replaced by its Stirling approximation. The second formula also indicates that we can approximate $P(\mu, \phi)$ by $\phi P(\mu/\phi, 1)$, both having the same means and variances, but different sample spaces. $P(\mu, \phi)$ is supported on the integers $\{0, 1, \ldots\}$, but $\phi P(\mu/\phi, 1)$ is on $\{0, \phi, 2\phi, \ldots\}$, so some interpolation is needed if we want to match the probabilities on the integers. Also, since $\phi P(\mu/\phi, 1)$ is the quasi–distribution of overdispersed Poisson data, the QL and the unnormalized double exponential family approach should have a close connection. This is made clear in the next section.

Figure 3.1 shows the different approximations of the double Poisson models $P(\mu = 3, \phi = 2)$ and $P(\mu = 6, \phi = 2)$. The approximations are very close for larger μ or larger y.

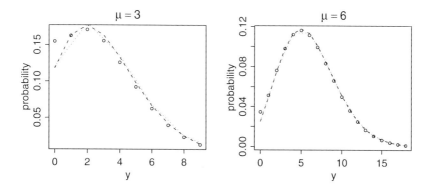

Figure 3.1 *Double Poisson models* $P(\mu = 3, \phi = 2)$ *and* $P(\mu = 6, \phi = 2)$: *the points are computed using formula (3.7), the dashed line using formula (3.8), and the dotted line is computed by interpolating the density of* $\phi P(\mu/\phi, 1)$ *on the odd values and scaling it so that it sums to one.*

Using a similar derivation, the approximate density of the overdispersed binomial $B(n, p, \phi)$ is

$$f(y) = \phi^{-1}\frac{(n/\phi)!}{(y/\phi)![(n-y)/\phi]!}p^{y/\phi}(1-p)^{(n-y)/\phi},$$

for $y = 0, \ldots, n$. The standard binomial and Poisson models have a well known relationship: if y_1 and y_2 are independent Poisson $P(\mu_1, 1)$ and $P(\mu_2, 1)$, the sum $n = y_1 + y_2$ is $P(\mu_1 + \mu_2, 1)$, and conditional on the sum, y_1 is binomial with parameters n and probability $p = \mu_1/(\mu_1 + \mu_2)$. We can show that the overdispersed case gives the same result: if y_1 and y_2 are independent Poisson $P(\mu_1, \phi)$ and $P(\mu_2, \phi)$, the sum $n = y_1 + y_2$ is $P(\mu_1 + \mu_2, \phi)$, and conditional on the sum, the component y_1 is overdispersed binomial $B(n, p, \phi)$ (Lee and Nelder, 2000a).

3.5 Extended quasi–likelihood

Although the standard QL formulation provides consistent estimators for the mean parameters provided the assumed first two-moment conditions hold, it does not include any likelihood-based method for estimating ϕ. Following Wedderburn's original paper, we can use the method of moments, giving

$$\mathrm{var}\left(\frac{y_i - \mu_i}{V(\mu_i)^{1/2}}\right) = \phi,$$

so we expect a consistent estimate

$$\widehat{\phi} = \frac{1}{n-p}\sum_i \frac{(y_i - \mu_i)^2}{V(\mu_i)},$$

where μ_i is evaluated using the estimated parameters, and p is the number of predictors in the model. Alternatively, we might consider the so-called pseudo-likelihood

$$PL(\phi) = -\frac{n}{2}\log\{2\pi\phi V(\widehat{\mu}_i)\} - \frac{1}{2\phi}\sum_i \frac{(y_i - \widehat{\mu}_i)^2}{V(\widehat{\mu}_i)},$$

where $\widehat{\mu}_i$ is computed using the QL estimate. The point estimate of ϕ from the PL is the same as the method-of-moments estimate. In effect, it assumes that the Pearson residuals

$$r_{pi} \equiv \frac{y_i - \widehat{\mu}_i}{V(\widehat{\mu}_i)^{1/2}}$$

are normally distributed. Note that the PL cannot be used to estimate the regression parameters, so that if we use it in conjunction with the quasi–likelihood, we are employing two distinct likelihoods.

However, if we want to use the GLM family (3.5) directly, estimation of ϕ needs an explicit $c(y_i, \phi)$. Nelder and Pregibon (1987) defined an extended quasi–likelihood (EQL) that overcomes this problem. The contribution of y_i to the EQL is

$$Q_i(\mu_i, \phi; y_i) = -\frac{1}{2}\log(2\pi\phi V(y_i)) - \frac{1}{2\phi}d(y_i, \mu_i)$$

and the total is denoted by $q^+ = \sum_i Q_i$, where $d(y_i, \mu_i)$ is the deviance function defined by

$$d_i \equiv d(y_i, \mu_i) = 2 \int_{\mu_i}^{y_i} \frac{y_i - u}{V(u)} du.$$

In effect, EQL treats the deviance statistic as $\phi\chi_1^2$-variate, a gamma variate with mean ϕ and variance $2\phi^2$. This is equivalent to assuming that the deviance residual

$$r_{d_i} \equiv \text{sign}(y_i - \mu_i)\sqrt{d_i}$$

is normally distributed. For one-parameter exponential families (3.3) the deviance residual has been shown to be the best normalizing transformation (Pierce and Schafer, 1986). Thus, we can expect the EQL to work well under GLM families.

The EQL approach allows a GLM for the dispersion parameter using the deviance as data. In particular, in simple problems with a single dispersion parameter, the estimated dispersion parameter is the average deviance

$$\widehat{\phi} = \frac{1}{n} \sum d(y_i, \mu_i),$$

which is analogous to the sample mean $\bar{d}$ for the parameter ϕ. Note that, in contrast with PL, the EQL is a function of both the mean and variance parameters. More generally, the EQL forms the basis for joint modelling of structured mean and dispersion parameters, both within the GLM framework.

To understand when the EQL can be expected to perform well, we can show that it is based on a saddlepoint approximation of the GLM family (3.5). The quality of the approximation varies, depending on the model and the size of the parameter. Assuming y is a sample from model (3.5), at fixed ϕ, the MLE of θ is the solution of

$$b'(\widehat{\theta}) = y.$$

Alternatively $\widehat{\mu} = y$ is the MLE of $\mu = b'(\theta)$. The Fisher information on θ based on y is

$$I(\widehat{\theta}) = b''(\widehat{\theta})/\phi. \tag{3.9}$$

From the equation (1.12) in Section 1.9, at fixed ϕ, the approximate density of $\widehat{\theta}_\phi$ is

$$p(\widehat{\theta}_\phi) \approx \{I(\widehat{\theta}_\phi)/(2\pi)\}^{1/2} \frac{L(\theta, \phi)}{L(\widehat{\theta}_\phi, \phi)}.$$

For simplicity let $\widehat{\theta} = \widehat{\theta}_\phi$. Since $\widehat{\mu} = b'(\widehat{\theta})$, the density of $\widehat{\mu}$, and hence of

y, is

$$
\begin{aligned}
p(y) \;=\; p(\widehat{\mu}) \;&=\; p(\widehat{\theta}) \left| \frac{\partial \widehat{\theta}}{\partial \widehat{\mu}} \right| \\
&=\; p(\widehat{\theta})\{b''(\widehat{\theta})\}^{-1} \\
&\approx\; \{2\pi\phi b''(\widehat{\theta})\}^{-1/2} \frac{L(\theta,\phi)}{L(\widehat{\theta},\phi)},
\end{aligned}
$$

from (3.9). The potentially difficult function $c(x,\phi)$ in (3.5) cancels out in the likelihood ratio term, so we end up with something simpler. The deviance function can be written as

$$
d(y,\mu) = 2\log \frac{L(\widehat{\theta},\phi = 1)}{L(\theta,\phi = 1)},
$$

and $\mu = b'(\theta)$ and $V(y) = b''(\widehat{\theta})$, so

$$
\log p(y) \approx -\frac{1}{2}\log\{2\pi\phi V(y)\} - \frac{1}{2\phi}d(y,\mu).
$$

Example 3.2: There is no explicit quasi–likelihood for the overdispersed Poisson model

$$
\frac{y\log\mu - \mu}{\phi} + c(y,\phi),
$$

since $c(y,\phi)$ is not available. Using

$$
V(y) = y
$$
$$
d(y,\mu) = 2(y\log y - y\log\mu - y + \mu)
$$

we get the extended quasi–likelihood

$$
q^+(\mu,\phi) = -\frac{1}{2}\log\{2\pi\phi y\} - \frac{1}{\phi}(y\log y - y\log\mu - y + \mu).
$$

Nelder and Pregibon (1987) suggest replacing $\log\{2\pi\phi y\}$ by

$$
\log\{2\pi\phi(y + 1/6)\}
$$

to make the formula work for $y \geq 0$. □

EQL turns out to be equivalent to Efron's (1986) unnormalized double exponential family. The loglihood contribution of y_i from an unnormalized double exponential density (3.6) is

$$
\begin{aligned}
\ell(\mu,\phi;y_i) \;&=\; -\frac{1}{2}\log\phi - \frac{1}{\phi}\{f(y_i;y_i) - f(\mu;y_i)\} + f(y_i,y_i) \\
&=\; -\frac{1}{2}\log\phi - \frac{1}{2\phi}d(y_i,\mu_i) + f(y_i,y_i),
\end{aligned}
$$

exactly the EQL up to some constant terms.

Bias can occur with EQL from ignoring the normalizing constant c_i, defined by

$$\int_y c_i \exp(Q_i)dy = 1.$$

For the overdispersed Poisson model with $V(\mu_i) = \phi\mu_i$, Efron (1986) showed that

$$c_i^{-1} \approx 1 + \frac{(\phi-1)}{12\mu_i}\left(1 + \frac{\phi}{\mu_i}\right).$$

This means that $c_i \approx 1$ if $\phi \approx 1$ or if the means μ_i are large enough; otherwise the EQL can produce a biased estimate. For the overdispersed binomial model, the mean is $\mu_i = n_i p_i$, the variance is $\phi n_i p_i(1-p_i)$, and the normalizing constant is

$$c_i \approx 1 + \frac{(\phi-1)}{12n_i}\left(1 - \frac{1}{p_i(1-p_i)}\right),$$

so the approximation is good if ϕ is not too large or if n_i is large enough.

Example 3.3: We consider the Poisson-gamma example, where we can compare the exact likelihood, EQL and pseudo-likelihood. Suppose, conditionally on u, the response $y|u$ is Poisson with mean u, and u itself is gamma with density

$$f(u) = \frac{1}{\Gamma(\nu)}\frac{1}{\alpha}\left(\frac{u}{\alpha}\right)^{\nu-1}\exp(-u/\alpha),\ u > 0,$$

so that

$$\begin{aligned} E(y) &= \mu = \alpha\nu \\ \mathrm{var}(y) &= \alpha\nu + \alpha^2\nu = \mu(1+\alpha). \end{aligned}$$

Thus y is an overdispersed Poisson model with $\phi = 1+\alpha$. The exact distribution of y is not in the exponential family. In fact, we can show that it is the negativebinomial distribution (Pawitan, 2001, Section 4.5).

Now, suppose we observe independent $y_1,\ldots,y_n$ from the above model. The true loglihood is given by

$$\begin{aligned} \ell(\mu,\alpha;y) &= \sum_i\left\{\log\Gamma\left(y_i + \frac{\mu}{\alpha}\right) - \log\Gamma\left(\frac{\mu}{\alpha}\right)\right\} \\ &- \sum_i\left\{\frac{\mu}{\alpha}\log\alpha + \left(y_i + \frac{\mu}{\alpha}\right)\log\left(1 + \frac{1}{\alpha}\right) + \log y_i!\right\}. \end{aligned}$$

Using the deviance derived in Example 3.2, the EQL is

$$q^+(\mu,\alpha) = -\frac{n}{2}\log\{2\pi(1+\alpha)(y+\frac{1}{6})\} - \frac{1}{1+\alpha}\sum_i(y_i\log y_i - y_i\log\mu - y_i + \mu),$$

and the pseudo-likelihood is

$$PL(\mu,\alpha) = -\frac{n}{2}\log\{2\pi\mu(1+\alpha)\} - \frac{1}{2(1+\alpha)}\sum_i\frac{(y_i-\mu)^2}{\mu}.$$

In all cases, the mean μ is estimated by the sample mean.

To be specific, suppose we observe the number of accidents in different parts of the country:

0 0 0 0 0 0 0 1 1 1 1 1 1 1 2 2 2 2 3 3

3 3 4 4 5 5 6 6 7 7 7 8 9 10 10 14 15 20

The average is $\bar{y} = 4.35$ and the variance is 21.0, clearly showing overdispersion. Figure 3.2 shows that the three likelihoods are quite similar in this example. The estimates of α using the true likelihood, the EQL and PL are 4.05 ± 0.82, 3.43 ± 1.01 and 3.71 ± 0.95, respectively. In this example, the dispersion effect is quite large, so the EQL will be biased if the mean is much lower than 4. □

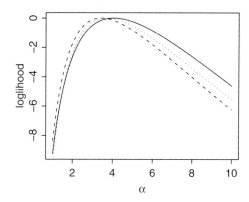

Figure 3.2 *The true negativebinomial likelihood of the dispersion parameter α (solid curve), the corresponding extended quasi–likelihood (dashed) and the pseudo–likelihood (dotted).*

EQL versus PL

The use of a quasi–distribution with a mismatched sample space gives consistent QL estimators for the location parameters β, because the estimating equations from the QL are from the method of moments based upon first-moment assumptions. However, it may not give consistent estimators for the dispersion parameters. We can make the EQL or the equivalent likelihood from the unnormalized double exponential family into a true likelihood by normalizing on a desirable sample space, as in (3.6). In such cases, the normalizing constant is hard to evaluate and depends upon the mean parameter. Furthermore, for this normalized two-parameter family, μ and ϕ are no longer the mean and dispersion parameters. Thus, we prefer the use of the unnormalized form.

The PL estimator can be viewed as the method-of-moments estimator using the Pearson residual sum of squares, while the EQL uses the deviance residual sum of squares. Because the deviance residuals are the best normalizing transformations (Pierce and Schafer, 1986) under the exponential family their use gives efficient (also consistent) estimators (Ha and Lee, 2005a), while the use of the PL can result in loss of efficiency. However, PL gives consistent estimators for a wider range of models because its expectation is based upon assumptions about the first two moments only, and not necessarily on an exponential family distributional assumption. Thus, under the model similar to the negative binomial model, which is not in the exponential family, the use of the unnormalized EQL gives inconsistent estimators, while the use of PL gives consistent but inefficient estimators. We also note that since the negativebinomial model is not in the exponential dispersion family, some loss of efficiency is to be expected from the use of EQL. Using more extensive simulations for various models, Nelder and Lee (1992) showed that in finite samples, EQL estimates often have lower mean-squared errors than PL estimates.

3.6 Joint GLM of mean and dispersion

Suppose we have two interlinked models for the mean and dispersion based on the observed data y and the deviance d:

$$E(y_i) = \mu_i, \ \eta_i = g(\mu_i) = x_i^t \beta, \ \text{var}(y_i) = \phi_i V(\mu_i)$$
$$E(d_i) = \phi_i, \ \xi_i = h(\phi_i) = g_i^t \gamma, \ \text{var}(d_i) = 2\phi_i^2$$

where g_i is the model matrix used in the dispersion model, which is a GLM with a gamma variance function. Now the dispersion parameters are no longer constant, but can vary with the mean parameters.

One key implication is that the dispersion values are needed in the IWLS algorithm for estimating the regression parameters, and that these values have a direct effect on the estimates of the regression parameters. The EQL q^+ yields a fitting algorithm, which can be computed iteratively using two interconnected IWLS:

(a) Given $\widehat{\gamma}$ and the dispersion estimates $\widehat{\phi}_i$s, use IWLS to update $\widehat{\beta}$ for the mean model.

(b) Given $\widehat{\beta}$ and the estimated means $\widehat{\mu}_i$s, use IWLS to update $\widehat{\gamma}$ with the deviances as data.

(b) Iterate Steps 1 and 2 until convergence.

For the mean model in the first step, the updating equation is

$$(X^t \Sigma^{-1} X)\beta = X^t \Sigma^{-1} z,$$

where

$$z_i = x_i^t \beta + \frac{\partial \eta_i}{\partial \mu_i}(y_i - \mu_i)$$

is the adjusted dependent variable and Σ is diagonal with elements

$$\Sigma_{ii} = \phi_i (\partial \eta_i / \partial \mu_i)^2 V(\mu_i).$$

As a starting value, we can use $\phi_i \equiv \phi$, so no actual value of ϕ is needed. Thus, this GLM is specified by a response variable y, a variance function $V()$, a link function $g()$, a linear predictor $X\beta$ and a prior weight $1/\phi$.

For the dispersion model, first compute the observed deviances $d_i = d(y_i, \widehat{\mu}_i)$ using the estimated means. For the moment, let $d_i^* = d_i/(1-q_i)$ with $q_i = 0$. For the REML adjustment, we use the GLM leverage for q_i described in the next section. The updating formula for $\widehat{\gamma}$ is

$$(G^t \Sigma_d^{-1} G)\gamma = G^t \Sigma_d^{-1} z_d,$$

where the dependent variables are defined as

$$z_{di} = g_i^t \gamma + \frac{\partial \xi_i}{\partial \phi_i}(d_i^* - \phi_i)$$

and Σ_d is diagonal with elements

$$\Sigma_{dii} = 2(\partial \xi_i / \partial \phi_i)^2 \phi_i^2.$$

This GLM is characterized by a response d, a gamma error, a link function $h()$, a linear predictor $G\gamma$ and a prior weight $(1-q)/2$. At convergence we can compute the standard errors of $\widehat{\beta}$ and $\widehat{\gamma}$. If we use the GLM deviance, this algorithm yields estimators using the EQL, while with the Pearson deviance it gives those from the PL.

The deviance components d^* become the responses for the dispersion GLM. Then the reciprocals of the fitted values from the dispersion GLM provide prior weights of the next iteration for the mean GLM; these connections are marked in Table 3.2. The resulting see-saw algorithm is very fast to converge. This means that all the inferential tools used for GLMs in Chapter 2 can be used for the GLMs for the dispersion parameters. For example, the model checking techniques for GLMs can be applied to check the dispersion model.

3.6.1 REML procedure for QL models

In estimating the dispersion parameters, if the size of β is large relative to the sample size, the REML procedure is useful in reducing bias. Be-

Table 3.2 *GLM attributes for joint GLMs.*

Component	β (fixed)	γ (fixed)
Response	y	d^*
Mean	μ	ϕ
Variance	$\phi V(\mu)$	ϕ^2
Link	$\eta = g(\mu)$	$\xi = h(\phi)$
Linear predictor	$X\beta$	$G\gamma$
Deviance	d	$\Gamma(d^*, \phi)$
Prior weight	$1/\phi$	$(1-q)/2$

$d = 2\int_{\hat{\mu}}^{y}(y-s)/V(s)\,ds,$
$d^* = d/(1-q),$
$\Gamma(d^*, \phi) = 2\{-\log(d^*/\phi) + (d^* - \phi)/\phi\}.$

This gives the EQL procedure if $q = 0$, and the REML procedure if q is the GLM leverage (Lee and Nelder, 1998).

cause $E(\partial^2 q^+/\partial\beta\partial\phi_i) = 0$, Lee and Nelder (1998) proposed the adjusted profile loglihood 1.14 in Section 1.9,

$$p_\beta(q^+) = [q^+ - \{\log\det(\mathcal{I}(\hat{\beta}_\gamma)/2\pi)\}/2]|_{\beta=\hat{\beta}_\gamma},$$

where $\mathcal{I}(\hat{\beta}_\gamma) = X^t\Sigma^{-1}X$ is the expected Fisher information, $\Sigma = \Phi W^{-1}$, $W = (\partial\mu/\partial\eta)^2 V(\mu)^{-1}$, and $\Phi = \text{diag}(\phi_i)$. In GLMs with the canonical link, satisfying $d\mu/d\theta = V(\mu)$, the observed and expected information matrices are the same. In general they are different. For confidence intervals, the use of observed information is better because it has better conditional properties; see Pawitan (2001, Section 9.6); but the expected information is computationally easier to implement.

The interconnecting IWLS algorithm is as shown earlier except for some modification to the adjusted deviance

$$d_i^* = d_i/(1-q_i)$$

where q_i is the ith diagonal element of

$$X(X^t\Sigma^{-1}X)^{-1}X^t\Sigma^{-1}.$$

The adjusted deviance also leads to a standardized deviance residual

$$r_{di} = \text{sign}(y_i - \mu_i)\sqrt{d_i^*/\phi_i}$$

can be used for model checkings that assuming that they follow the standard normal.

Suppose that the responses y have a normal distribution, i.e., $V(\mu) = 1$. If the β were known, each $d_i^* = (y_i - x_i\beta)^2 = d_i$ would have a prior weight $1/2$, which is reciprocal of the dispersion parameter. This is because

$$E(d_i^*) = \phi_i \text{ and } \text{var}(d_i^*) = 2\phi_i^2$$

and 2 is the dispersion for the $\phi_i\chi_1^2$ distribution, a special case of the gamma. With β unknown, the responses

$$d_i^* = (y_i - x_i\hat{\beta})^2/(1 - q_i)$$

would have a prior weight $(1 - q_i)/2$ because

$$E(d_i^*) = \phi_i \text{ and } \text{var}(d_i^*) = 2\phi_i^2/(1 - q_i).$$

Another intuitive interpretation would be that d_i^*/ϕ_i has approximately χ^2 distribution with $1 - q_i$ degrees of freedom instead of 1, because they have to be estimated. For normal models, our method provides the ML estimators for β and the REML estimators for ϕ. For the dispersion link function $h()$, we usually take the logarithm.

3.6.2 REML method for JGLMs allowing true likelihood

The REML algorithm using EQL gives a unified framework for joint GLMs (JGLMs) with an arbitrary variance function $V()$. However, since the EQL is an approximation to the GLM likelihood, we use the true likelihood for that variance function, if it exists. For example, suppose that the y component follows the gamma GLM such that $E(y) = \mu$ and $\text{var}(y) = \phi\mu^2$; we have

$$-2\log L = \sum\{d_i/\phi_i + 2/\phi_i + 2\log(\phi_i)/\phi_i + 2\log\Gamma(1/\phi_i)\},$$

where $d_i = 2\int_{\mu_i}^{y_i}(y_i - s)/s^2 ds = 2\{(y - \mu)/\mu - \log y/\mu\}$. The corresponding EQL is

$$-2\log q^+ = \sum\{d_i/\phi_i + \log(2\pi\phi_i y_i^2)\}.$$

Note that $\log f(y)$ and $\log q(y)$ are equivalent up to the Stirling approximation

$$\log\Gamma(1/\phi_i) \approx -\log\phi_i/\phi_i + \log\phi_i/2 + \log(2\pi)/2 - 1/\phi_i.$$

Thus, the EQL can give a poor approximation to the gamma likelihood when the value of ϕ is large. It can be shown that $\partial p_\beta(L)/\partial\gamma_k = 0$ leads to the REML method with

$$q_i^* = q_i + 1 + 2\log\phi_i/\phi_i + 2\text{dg}(1/\phi_i)/\phi_i,$$

where dg() is the digamma function.

Example 3.4: Geissler's data on sex ratio consisted of the information on the number of boys in families of size $n = 1, \dots, 12$. In total there were 991,958 families. Families of size $n = 12$ are summarized in Table 3.3. The expected frequencies are computed based on a standard binomial model, giving an estimated probability of 0.5192 for a boy birth. It is clear that there were more families than expected with few or many boys, indicating some clustering within the families or an overdispersion effect.

Table 3.3 *Distribution of number of boys in a family of size 12.*

No. of boys k	0	1	2	3	4	5	6
Observed n_k	3	24	104	286	670	1033	1343
Expected e_k	1	12	72	258	628	1085	1367

No. of boys k	7	8	9	10	11	12
Observed n_k	1112	829	478	181	45	7
Expected e_k	1266	854	410	133	26	2

Hence, for a family of size n, assume that the number of boys is an overdispersed binomial with probability p_n and dispersion parameter ϕ_n. We model the mean and dispersion by

$$\log\left(\frac{p_n}{1 - p_n}\right) = \beta_0 + \beta_1 n$$
$$\log \phi_n = \gamma_0 + \gamma_1 n;$$

Lee and Nelder (2000a) analysed this joint model and showed that the linear dispersion model is not appropriate. In fact, the dispersion parameter decreases as the family size grows to 8, then it increases slightly. We can fit a non-parametric model

$$\log \phi_n = \gamma_0 + \gamma_n,$$

with the constraint $\gamma_1 = 0$. For the mean parameter, Lee and Nelder (2000a) tried a more general mean model

$$\log\left(\frac{p_n}{1 - p_n}\right) = \beta_0 + \beta_n$$

and they found a decrease of deviance 2.29 with nine degrees of freedom, so the linear model is a good fit. For the linear logistic mean model and general dispersion model, they obtained the following estimates:

$$\widehat{\beta}_0 = 0.050 \pm 0.003, \ \widehat{\beta}_1 = 0.0018 \pm 0.0004$$

$$\widehat{\gamma}_0 = 0.43 \pm 0.003, \ \widehat{\gamma}_2 = -0.007 \pm 0.005, \widehat{\gamma}_3 = -0.09 \pm 0.005$$

$$\widehat{\gamma}_4 = -0.17 \pm 0.005, \ \widehat{\gamma}_5 = -0.21 \pm 0.006, \ \widehat{\gamma}_6 = -0.23 \pm 0.006$$

$$\widehat{\gamma}_7 = -0.24 \pm 0.007, \ \widehat{\gamma}_8 = -0.24 \pm 0.008, \ \widehat{\gamma}_9 = -0.23 \pm 0.009$$

$$\widehat{\gamma}_{10} = -0.23 \pm 0.012, \ \widehat{\gamma}_{11} = -0.20 \pm 0.014, \ \widehat{\gamma}_{12} = -0.20 \pm 0.018.$$

Thus it appears that for increasing family size, the rate of boy births increases slightly, but the overdispersion effect decreases. □

3.7 Joint GLMs for quality improvement

The Taguchi method (Taguchi and Wu, 1985) has been widely used for the analysis of quality improvement experiments. It begins by defining summarising quantities, called signal-to-noise ratios (SNRs), depending upon the aim of an experiment, e.g., maximising mean, minimising variance while controlling mean, etc. This SNR then becomes the response to be analysed. However, there are several reasons why this proposal should be rejected.

- The analysis process should consist of two main activities: the first is model selection, which aims to find parsimonious well-fitting models for the basic responses measured, and the second is model prediction, where the output from the primary analysis is used to derive summarising quantities of interest and their uncertainties (Lane and Nelder, 1982). In this formulation, it is clear that SNRs are quantities of interest belonging to the prediction phase, and that Taguchi's proposal inverts the two phases of analysis. Such an inversion makes no inferential sense.

- Suppose that model selection can be skipped because the model is known from past experience. The SNRs are still summarising statistics and the use of them as response variables is always likely to be a relatively primitive form of analysis. Such data reduction would be theoretically justifiable only if it constituted a sufficiency reduction.

- In the definition of SNRs, there appears to be an assumption that if a response y has mean μ, $f(y)$ will be a good estimate of $f(\mu)$. In one of the SNRs the term $10 \log_{10}(\bar{y}^2/s^2)$ appears as an estimate of $10 \log_{10}(\mu^2/\sigma^2)$. As a predictive quantity we would use $10 \log_{10} \hat{\mu}^2/\hat{\sigma}^2$, where $\hat{\mu}$ and $\hat{\sigma}^2$ are derived from a suitable model. Even then it is a moot point whether this revised version is useful for making inferences. What is certain, however, is that SNRs have no part to play in model selection.

To overcome these drawbacks of the original Taguchi method, Box (1988) proposed the use of linear models with data transformation. He regarded the following two criteria as very important:

- **Separation:** the elimination of any unnecessary complications in the model due to functional dependence between the mean and variance (or the elimination of *cross-talk* between location and dispersion effects)
- **Parsimony:** the provision of the simplest additive models

A single data transformation may fail to satisfy the two criteria simultaneously if they are incompatible. Shoemaker *et al.* (1988) suggested that the separation criterion is the more important. In GLMs the variance of y_i is the product of two components; $V(\mu_i)$ expresses the part of the variance functionally dependent on the mean μ_i, while ϕ_i expresses the variability independent of the mean involved. Under a model similar to a gamma GLM, satisfying $\sigma_i^2 = \phi_i \mu_i^2$, the maximisation of the $SNR_i = 10 \log_{10}(\mu_i^2/\sigma_i^2) = -10 \log_{10} \phi_i$ is equivalent to minimising the dispersion ϕ_i, which Leon *et al.* (1987) called performance-measure-independent-of-mean adjustment (PERMIA). This means that the Taguchi method implicitly assumes the mean-variance relationship $\sigma_i^2 = \phi_i \mu_i^2$, without model checking. Thus, we identify a suitable mean-variance relationship $V()$ and then minimize the dispersion under a selected model. The definition of PERMIA depends upon the identification of $V()$ and whether it is meaningful as a dispersion. With GLMs, the separation criterion reduces to the correct specification of the variance function $V()$. Furthermore, these two criteria can be met separately; i.e., the parsimony criterion for both mean and dispersion models can be met independently of the separation criterion by choosing a suitable link function together with covariates.

Suppose that we want to minimize variance among items while holding the mean on target. For minimizing variance, Taguchi classifies the experimental variables into *control* variables that can be easily controlled and manipulated and *noise* variables that are difficult or expensive to control. The Taguchi method aims to find a combination of the control variables at which the system is insensitive to variance caused by the noise variables. This approach is called *robust design*. The term *design* refers not to statistical experimental design, but to the selection of the best combination of the control variables, which minimizes the variance $\phi_i V(\mu_i)$. Given μ_i on target the variance is minimized by minimizing the PERMIA (ϕ_i). Control variables for the mean model, to bring it on target, must not appear in the dispersion model; otherwise adjustment of the mean will alter the dispersion.

For the analysis of quality improvement experiments, the REML procedure should be used because the number of β is usually large compared with the small sample sizes. Also the use of EQL is preferred because it has better finite sampling properties than the pseudo-likelihood method.

3.7.1 Example: injection moulding data

Engel (1992) presented data from an injection moulding experiment, shown in Table 3.5. The responses were percentages of shrinkage of products made by injection moulding. There are seven controllable factors (A through G), listed in Table 3.4, in a 2^{7-4} fractional factorial design. At each setting of the controllable factors, four observations were obtained from a 2^{3-1} fractional factorial with three noise factors (M through O).

Table 3.4 *Factors in experiment.*

Control variables	Noise variables
A: cycle time	M: percentage regrind
B: mould temperature	N: moisture content
C: cavity thickness	O: ambient temperature
D: holding pressure	
E: injection speed	
F: holding time	
G: gate size	

Engel's model included
$$V(\mu) = 1$$
and
$$\eta = \mu = \beta_0 + \beta_A A + \beta_D D + \beta_E E \quad \text{and} \quad \log \phi = \gamma_0 + \gamma_F F. \quad (3.10)$$

The high and low levels of each factor are coded as 1 and -1 respectively. Using a normal probability plot based upon the whole data, with the identity link for the mean, the fifth and sixth observations appear as outliers with residuals of opposite sign; this led Steinberg and Bursztyn (1994) and Engel and Huele (1996) to suspect that the two observations might have been switched during the recording.

After interchanging the two points, they recommended the mean model
$$\eta = \mu = \beta_0 + \beta_A A + \beta_D D + \beta_G G + \beta_{C \cdot N} C \cdot N + \beta_{E \cdot N} E \cdot N. \quad (3.11)$$

Table 3.5 *Experimental data from injection moulding experiment, with cell means $\bar{y}_{i.}$ and standard deviations s_i.*

Controllable factors							Percentage shrinkage for noise factors (M,N,O)			
A	B	C	D	E	F	G	$(1,1,1)$	$(1,2,2)$	$(2,1,2)$	$(2,2,1)$
1	1	1	1	1	1	1	2.2	2.1	2.3	2.3
1	1	1	2	2	2	2	0.3	2.5	2.7	0.3
1	2	2	1	1	2	2	0.5	3.1	0.4	2.8
1	2	2	2	2	1	1	2.0	1.9	1.8	2.0
2	1	2	1	2	1	2	3.0	3.1	3.0	3.0
2	1	2	2	1	2	1	2.1	4.2	1.0	3.1
2	2	1	1	2	2	1	4.0	1.9	4.6	2.2
2	2	1	2	1	1	2	2.0	1.9	1.9	1.8

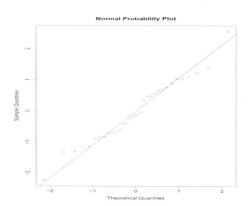

Figure 3.3 *The normal probability plot for mean model (3.11) without switching two points.*

Because the factor F is confounded with C·E they suspected that the dispersion effect F in (3.10) was an artifact caused by ignoring the interactions C·N and E·N in the mean model (3.11). Engel and Huele (1996) found that the factor A, is needed for the dispersion factor (not F).

Because $V(\mu) = 1$, the EQL becomes the normal likelihood. The ML method underestimates the standard errors and so gives larger absolute t-values in both mean and dispersion analysis, compared with the REML analysis, so we use only the REML analysis. The normal probability plot of residuals for Engel and Huele's mean model (3.11) is in Figure 3.4. This shows a jump in the middle, implying that some significant factors may be missing.

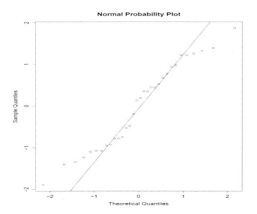

Figure 3.4 *The normal probability plot for mean model (3.11) after switching two points.*

Note that model (3.11) breaks the marginality rules in Chapter 2 which require interactions to have their corresponding main effects included in the model (Nelder, 1994). Lee and Nelder (1997) found that if we add all the necessary main effects in (3.11) to give the mean model with $\eta = \mu$, where

$$\eta = \beta_0 + \beta_A A + \beta_C C + \beta_D D + \beta_G G + \beta_E E + \beta_N N + \beta_{C \cdot N} C \cdot N + \beta_{E \cdot N} E \cdot N \tag{3.12}$$

and dispersion model

$$\log \phi = \gamma_0 + \gamma_A A + \gamma_F F, \tag{3.13}$$

the jump disappears, and F remains a significant dispersion factor, regardless of whether the two observations are interchanged or not. Furthermore, with the log link $\eta = \log \mu$, the two suspect observations no longer appear to be outliers: see the normal probability plot of model (3.12) in Figure 3.5. From now on we use the original data.

From Table 3.2 this dispersion model can be viewed as a gamma GLM and we can use all the GLM tools for inferences. As an example, model-checking plots for the dispersion GLM (3.13) are in Figure 3.6. Further extended models can be made by having more component GLMs as we shall see.

The results with the log link are in Table 3.6. The effect of the dispersion factor F is quite large so that the observations with the lower level of F

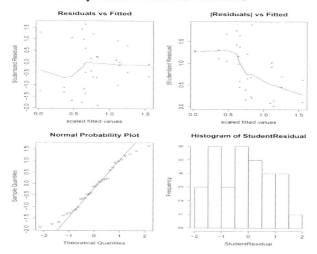

Figure 3.5 *The model checking plots for mean model (3.12).*

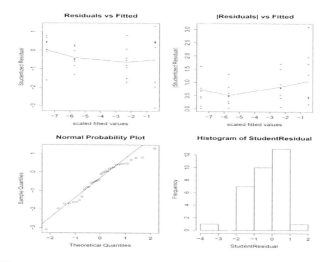

Figure 3.6 *The model checking plots for dispersion model (3.13).*

have weights exp(2*2.324) =104 times as large as those with the higher level; it is almost as if the number of observations is reduced by half, i.e., restricted to those at the lower level of F. In consequence, A, B and C are almost aliased to G, D and E respectively in the mean model, so that parameter estimates of factors appearing together with near-aliased factors are unstable, with larger standard errors.

The assertion that observations 5 and 6 are suspect is sensitive to the

Table 3.6 *Results from mean model (3.12) and dispersion model (3.13).*

	Mean model				Dispersion model		
	Estimate	Standard error	t test result		Estimate	Standard error	t test result
Constant	0.800	0.010	82.11	Constant	-2.849	0.300	-9.51
A	0.149	0.035	4.20	A	-0.608	0.296	-2.05
C	0.069	0.031	2.22	F	2.324	0.300	7.76
D	-0.152	0.010	-15.64				
E	0.012	0.032	0.37				
G	-0.074	0.035	-2.13				
N	-0.006	0.008	-0.80				
$C{\cdot}N$	0.189	0.034	5.52				
$E{\cdot}N$	-0.173	0.034	-5.07				

assumption made about the link function in the model for the mean. The apparently large effect of F in the dispersion model may not be caused by interactions in the mean model; F is aliased with C·E, but C·N and E·N are large in the mean model, and this says nothing about the form of the C·E response. In fact F (or C·E) is not required for the mean. Finally, we note that the ability of a large dispersion effect to produce near-aliasing of effects in the mean model has not been commented upon in other analyses of this data set.

CHAPTER 4

Extended likelihood inferences

In the previous three chapters we reviewed likelihood inferences about fixed parameters. There have been several attempts to extend Fisher's likelihood beyond its use in parametric inference to inference from more general models that include unobserved random variables. Special cases are inferences for random effect models, prediction of unobserved future observations and missing data problems. We want to extend classical likelihood inferences about fixed parameters to those about random parameters or combinations of fixed and random parameters. This chapter aims to illustrate that such an extension is possible and useful for statistical modelling and inference.

Classical likelihood and its extensions that we have discussed so far are defined for fixed parameters. We may say confidently that we understand their properties well, and there is a reasonable consensus about their usefulness. Here we shall discuss the concept of extended and h-likelihoods for inferences about unobserved random variables in more general models than we covered earlier. The need for a likelihood treatment of random parameters is probably best motivated by the specific applications that we shall describe in the following chapters.

Statisticians have disagreed on a general definition of likelihood that also covers unobserved random variables, e.g., Bayarri *et al.* (1988). Is there a theoretical basis for choosing a particular form of general likelihood? We can actually ask a similar question about the classical likelihood, and the answer seems to be provided by the principle (Birnbaum, 1962) that the likelihood contains all the evidence about a (fixed) parameter. Bjørnstad (1996) established the extended likelihood principle, showing that a particular definition of general likelihood contains all the evidence about both fixed and random parameters. We shall use this particular form as the basis for defining extended likelihood and h-likelihood.

Lee and Nelder (1996) introduced the h-likelihood for inferences in hierarchical GLMs, but because it is fundamentally different from classical likelihood, it has generated some controversies. One key property of likelihood inference that people expect is an invariance with respect to transformations. Here it will be important to distinguish extended likelihood from h-likelihood. We shall give examples where a blind optimization of the extended likelihood for estimation lacks invariance, so that different scales of the random parameters can lead to different estimates. The dependence on scale makes the extended likelihood immediately open to criticism. In fact, this has been the key source of the controversies. This problem is resolved for the h-likelihood defined as an extended likelihood for a *particular scale* of the random parameters with special properties, i.e., it is not defined on an arbitrary scale, so that transformation is not an issue as we shall show.

Another source of the controversies stems from the fact that MLEs of fixed unknown parameters are generally consistent and asymptotically normal, while MLEs of random unknowns are often neither consistent nor asymptotically normal. Asymptotics of MLEs of random parameters are not well established yet. This chapter outlines how to establish an asymptotic optimality of random parameter estimators and its finite sample adjustment.

As a reminder, we use $f_\theta()$ to denote probability density functions of random variables with fixed parameters θ; the arguments within the brackets determine what the random variable is, and it can be conditional or unconditional. Thus, $f_\theta(y,v)$, $f_\theta(v)$, $f_\theta(y|v)$ and $f_\theta(v|y)$ correspond to different densities even though we use the same basic notation $f_\theta()$. Similarly, the notation $L(a;b)$ denotes the likelihood of parameter a based on data or model b, where a and b can be of arbitrary complexity. For example, $L(\theta;y)$ and $L(\theta;v|y)$ are likelihoods of θ based on different models. The corresponding loglihood is denoted by $\ell()$.

4.1 Two kinds of likelihoods

4.1.1 Fisher's likelihood

The classical likelihood framework has two types of objects: a random outcome but observed y and an unknown parameter θ, and two related processes on them:

- *Data generation:* Generate an instance of the data y from a probability function with fixed parameters θ

$$f_\theta(y).$$

- *Parameter inference:* Given the data y, make statements about the unknown fixed θ in the stochastic model by using the likelihood

$$L(\theta;y).$$

The connection between these two processes is

$$L(\theta; y) \equiv f_\theta(y),$$

where L and f are algebraically identical, but on the left-hand side y is fixed while θ varies, while on the right-hand side θ is fixed while y varies. The function $f_\theta(y)$ summarizes, for fixed θ, where y will occur if we generate it from $f_\theta(y)$, while $L(\theta; y)$ shows the distribution of information as to where θ might be, given a fixed data set y. Since θ is a fixed number, the information is interpreted in a qualitative way.

Fisher's likelihood framework has been fruitful for inferences about fixed parameters. However, a new situation arises when a mathematical model involves random quantities at more than one level. Consider the simplest example of a two-level hierarchy with the model

$$y_{ij} = \beta + v_i + e_{ij},$$

where $v_i \sim N(0, \lambda)$ and $e_{ij} \sim N(0, \phi)$ with v_i and e_{ij} uncorrelated. This model leads to a specific multivariate distribution. Classical analysis of this model concentrates on estimation of the parameters β, λ and ϕ. From this view, it is straightforward to write the likelihood from the multivariate normal distribution and to obtain estimates by maximizing it. However, in many recent applications the main interest is often the estimation of $\beta + v_i$. These applications are often characterized by a large number of parameters. Although the v_is are thought of as having been obtained by sampling from a population, once a particular sample has been obtained they are fixed quantities and the likelihood based upon the marginal distribution provides no information on them.

4.1.2 Extended likelihood

There have been many efforts to generalize the likelihood, e.g., Lauritzen (1974), Butler (1986), Bayarri *et al.* (1987), Berger and Wolpert (1988) and Bjørnstad (1996), where the desired likelihood must deal with three types of objects: unknown parameters θ, unobservable random quantities v and observed data y. The previous two processes now take the forms:

- *Data generation:* (i) Generate an instance of the random quantities v from a probability function $f_\theta(v)$ and then with v fixed, (ii) generate an instance of the data y from a probability function $f_\theta(y|v)$. The combined stochastic model is given by the product of the two probability functions

$$f_\theta(v)f_\theta(y|v) = f_\theta(y, v). \tag{4.1}$$

- *Parameter inference:* Given the data y, we can (i) make inferences about θ by using the marginal likelihood $L(\theta; y)$, and (ii) given θ, make inferences about v by using a conditional likelihood of the form

$$L(\theta, v; v|y) \equiv f_\theta(v|y) \equiv P_\theta(v|y).$$

The conditional likelihood is called the predictive probability (or density) $P_\theta(v|y)$ to highlight its probability property,

$$\int L(\theta, v; v|y)dv = 1.$$

The extended likelihood of the unknown (θ, v) is defined by

$$L(\theta, v; y, v) \equiv L(\theta; y)L(\theta, v; v|y). \qquad (4.2)$$

The connection between these two processes is given by

$$L(\theta, v; y, v) \equiv f_\theta(y, v), \qquad (4.3)$$

so the extended likelihood matches the definition used by Butler (1986), Berger and Wolpert (1988) and Bjørnstad (1996). On the left-hand side y is fixed while (θ, v) vary, while on the right-hand side θ is fixed while (y, v) vary. In the extended likelihood framework the v values appear in data generation as random instances and in parameter estimation as unknowns. In the original framework there is only one kind of random object y, while in the extended framework there are two kinds of random objects, so that there may be several likelihoods, depending on how these objects are used.

The word *predictor* is often used for the estimate of random variables. For the prediction of unobserved future observations, we believe it is the right one. However, for estimation of unknown but realized values of random variables, the word *estimator* is more appropriate. Thus, for unknown fixed (realized) value of v once the data y are given, a concept like best linear unbiased estimators (BLUEs) is relevant, while for inferences about future random effects a concept like best linear unbiased predictors (BLUPs) would be appropriate. However, in literature these two very different statistical practices have not been clearly distinguished, causing widespread confusion. Thus, in this book for estimation we use the conditional likelihood $L(\theta, v; v|y)$, while for prediction we use the predictive probability $P_\theta(V|y)$. We sometimes use capital V to indicate a not yet realized (or future) random variable.

The *h-likelihood* is a special kind of extended likelihood, where the scale of the random parameter v is specified to satisfy a certain condition as we shall discuss in Section 4.5.

4.1.3 Classical and extended likelihood principles

Two theoretical principles govern what we can do with an extended likelihood. The *classical likelihood principle* of Birnbaum (1962) discussed in Section 1.1 states that the first term in (4.2), the marginal likelihood $L(\theta; y)$ carries all the (relevant experimental) information in the data about the fixed parameters θ, so that $L(\theta; y)$ should be used for inferences about θ. Bjørnstad (1996) proved an *extended likelihood principle* stating that the extended likelihood $L(\theta, v; y, v)$ carries all the information in the data about the unobserved quantities θ and v. Thus, when θ is known, from (4.2) the second term $L(\theta, v; v|y) \equiv P_\theta(v|y)$ in (4.2) must carry the information in the data about the random parameter.

When v is absent, $L(\theta, v; y, v)$ reduces to $L(\theta; y)$ and the extended likelihood principle reduces to the classical likelihood principle. When θ is absent, $L(\theta, v; y, v)$ leads to the predictive probability $P_\theta(v|y)$. This looks similar to the Bayesian posterior probability $P(\theta|y)$, where in the absence of v,

$$f_\theta(y)\pi(\theta) = P(y|\theta)\pi(\theta) = P(\theta|y)P(y),$$

$P(y|\theta) = f_\theta(y)$, $\pi(\theta)$ is a prior of hyperparameter θ and $P(y) = \int P(y|\theta)\pi(\theta)d\theta$. In the Bayesian approach, $P(y|\theta) = f_\theta(y)$ is the model, while in the extended likelihood framework both $f(y|v)$ and $f(v)$ belong to the model by which the data are generated. Thus, the model assumption $f(v)$ should be checkable or at least validated via the data.

When both parameters are unknown, the extended likelihood principle does not tell us how inference should be done for each component parameter. However, the classical likelihood principle still holds for the fixed parameter, so we have $L(\theta; y)$ as the whole carrier of information for θ alone. This is an unusual situation in likelihood theory where, in contrast to the classical fixed-parameter world, we now have a proper and unambiguous likelihood for a component parameter θ. This means that the second term $L(\theta, v; v|y)$ $(=P_\theta(v|y))$ in (4.2) does not carry any information about the fixed parameter θ, so generally it cannot be used for inference about θ. *It also means that, without special justification, we cannot estimate θ by joint maximization of the extended likelihood $L(\theta, v; y, v)$ with respect to (θ, v).* Doing so will violate the classical likelihood principle and make the analysis open to contradictions. As is shown in the examples below, such contradictions are easy to construct. If we seem to over-emphasize this point, it is because it has been the source of controversies surrounding the use of extended likelihood for random parameters.

Example 4.1: Consider an example adapted from Bayarri *et al.* (1988).

There is a single fixed parameter θ, a single unobservable random quantity U and a single observable quantity Y. The unobserved random variable U has an exponential density

$$f_\theta(u) = \theta \exp(-\theta u) \quad \text{for } u > 0, \theta > 0,$$

and given u, the observable outcome Y also has an exponential density

$$f_\theta(y|u) = f(y|u) = u \exp(-uy) \quad \text{for } y > 0, u > 0,$$

free of θ. Then we can derive these likelihoods:

- The marginal likelihood

$$L(\theta; y) = f_\theta(y) = \int_0^\infty f(y|u) f_\theta(u) du = \theta/(\theta + y)^2,$$

  which gives the (marginal) MLE $\hat{\theta} = y$. However, this classical likelihood is uninformative about the unknown value of u of U.

- The conditional likelihood

$$L(\theta, u; y|u) = f(y|u) = u \exp(-uy),$$

  which is uninformative about θ and loses the relationship between u and θ reflected in $f_\theta(u)$. This likelihood carries only the information about u in the data y. This gives, if maximized, $\hat{u} = 1/y$.

- The extended likelihood

$$L(\theta, u; y, u) = f(y|u) f_\theta(u) = u\theta \exp\{-u(\theta + y)\},$$

  which yields, if jointly maximized with respect to θ and u, useless estimators $\hat{\theta} = \infty$ and $\hat{u} = 0$.

- The predictive probability

$$P_\theta(u|y) = L(\theta, u; u|y) = u(\theta + y)^2 \exp\{-u(\theta + y)\}$$

  carries the combined information concerning u from $f_\theta(u)$ and $f(y|u)$. If θ is known, this is useful for inference about u. However, if θ is not known, joint maximization yields again the useless estimators $\hat{\theta} = \infty$ and $\hat{u} = 0$.

This example shows that different likelihoods carry different information, and that a naive joint inference of (θ, v) from an extended likelihood potentially violating the classical likelihood principle can be treacherous. □

Example 4.2: This is an example with pretend-missing data from Little and Rubin (2002): Suppose $y = (y_{obs}, y_{mis})$ consists of n iid variates from $N(\mu, \sigma^2)$, where $y_{obs} = (y_1, \ldots, y_k)$ consists of k observed values and $y_{mis} = (y_{k+1}, \ldots, y_n)$ represents $(n - k)$ missing values. Simply as a thought experiment, we can always add such (useless) missing data to any iid sample, so we know that they should not change the estimation of μ and σ^2 based on y_{obs}. Since

$$f_\theta(y) = f_\theta(y_{obs}) f_\theta(y_{mis})$$

where $\theta = (\mu, \sigma^2)$, $f_\theta(y_{obs}) = \Pi_{i=1}^{k} f_\theta(y_i)$ and $f_\theta(y_{mis}) = \Pi_{i=k+1}^{n} f_\theta(y_i)$, we have

$$\log L(\mu, \sigma^2, y_{mis}; y_{obs}, y_{mis}) = \log f_\theta(y_{obs}) + \log f_\theta(y_{mis}) \qquad (4.4)$$

$$= -\frac{n}{2} \log \sigma^2 - \frac{1}{2\sigma^2} \sum_{i=1}^{k} (y_i - \mu)^2 - \frac{1}{2\sigma^2} \sum_{i=k+1}^{n} (y_i - \mu)^2.$$

For $i = k+1, ..., n$ we have $\partial \log L / \partial y_i = -(y_i - \mu)/\sigma^2$ to give $\widehat{y}_i = \mu$. This means that the joint maximization of (μ, σ^2, y_{mis}) gives

$$\widehat{\mu} = \frac{1}{k} \sum_{i=1}^{k} y_i$$

$$\widehat{\sigma}^2 = \frac{1}{n} \sum_{i=1}^{k} (y_i - \bar{y})^2,$$

with a correct estimate of the mean and a wrong estimate of the variance. It is clear that the second term $\log f_\theta(y_{mis}) = \log L(\mu, \sigma^2, y_{mis}; y_{mis}|y_{obs})$ in (4.4) leads us to the wrong estimate of σ^2. $\square$

Example 4.3: Consider a one-way layout model

$$y_{ij} = \mu + v_i + e_{ij}, \quad i = 1, \ldots, n; \ j = 1, \ldots, m$$

where conditional on v_i, y_{ij} is $N(\mu + v_i, 1)$, and $v_1, \ldots, v_n$ are iid $N(0, 1)$. The extended likelihood, with μ and $v = (v_1, \ldots, v_n)$ as the unknown parameters, is

$$\log L(\mu, v; y, v) = -\frac{1}{2} \sum_{ij} (y_{ij} - \mu - v_i)^2 - \frac{1}{2} \sum_i v_i^2.$$

By jointly maximizing this likelihood, we get the standard MLE $\widehat{\mu} = \bar{y}_{..}$. But suppose we reparametrize the random effects, by assuming $v_i = \log u_i$. Allowing the Jacobian, we now get

$$\log L(\mu, u; y, u) = -\frac{1}{2} \sum_{ij} (y_{ij} - \mu - \log u_i)^2 - \frac{1}{2} \sum_i (\log u_i)^2 - \sum_i \log u_i,$$

and, by joint maximization, obtain $\widehat{\mu} = \bar{y}_{..} + 1$.

Hence the estimate of the fixed parameter μ is not invariant with respect to reparameterization of the random parameters. This example shows again how an improper use of the extended likelihood by a naive joint optimization of fixed and random parameters leads to a wrong result. However, it also suggests that sometimes a joint optimization gives the correct MLE, i.e., the MLE that would have resulted from the marginal likelihood. In Section 4.5 we shall see that the scale of the random parameters determines whether we can or cannot perform joint maximization of the extended likelihood. $\square$

4.2 Wallet game and extended likelihood

The so-called wallet game (Gardner, 1982; Pawitan and Lee, 2016) is a well-known paradox that shows the logical need for the extended likelihood concept, if we want to (i) allow the sense of uncertainty associated with a realized but still unobserved random outcome, while (ii) at the same time avoid the probability-related paradox. The setup is as follows:

> Two people, equally rich or equally poor, meet to compare the contents of their wallets. Each is ignorant of the contents of the two wallets. Here is the game: whoever has less money receives the contents of the wallet of the other. One of the two men can reason: "I have a fixed amount x_1 in my wallet; either I lose that or I win an amount $x_2 > x_1$ with 50-50 chance. Therefore the game is favorable to me." The other man can reason in exactly the same way. In fact, by symmetry, the game is fair. Where is the mistake in the reasoning of each man?

We need to emphasize: the issue is *not* to show that the game is actually fair — that is easy to do, see below —, but to point where the reasoning fails. Because of its close connection we also describe the so-called exchange paradox (see also Pawitan, 2001, Chapter 2):

> Unknown θ dollars are put in one envelope and 2θ dollars in another. You are to pick one envelope at random, open it and then decide whether you would exchange it with the other envelope. You pick one (randomly), open it and see the outcome $x = 100$ dollars. You reason that if Y is the content of the other envelope, Y is either 50 or 200 with 50-50 chance; if you exchange it you expect to get $(50+200)/2 = 125$, which is more than your current 100. So, you would exchange the envelope, wouldn't you? Since the reasoning above holds for any value of x, which means that you actually *do not need to open the envelope* in the first place and you would still want to exchange it!

The explanation of the exchange paradox relies on the classical likelihood, while that of the wallet game relies on the extended likelihood. We start with the former. Let the *fixed unknown* amounts in the two envelopes of the exchange paradox be θ and 2θ, and we observe an amount (data) $x = 100$ from a randomly chosen envelope. The *classical* likelihood of θ is then

$$
\begin{aligned}
L(\theta = 100) &= P(X = 100|\theta = 100) \\
&= P(X = \theta|\theta = 100) = 0.5 \\
L(\theta = 50) &= P(X = 100|\theta = 50) \\
&= P(X = 2\theta|\theta = 50) = 0.5,
\end{aligned}
$$

which means the data $x = 100$ cannot indicate any preference over two possible values of θ. Hence, the other envelope contains either 50 (if $\theta = 50$) or 200 (if $\theta = 100$) with equal likelihood, not probability. The paradox occurs if we wrongly take an average using likelihood as weights; we do not have any theory that justifies such an operation.

Now back to the wallet game. Let X_1 and X_2 be the random amounts of money in the two wallets; we assume that they are iid samples from a positive-valued distribution F. For convenience, assume F is continuous and EX_1 exists. First, we note again that, unlike in the exchange paradox, in this game we do not observe any data. Second, it is obvious from this setup that the game is fair, i.e., on average each person would receive EX_1 when he or she plays the game repeatedly.

Now consider a specific realization $X_1 = x_1$ and $X_2 = x_2$ (*but still unobserved to both players*). You are thinking that either $x_1 < x_2$ or $x_1 > x_2$. If the former occurs you gain x_2, if the latter you lose x_1. To emphasize, since x_1 is fixed, you think that your loss is limited to x_1, but if you win your gain is x_2, which by construction is greater than x_1. You are of course uncertain whether you will win or lose. If we account for this uncertainty by attaching probability 0.5 to those specific events, the expected gain minus loss is $0.5(x_2 - x_1) > 0$, hence the paradox. Only a probability reasoning allows this final averaging step.

Now let $V \equiv I(X_1 < X_2)$, so V is a random 0 to 1 Bernoulli random variable with $P(V = 1) = P(X_1 < X_2) = 0.5$. The specific realization of $v = 0$ or 1 is unobserved, so we are in a state of uncertainty. The paradox clearly arises from the use of probability to represent this uncertainty. The question is what is this sense of uncertainty whether as to the realized v is 0 or 1? It cannot be a probability, since that it must allow an averaging step that would lead to the paradox. It cannot be the classical likelihood, since v is a realized random variable. In fact in this problem, it is given by the extended likelihood, i.e.,

$$L(v = 1) \equiv P(V = 1) = 0.5, \qquad (4.5)$$

and $L(v = 0) \equiv P(V = 0) = 0.5$, so the specific realizations of v are equally likely. As in the exchange paradox, we cannot take expectation with likelihood as weight, so we have no rational basis to believe the game is favorable to us. Again the paradox arises because we wrongly take an average using likelihood as weights.

To see that (4.5) follows from (4.3), there is no data y nor unknown parameter θ, or equivalently for y we can simply generate an independent toss of a fair coin, so $f_\theta(y|v)$ is constant with respect to v and θ, while $p_\theta(v)$ is given by (4.5).

The key logical difference between the wallet game and the exchange paradox is that in the former the unknown parameter v is a (discrete) random unobservable, thus requiring the extended likelihood, while in the latter the parameter is fixed, so the classical likelihood suffices. The use of discrete random effects is discussed in Chapter 13 on hypothesis testing.

4.3 Inference about the fixed parameters

To keep the distinctions clear between extended and classical likelihoods, we use the following notation:

$$
\begin{aligned}
\ell_e(\theta, v) &= \log L(\theta, v; y, v) \\
\ell(\theta) &= \log L(\theta; y).
\end{aligned}
$$

From the previous discussion

$$\ell_e(\theta, v) = \ell(\theta) + \log f_\theta(v|y). \qquad (4.6)$$

The use of $\ell(\theta)$ for fixed parameters is the classical likelihood approach and the use of $\log f_\theta(v|y)$ for random parameters is the empirical Bayes approach. From the data generation formulation, the marginal distribution is obtained via an integration, so that

$$L(\theta; y) \equiv f_\theta(y) = \int f_\theta(v, y) dv = \int L(\theta, v; y, v) dv. \qquad (4.7)$$

However, for the non-normal models that we consider in this book, the integration required to obtain the marginal likelihood is often intractable.

One method of obtaining the marginal MLE for fixed parameters θ is the so-called EM (expectation-maximization) algorithm (Dempster et $al.$, 1977). This exploits the property (4.6) of extended loglihood, where, under appropriate regularity conditions,

$$
\begin{aligned}
\frac{\partial}{\partial \theta} E(\ell_e|y) &= \partial \ell / \partial \theta + E(\partial \log f_\theta(v|y)/\partial \theta|y) \\
&= \partial \ell / \partial \theta. \qquad (4.8)
\end{aligned}
$$

This means that the optimization of $\ell(\theta)$ is equivalent to the optimization of $E(\ell_e|y)$. The E step in the EM algorithm involves finding $E(\ell_e|y)$ analytically, and the M step maximizes it. The last equality in (4.8) follows from

$$
\begin{aligned}
E(\partial \log f_\theta(v|y)/\partial\theta|y) &= \int \frac{\partial f_\theta(v|y)/\partial\theta}{f_\theta(v|y)} f_\theta(v|y) dv \\
&= \int \partial f_\theta(v|y)/\partial\theta dv \\
&= \frac{\partial}{\partial\theta} \int f_\theta(v|y) dv = 0,
\end{aligned}
$$

as the last integral is equal to one.

The EM algorithm is known to have slow convergence and, for non-normal models, it is often hard to evaluate the conditional expectation $E(\ell_e|y)$ analytically. Alternatively, simulation methods, such as Monte Carlo EM (Vaida and Meng, 2004) and Gibbs sampling (Gelfand and Smith, 1990) can be used to evaluate the conditional expectation, but these methods are computationally intensive. Another approach, numerical integration via Gauss-Hermite quadrature (Crouch and Spiegelman, 1990), can be directly applied to obtain the MLE, but this also becomes computationally intractable as the number of integrals increases.

Ideally, we should not need to evaluate an analytically difficult expectation step, nor use the computationally intensive methods required for Monte Carlo EM, MCMC or numerical integration. Rather, we should be able to obtain necessary estimators by directly maximizing appropriate quantities derived from the extended likelihood, and compute their standard error estimates from its second derivatives. Later we shall show how to implement inferential procedures using $\ell_e(\theta, v)$.

In the extended likelihood framework, the proper likelihood for inferences about fixed parameters θ — the marginal likelihood $\ell(\theta)$ — can be obtained from $\ell_e(\theta, v)$ by integrating out the random parameters in (4.7). However, for general models that we consider in this book, the integration to obtain the marginal likelihood is mostly intractable. For such cases, the marginal loglihood can be approximated by the Laplace approximation (1.19)

$$
\ell(\theta) \approx p_v(\ell_e) = [\ell_e - \frac{1}{2}\log\det\{D(\ell_e, v)/(2\pi)\}]|_{v=\widehat{v}_\theta} \qquad (4.9)
$$

where $D(\ell_e, v) = -\partial^2 \ell_e/\partial v^2$ and $\widehat{v}_\theta = \hat{v}(\theta)$ solves $\partial \ell_e/\partial v = 0$ for fixed θ. This approximation is the adjusted profile loglihood (1.14) to the integrated loglihood $\ell(\theta)$ as shown in (1.20). Throughout this book we shall study the various forms of adjusted profile loglihoods $p_*()$ that can be used for statistical inference; these functions $p_*()$ may be regarded as derived loglihoods for various subsets of parameters. We use either $\widehat{v}_\theta$ or $\hat{v}(\theta)$ for convenience to emphasise the estimator $\hat{v}$ is a function of θ.

The use of the Laplace approximation has been suggested by many authors (see, e.g., Tierney and Kadane, 1986) and is highly accurate when $\ell_e(\theta, v)$ is approximately quadratic in v for fixed θ. When the Laplace approximation fails, e.g., non-normal data where there is too little information on the random effects, we may expect some bias in the estimation of θ that persists for large samples. In this case, a higher-order Laplace approximation may be considered.

4.4 Inference about the random parameters

First let us consider the case when θ is known, so that the information about v is contained in $L(\theta, v; v|y)$. In statistical inference estimation, confidence intervals and hypothesis testing are important. How do we estimate v from $L(\theta, v; v|y)$? One option is to use the conditional mean $E_\theta(v|y)$, but this will often require analytical derivation. Instead of a sample mean, the standard maximum likelihood (ML) approach uses the mode as an estimator. We shall use the MLE from $L(\theta, v; v|y)$ which is the mode of $P_\theta(v|y)$. Due to the similarity of the predictive probability $P_\theta(v|y)$ to the Bayesian posterior, such an estimate is known in some areas such as statistical image processing as the maximum *a posteriori* (MAP) estimate. In other wide areas, such as non–parametric function estimation and smoothing, generalized additive modelling, this is known as the penalized likelihood method (e.g., Green and Silverman, 1994).

One significant advantage of the ML approach over the conditional mean approach is that, *for fixed θ*, maximizing $L(\theta, v; v|y)$ with respect to v is equivalent to maximizing the extended likelihood $L(\theta, v; y, v)$. In *all* applications of the extended likelihood of which we are aware, the statistical models are explicitly stated in the form

$$f_\theta(y|v)f_\theta(v) = L(\theta, v; y, v),$$

which means that $L(\theta, v; y, v)$ is immediately available. By contrast, with the conditional mean approach we have to find various potentially complicated functions due to the integration steps:

$$
\begin{aligned}
f_\theta(y) &= \int f_\theta(y|v)f_\theta(v)dv \\
P_\theta(v|y) &= \frac{f_\theta(y|v)f_\theta(v)}{f_\theta(y)} \\
E_\theta(v|y) &= \int v f_\theta(v|y)dv.
\end{aligned}
$$

The conditional density of $v|y$ is explicitly available only for the so-called conjugate distributions. The Bayesians have recently developed a massive collection of computational tools, such as the Gibbs sampling or Markov chain Monte Carlo methods, to evaluate these quantities. There is, however, a computational and philosophical barrier to these methods, as they require fully Bayesian models, thus needing priors for the fixed parameters; also, the iterative steps in the algorithm are not always immediate and convergence is often an issue.

As shown in the examples above, when θ is unknown, joint maximization of (θ, v) from the extended likelihood is generally not justified; we can do that only under special conditions as we shall discuss in Section 4.5. In general, θ can first be estimated from the marginal likelihood. To summarize, the safe option in the use of extended likelihood is as follows:

- For inferences about the fixed parameters, use the classical likelihood approach based on the marginal likelihood $L(\theta; y)$.
- Given the estimated fixed parameters, use the MLE from the extended likelihood for an estimate for random parameters.

This procedure is already in heavy use, such as in the analogous MAP and penalized likelihood methods. We note that the resulting random parameter estimate depends upon the scale of random parameters used in defining the extended likelihood. This is a recognized problem in the penalized likelihood method, for which there are no clear guidelines.

4.5 Canonical scale, h-likelihood and joint inference

If we insist on a rigid separation of the fixed and random parameters, the extended likelihood framework will be no richer than the empirical Bayes framework. However, for certain general classes of models, we can exploit the extended likelihood to give a joint inference — not just maximization — for some fixed and random parameters. Here we have to be careful, since we have shown that a naive use of the extended likelihood involving the fixed parameters violates the classical likelihood principle and can lead to contradictions.

A key property to keep in extending the likelihood inferences is the invariance of likelihood ratio with respect transformations of the data or parameterizations. Care is necessary in the transformation of random parameters because of the Jacobian term caused by the transformation, leading to different parameter estimators. We now derive a condition that allows a joint inference from the extended likelihood, maintaining the invariance property. Let θ_1 and θ_2 be an arbitrary pair of values of fixed parameter θ. The evidence about these two parameter values is contained in the likelihood ratio

$$\frac{L(\theta_1; y)}{L(\theta_2; y)}.$$

Suppose there exists a scale v, such that the likelihood ratio is preserved in the following sense

$$\frac{L(\theta_1, \widehat{v}_{\theta_1}; y, v)}{L(\theta_2, \widehat{v}_{\theta_2}; y, v)} = \frac{L(\theta_1; y)}{L(\theta_2; y)}, \qquad (4.10)$$

where $\widehat{v}_{\theta_1}$ and $\widehat{v}_{\theta_2}$ are the MLEs of v for θ at θ_1 and θ_2, so that $\widehat{v}_\theta$ is *information-neutral* concerning θ. Alternatively, (4.10) is equivalent to

$$\frac{L(\theta_1, \widehat{v}_{\theta_1}; v|y)}{L(\theta_2, \widehat{v}_{\theta_2}; v|y)} = 1,$$

which means that neither the likelihood component $L(\theta, \widehat{v}_\theta; v|y)$ nor $\widehat{v}_\theta$ carries any information about θ, as required by the classical likelihood principle. We shall call such a v-scale the *canonical scale* of the random parameter, and we make an explicit definition to highlight this special situation.

Definition 4.1 *If the parameter v in $L(\theta, v; y, v)$ is canonical, we call L an h-likelihood.*

If such a scale exists, the definition of h-likelihood is immediate. However, in an arbitrary statistical problem no canonical scale may exist, and we shall study how to extend its definition (Chapter 6).

In definition 4.1 h-likelihood appears as a special kind of extended likelihood: to call $L(\theta, v; y, v)$ an h-likelihood assumes that v is canonical, and we shall use the notation $H(\theta, v)$ to denote h-likelihood and $h(\theta, v)$ the h-loglihood. The h-loglihood can be treated like an ordinary loglihood, where, for example, we can take derivatives and compute Fisher information for both parameters (θ, v). In an arbitrary statistical problem, the canonical scale may not be obvious. However, it is easy to check whether a particular scale is canonical; see below. For some classes of models considered in this book, the canonical scale may be easily recognized. A canonical scale has many interesting properties that make it the most convenient scale to use. The most useful results are summarized in the following.

Let $I_m(\widehat{\theta})$ be the observed Fisher information of the MLE $\widehat{\theta}$ from the marginal likelihood $L(\theta; y)$ and let the partitioned matrix

$$I_h^{-1}(\widehat{\theta}, \widehat{v}) = \begin{pmatrix} I^{11} & I^{12} \\ I^{21} & I^{22} \end{pmatrix}$$

be the inverse of the observed Fisher information matrix of $(\widehat{\theta}, \widehat{v})$ from the h-likelihood $H(\theta, v; y, v)$, where I^{11} corresponds to the θ part. Then

- The MLE $\widehat{\theta}$ from the marginal likelihood $L(\theta; y)$ coincides with the $\widehat{\theta}$ from the joint maximizer of the h-likelihood $L(\theta, v; y, v)$.
- The information matrices for $\widehat{\theta}$ from the two likelihoods also match in the sense that
$$I_m^{-1} = I^{11}.$$
This means that (Wald-based) inference on the fixed parameter θ can be obtained directly from the h-likelihood framework.
- I^{22} yields an estimate of $\text{var}(\widehat{v} - v)$. If $\widehat{v} = E(v|y)|_{\theta=\widehat{\theta}}$, this estimates
$$\text{var}(\widehat{v} - v) \geq E\{var(v|y)\},$$
accounting for the inflation of variance caused by estimating θ.

We now outline the proof of these statements. From the definition of a canonical scale for v and for free choice of θ_1, the condition (4.10) is equivalent to a marginal likelihood $L(\theta; y)$ proportional to the profile likelihood $L(\theta, \widehat{v}_\theta; y, v)$. The first statement follows immediately. The second part follows from Pawitan (2001, Section 9.11), where it is shown that the curvature of the profile likelihood of θ obtained from a joint likelihood of (θ, v) is $(I^{11})^{-1}$. The third part can be shown similarly by using the derivation in Section 5.4.1. The last two properties hold for general models; see derivations in Lee *et al.* (2011), Ha *et al.* (2016) and Paik *et al.* (2015).

Now suppose v is canonical; a nonlinear transform $u \equiv v(u)$ will change the extended likelihood in a nontrivial way to give
$$L(\theta, u; y, u) = f_\theta(y|v(u))f(v(u))|J(u)|. \qquad (4.11)$$
Because of the Jacobian term , $|J(u)|$, u cannot be canonical. This means that, up to linear transformations, the canonical scale v is unique. Thus, from the above results, joint inference of (θ, v) is possible only through the h-likelihood. Moreover, inferences from h-likelihood can be treated like inferences from ordinary likelihood. We can now recognize that all the supposed counter-examples about the h-likelihood involved the use of non-canonical scales and joint maximization: for more discussion see Lee and Nelder (2005).

Definition 4.1 of the h-likelihood is too restrictive because the canonical scale is assumed to work for the full set of fixed parameters. As an extension, suppose there are two subsets of the fixed parameters (θ, ϕ) such that
$$\frac{L(\theta_1, \phi, \widehat{v}_{\theta_1, \phi}; y, v)}{L(\theta_2, \phi, \widehat{v}_{\theta_2, \phi}; y, v)} = \frac{L(\theta_1, \phi; y)}{L(\theta_2, \phi; y)}, \qquad (4.12)$$

but

$$\frac{L(\theta, \phi_1, \widehat{v}_{\theta,\phi_1}; y, v)}{L(\theta, \phi_2, \widehat{v}_{\theta,\phi_2}; y, v)} \neq \frac{L(\theta, \phi_1; y)}{L(\theta, \phi_2; y)}. \qquad (4.13)$$

In this case the scale v is information-neutral only for θ and not for ϕ, so that joint inference using the h-likelihood is possible only for (θ, v), with ϕ needing a marginal likelihood. For example, joint maximization of (θ, v) gives the marginal MLE for θ, but not that for ϕ, as we shall see in an example below.

From (4.9), the marginal loglihood $\log L(\theta, \phi; y)$ is given approximately by the adjusted profile likelihood

$$p_v(h) = [h - \frac{1}{2}\log\det\{D(h, v)/(2\pi)\}]|_{v=\widehat{v}}.$$

In this case $D(h, v)$ is a function of ϕ, but not θ.

4.5.1 Checking whether a scale is canonical

There is no guarantee in an arbitrary problem that a canonical scale exists, and even if it exists it may not be immediately obvious what it is. However, as stated earlier, there are large classes of models in this book where canonical scales are easily identified.

In general, for v to be canonical, the profile likelihood of θ from the extended likelihood must be proportional to the marginal likelihood $L(\theta)$. As noted earlier, if a canonical scale exists, it is unique up to linear transformations. The marginal loglihood ℓ is approximated by the adjusted profile loglihood $p_v(\ell_e)$, and we often find that v is canonical if the adjustment term $I(\widehat{v}_\theta)$ in $p_v(\ell_e)$ is free of θ. If the fixed parameters consist of two subsets (θ, ϕ), then v is canonical for θ if $I(\widehat{v}_{\theta,\phi})$ is free of θ. In practice, checking this condition is straightforward. However, there is no guarantee that this condition is sufficient for the canonical scale, but we have found it useful for finding a true canonical scale.

Example 4.4: Continuing Example 4.1, consider the scale $v = \log u$ for the random parameter, so that

$$f_\theta(v) = e^v \theta \exp(-e^v \theta)$$

and the extended likelihood is given by

$$L(\theta, v; y, v) = e^{2v} \theta \exp\{-e^v(\theta + y)\},$$

or

$$\log L = 2v + \log\theta - e^v(\theta + y),$$

and we obtain

$$\widehat{u}_\theta = \exp\widehat{v}_\theta = E\{u|y\} = \frac{2}{\theta + y},$$

and, up to a constant term, the profile loglihood is equal to the marginal loglihood:

$$
\begin{aligned}
\log L(\theta, \widehat{v}_\theta) &= 2\log\left(\frac{2}{\theta + y}\right) + \log\theta - 2 \\
&= \log L(\theta; y) + \text{constant},
\end{aligned}
$$

so $v = \log u$ is the canonical scale for the extended likelihood. By taking the derivative of the h-loglihood $h(\theta, v) \equiv \log L(\theta, v)$ with respect to θ and setting it to zero we obtain

$$
\frac{1}{\theta} - e^v = 0
$$

or

$$
\widehat{\theta} = y,
$$

exactly the marginal MLE from $L(\theta; y)$. Its variance estimator is

$$
\widehat{\text{var}(\widehat{\theta})} = -\{\partial^2 \log L(\theta; y)/\partial\theta^2|_{\theta=\widehat{\theta}}\}^{-1} = 2y^2.
$$

Note here that

$$
\begin{aligned}
&\begin{pmatrix} I_{11} = -\partial^2 h/\partial\theta^2|_{\theta=\widehat{\theta}, v=\widehat{v}} & I_{12} = -\partial^2 h/\partial\theta\partial u|_{\theta=\widehat{\theta}, v=\widehat{v}} \\ I_{21} = I_{12} & I_{22} = -\partial^2 h/\partial u^2|_{\theta=\widehat{\theta}, v=\widehat{v}} \end{pmatrix} \\
&= \begin{pmatrix} 1/y^2 & 1 \\ 1 & (y+\widehat{\theta})^2/2 = 2y^2 \end{pmatrix},
\end{aligned}
$$

so that $I^{11} = 2y^2$, matching the estimated variance of the marginal MLE.

Here I_{22} is free from θ to indicate v is canonical and $1/I_{22} = 1/(2y^2)$ is estimating $\text{var}(u|y) = 2/(y+\theta)^2$, while $I^{22} = 1/y^2$ takes account of the estimation of θ when estimating random parameters.

Let $w = \theta u$, so that $E(w) = 1$. It follows that

$$
\widehat{w} = 2\widehat{\theta}/(y + \widehat{\theta}) = \widehat{E(w|y)} = \widehat{\theta}E(u|y) = 1.
$$

Now we have

$$
\widehat{\text{var}(\widehat{w} - w)} = 1 = \text{var}(w),
$$

which reflects the variance increase caused by estimating θ; note that

$$
\widehat{\text{var}(w|y)} = 2\theta^2/(y+\theta)^2|_{\theta=\widehat{\theta}} = 1/2.
$$

Thus,

$$
\widehat{\text{var}(u|y)} = \widehat{\text{var}(w/\theta|y)} = 2/(y+\theta)^2|_{\theta=\widehat{\theta}} = 1/(2y^2)
$$

and

$$
\widehat{\text{var}(\widehat{u} - u)} = \widehat{\text{var}(\widehat{w} - w)}/\theta^2 = 1/\widehat{\theta}^2 = 1/y^2 = I^{22}. \ \square
$$

Example 4.5: In the missing data problem in Example 4.2, it is readily seen that for fixed (μ, σ^2), the observed Fisher information for the missing data $y_{mis,i}$, for $i = k+1, \ldots, n$, is

$$
I(\widehat{y}_{mis,i}) = 1/\sigma^2,
$$

so y_{mis} is a canonical scale that is information-neutral for μ, but not for σ^2. This means that the h-likelihood can be used to estimate (μ, y_{mis}) jointly, but that σ^2 must be estimated using the marginal likelihood. It can be shown that $I^{11} = \sigma^2/k$ is as a variance estimate for $\hat{\mu}$ and $I^{1+i,1+i} = (1 + 1/k)\sigma^2$ as an estimate of $\text{var}(y_{mis,i} - \hat{y}_{mis,i}) = \text{var}(y_{mis,i} - \hat{\mu}) = (1+1/k)\sigma^2$; both are proper estimates. In this case, the adjusted profile likelihood is given by

$$p_{y_{mis}}(h) = -(n/2)\log\sigma^2 - \sum_{i=1}^{k}(y_i - \mu)^2/2\sigma^2 + ((n-k)/2)\log\sigma^2$$

and is equal to the marginal loglihood $\ell(\mu, \sigma^2)$; it leads to the correct MLE of σ^2, namely

$$\hat{\sigma}^2 = \frac{1}{k}\sum_{i=1}^{k}(y_i - \bar{y})^2.$$

The estimate of its variance can be obtained from the Hessian of $p_{y_{mis}}(h)$.

4.5.2 Transformation of the canonical scale

With ordinary likelihoods we deal with transformation of parameters via the invariance principle set out in Chapter 1. If an h-likelihood $L(\theta, v(u); y, v)$ with canonical scale v is to be treated like an ordinary likelihood, something similar is needed. Thus, to maintain invariant h-likelihood inferences between equivalent models generated by monotone transformations $u = u(v)$, we shall define

$$\begin{aligned} H(\theta, u) &\equiv H(\theta, v(u)) &\text{(4.14)} \\ &= L(\theta, v(u); y, v) \\ &= f_\theta(y|v(u))f_\theta(v(u)), \end{aligned}$$

which is *not* the same as the extended likelihood

$$\begin{aligned} L(\theta, u; y, u) &= f_\theta(y|v(u))f_\theta(v(u))|J(u)| \\ &= H(\theta, u)|J(u)|. \end{aligned}$$

Here u is not canonical for its own extended likelihood $L(\theta, u; y, u)$, but by definition it is canonical for its h-likelihood $H(\theta, u)$.

The definition has the following consequence. Let $H(\theta, v; y, v)$ be the h-likelihood defined on a particular scale v; then for monotone transformation of $\phi = \phi(\theta)$ and $u = u(v)$ we have

$$\frac{H(\theta_1, v_1; y, v)}{H(\theta_2, v_2; y, v)} = \frac{H(\phi_1, u_1; y, v)}{H(\phi_2, u_2; y, v)} = \frac{H(\phi_1, u(v_1); y, v)}{H(\phi_2, u(v_2); y, v)}.$$

This means that the h-likelihood keeps the invariance property of the

MLE:

$$\hat{\phi} = \phi(\hat{\theta})$$
$$\hat{u} = u(\hat{v}),$$

i.e., ML estimation is invariant with respect to both fixed and random parameters. The invariance property is kept by determining the h-likelihood in a particular scale. This is in contrast to the penalized likelihood, the maximum *a posteriori* or the empirical Bayes estimators, where transformation of the parameter may require non–trivial re-computation of the estimate. In general, joint inferences about (θ, u) from the h-likelihood $H(\theta, u)$ are equivalent to joint inferences about (θ, v) from $H(\theta, v)$. Furthermore, likelihood inferences about θ from $H(\theta, u)$ are equivalent to inferences from the marginal likelihood $L(\theta; y)$.

Example 4.6: Continuing Examples 4.1 and 4.4, we have shown earlier that the scale $v = \log u$ is canonical. To use the u-scale for joint estimation, we must use the h-loglihood

$$\begin{aligned} h(\theta, u) &\equiv \log H(\theta, v(u)) \\ &= \log L(\theta, \log u; y, \log u) \\ &= 2 \log u + \log \theta - u(y + \theta). \end{aligned}$$

In contrast, the extended loglihood is

$$\ell(\theta, u) = \log L(\theta, u; y, u) = \log u + \log \theta - u(y + \theta),$$

with a difference of $\log u$ due to the Jacobian term. It is now straightforward to produce meaningful likelihood inferences for both θ and u from $h(\theta, u)$. For known θ, setting $\partial h/\partial u = 0$ gives

$$\hat{u} = 2/(y + \theta) = E(u|y).$$

Then, the corresponding Hessian $-\partial^2 h/\partial u^2|_{u=\hat{u}} = 2/\hat{u}^2 = (y+\theta)^2/2$ gives as an estimate for $\text{var}(\hat{u} - u)$:

$$\text{var}(u|y) = 2/(y + \theta)^2.$$

If θ is unknown, as we expect, the joint maximization of $h(\theta, u)$ gives the MLE $\hat{\theta} = y$, and the random effect estimator

$$\hat{u} = E(u|y)|_{\theta=\hat{\theta}} = 1/y.$$

From the marginal loglihood

$$\ell(\theta) = \log L(\theta; y) = \log \theta - 2 \log(\theta + y),$$

we have the variance estimator of the MLE $\hat{\theta} = y$

$$\widehat{\text{var}}(\hat{\theta}) = -\{\partial^2 \ell/\partial \theta^2|_{\theta=\hat{\theta}}\}^{-1} = 2y^2.$$

From the extended likelihood we derive the observed Fisher information matrix

$$I(\widehat{\theta}, \widehat{u}) = \begin{pmatrix} 1/y^2 & 1 \\ 1 & \frac{1}{2y^2} \end{pmatrix},$$

which gives the variance estimator

$$\widehat{\mathrm{var}}(\widehat{\theta}) = 2y^2,$$

exactly the same as the one from the marginal loglihood.

We also have

$$\widehat{\mathrm{var}}(\widehat{u} - u) = 1/y^2 = \mathrm{var}(u)|_{\theta=\hat{\theta}},$$

which is larger than the plug-in estimate

$$\widehat{\mathrm{var}}(u|y) = 2/(y+\theta)^2|_{\theta=\hat{\theta}} = 1/(2y^2)$$

obtained from the variance formula when θ is known. The increase reflects the extra uncertainty caused by estimating θ.

Suppose that instead we use the extended likelihood $L(\theta, u; y, u)$. Here we can still use $p_u(\ell)$ for inferences about fixed parameters. The equation $\partial\ell/\partial u = 1/u - (\theta + y) = 0$ gives $\tilde{u} = 1/(\theta + y)$. From this we get

$$
\begin{aligned}
p_u(\ell) &= \log \tilde{u} + \log \theta - \tilde{u}(\theta + y) - \frac{1}{2}\log\{1/(2\pi\tilde{u}^2)\} \\
&= \log \theta - 2\log(\theta + y) - 1 + \frac{1}{2}\log 2\pi,
\end{aligned}
$$

which, up to a constant term, is equal to the marginal loglihood $\ell(\theta)$, to yield the same inference for θ. What happens is that

$$-\partial^2 \ell/\partial u^2|_{u=\tilde{u}} = 1/\tilde{u}^2 = (\theta + y)^2,$$

so that ℓ and $p_u(\ell)$ are no longer equivalent, and the joint maximization of $\ell(\theta, u)$ cannot give the MLE for θ. □

4.6 Prediction of random parameters

Estimation theory for MLEs of fixed parameters is well established, whereas that for random parameters is less so. Suppose that we are interested in predicting an unobservable random variable

$$r = r(v, \theta).$$

Let $t(y)$ be an unbiased predictor for r in the sense that

$$E(t(y)) = E(r).$$

Now we want to find an unbiased predictor which minimizes

$$E(t(y) - r)^2 = \mathrm{var}(t(y) - r).$$

Let $\delta = E(r|y)$. Because

$$
\begin{aligned}
E\{(t(y) - \delta)(\delta - r)\} &= EE\{(t(y) - \delta)(\delta - r)|y\} \\
&= E\{(t(y) - \delta)E(\delta - r|y)\} = 0,
\end{aligned}
$$

we have

$$
\begin{aligned}
E(t(y) - r)^2 &= E(t(y) - \delta)^2 + E(\delta - r)^2 + 2E\{(t(y) - \delta)(\delta - r)\} \\
&= E(t(y) - \delta)^2 + E(\delta - r)^2 \\
&= E(t(y) - \delta)^2 + E(\text{var}(r|y)) \\
&\geq E(\text{var}(r|y)).
\end{aligned}
$$

This means that $\text{var}(r|y)$ is the unavoidable variation in predicting random variable r. Thus, if $\text{var}(r|y) > 0$,

$$
t(y) - r = O_p(1)
$$

for any function $t(y)$, so that there is no consistent predictor for r.

In this book we distinguish the target of estimation (TOE) and target of prediction (TOP) . Here δ is the TOE and r the TOP . Let $\hat{\theta}$ be the MLE. We use the plug-in estimator

$$
\hat{r} = \hat{\delta} = E(r|y)|_{\theta = \hat{\theta}}
$$

as a predictor for r and as an estimator for δ. But, $\hat{r} - \delta$ can be asymptotically normal with zero mean, while $\hat{r} - r$ may not. For prediction for future random effects, we can estimate the predictive probability $P_\theta(R = r|y)$ consistently as we shall discuss in the next section, even though we cannot predict r consistently.

TOE is a more general concept than a parameter, because $g(\hat{\theta}) = \widehat{g(\theta)}$ for all transformations $g()$ but

$$
E(g(r)|y) \neq g(E(r|y)), \tag{4.15}
$$

unless $g()$ is a linear function or $var(r|y) = 0$. Thus, the TOEness of δ is not sufficient for $g(\delta)$ to be a TOE . In a previous section, we illustrated via examples that MLE from the h-loglihood of the canonical scale gives meaningful estimators of random parameters, which can be interpreted as estimators of the TOE on particular scales. For an estimation of random effects we should use the h-likelihood.

Example 4.7: Consider a linear model

$$
y_i = x_i \beta + e_i.
$$

Given the data $Y = y$, the MLE for the error $e_i = y_i - x_i \beta$ is the residual

$$
\hat{e}_i = y_i - x_i \hat{\beta}.
$$

Here the TOE δ for e_i is e_i itself because

$$\delta = E(e_i|y) = E(y_i - x_i\beta|y) = y_i - x_i\beta = e_i.$$

Thus, the MLE of e_i is

$$\hat{e}_i = \hat{E}(e_i|y) = E(e_i|y)|_{\beta=\hat{\beta}} = y_i - x_i\hat{\beta},$$

so that

$$\hat{e}_i - e_i = x_i(\hat{\beta} - \beta) = o_p(1)$$

is asymptotically normal for arbitrary distribution of e_i. Given y, the $100(1 - \alpha)\%$ confidence interval for realized value of e_i is $\hat{e}_i \pm z_{\alpha/2}\sqrt{\text{var}(x_i\hat{\beta})}$ with $z_{\alpha/2}$ being the appropriate value from the normal table. This interval is based on asymptotic normality. In the next sections, we discuss how to predict a future outcome $y_{n+1} = x_{n+1}\beta + e_{n+1}$, where asymptotic normality cannot be applied.

Now we discuss the general case when the consistency fails.

Example 4.8: Consider the model for paired data, with $i = 1, ..., m$ and $j = 1, 2$,

$$y_{ij} = \beta_0 + \beta_j + v_i + e_{ij}, \qquad (4.16)$$

where β_0 is the intercept, β_1 and β_2 are treatment effects for the two groups and the white noise $e_{ij} \sim N(0, \phi)$. For a moment, assume v_k is a fixed unknown. For a contrast we have the MLE of $v_k - v_t$ for $k \neq t$ to be

$$\widehat{v_k - v_t} = \bar{y}_{k\cdot} - \bar{y}_{t\cdot}.$$

where $\bar{y}_{i\cdot} = \sum_j y_{ij}/J$, and have

$$\text{var}(\widehat{v_k - v_t}) = 2\phi/J.$$

Thus, the MLE is no longer consistent, having

$$\text{var}(\widehat{v_k - v_t}) = \phi > 0.$$

The linear model theory gives correct variance estimation, but a correct confidence interval normality of e_{ij} is crucial. To have a correct confidence interval in general, we need the confidence density $c(\theta)$. In practice, the most convenient way of computing $c(\theta)$ or approximating $c(\theta)$ would be to use the bootstrap distribution of θ (Pawitan, 2001, Chapter 5); for more discussion see Xie and Singh (2013).

Now let $v_i \sim N(0, \lambda)$. For identifiability, we put a constraint $E(v_i) = 0$ upon individual random effects. Without such a constraint, the individual v_i is not estimable, although the contrasts are. Here the targets for v are given by

$$\delta_i = E(v_i|y) = \frac{2\lambda}{2\lambda + \phi}\{\bar{y}_{i\cdot} - \beta_0 - (\beta_1 + \beta_2)/2\}, \qquad (4.17)$$

so that, given dispersion parameters (λ, ϕ), the MLE for δ_i is given by

$$\begin{aligned}
\hat{v}_i &= \hat{\delta}_i = E(v_i|Y)|_{\beta=\hat{\beta}} = \frac{2\lambda}{2\lambda + \phi}\{\bar{y}_{i\cdot} - \hat{\beta}_0 - (\hat{\beta}_1 + \hat{\beta}_2)/2\} \\
&= \frac{2\lambda}{2\lambda + \phi}(\bar{y}_{i\cdot} - \bar{y}_{\cdot\cdot}),
\end{aligned}$$

where $\bar{y}_{i\cdot} = (y_{i1} + y_{i2})/2$ and $\bar{y}_{\cdot\cdot} = \sum_i \bar{y}_{i\cdot}/m$. Because

$$\hat{v}_i - \delta_i = \frac{2\lambda}{2\lambda + \phi}[\beta_0 - \hat{\beta}_0 + \{(\beta_1 + \beta_2) - (\hat{\beta}_1 + \hat{\beta}_2)\}/2] = o_p(1),$$

$\hat{v}_i - \delta_i$ is asymptotically normal with a zero mean as m goes to infinity. However, because $\mathrm{var}(v_i|y) > 0$, v_i cannot be consistently estimated by $\hat{v}_i$. As shown earlier, the h-loglihood provides a consistent estimator for $\mathrm{var}(\hat{v}_i - v_i)$. However, for interval estimations, normality assumptions, $v_i \sim N(0, \lambda)$ and $e_{ij} \sim N(0, \phi)$ are crucial. We discuss finding an interval estimation for v_i under general models in the following sections. $\square$

4.7 Prediction of future outcome

The nature of the prediction problem is to extract information from the data to be able to say something about a not-yet realized random quantity. In prediction problems, to get inferences about an unobserved future observation U, we usually have to deal with unknown fixed parameters θ. Here we use the capital letter for an unobserved future observation to emphasize that it is not fixed based on the data.

Suppose we observe $Y_1 = y_1, ..., Y_n = y_n$ from iid $N(\mu, \sigma^2)$, where μ is not known but σ^2 is, and denote the sample average by $\bar{y}$. Let $U = Y_{n+1}$ be an unobserved future observation (TOP). Then

$$Y_{n+1} - \bar{y} \sim N(0, \sigma^2(1 + 1/n))$$

from which we can get a correct $100(1 - \alpha)\%$ prediction interval for U as

$$\bar{y} \pm z_{\alpha/2}\sigma\sqrt{1 + 1/n}.$$

Now we want to investigate how to reach such an interval by using the predictive probability

$$P_\mu(U|y) = f_\mu(U|Y_1 = y_1, ..., Y_n = y_n) = P_\mu(U),$$

on observing the data $Y = y$. This predictive probability does not carry information about U in the data, which seems surprising as we *think* that knowing the data should tell us something about μ and hence the future U.

An *ad hoc* solution is simply to specify that U follows $N(\bar{x}, \sigma^2)$, which is a short way of saying that we want to estimate $P_\mu(U)$ by $P_{\bar{y}}(U)$, using $\hat{\mu} = \bar{y}$. Asymptotically, $P_{\bar{y}}(U)$ is the MLE for a function $P_\mu(U)$. This *plug-in (classical ML) solution* is in the same spirit as the empirical Bayes (EB) approach. The weakness of this approach is obvious: it does not account for the uncertainty of $\hat{\mu}$ (nuisance parameter estimator) in the prediction. This gives a prediction interval

$$\bar{x} \pm z_{\alpha/2}\sigma$$

which could be far from the correct solution if n is small. We need a finite sample adjustment.

Let us consider an h-loglihood solution to this problem. First, the h-loglihood is

$$\begin{aligned}
\ell(\mu, U) &= \log f_\mu(U, Y = y) \\
&= \log L(\mu; y) + \log P_\mu(U|y) \\
&= -[(n+1)\log\sigma^2 + \{\sum(y_i - \mu)^2 + (U - \mu)^2\}/\sigma^2]/2.
\end{aligned}$$

We see that the EB approach uses only the $P_\mu(U|y)$ component in the h-likelihood, but the $L(\mu; y)$ component has information on uncertainty in estimating μ. Thus, we should use the whole h-likelihood for a proper inference. Now we can show immediately that U is in the canonical scale, so we have the h-loglihood $h(\mu, U) = \ell(\mu, U)$ and we can have combined inference of (μ, U). The joint maximization gives solutions

$$\begin{aligned}
\hat{U} &= \hat{\mu} \\
\hat{\mu} &= (U + n\bar{y})/(n+1) = \bar{y},
\end{aligned}$$

so that the MLE is an optimal estimator for $\delta = E(U|y)$

$$\begin{aligned}
\hat{U} &= E(U|Y_1 = y_1, ..., Y_n = y_n)|_{\mu=\bar{y}} \\
&= E(U)|_{\mu=\bar{y}} \\
&= \bar{y}.
\end{aligned}$$

Because

$$-\partial^2 h/\partial U^2 = 1/\sigma^2, \quad -\partial^2 h/\partial U\partial\mu = -1/\sigma^2, \quad -\partial^2 h/\partial\mu^2 = (n+1)/\sigma^2,$$

the Hessian matrix gives a variance estimate

$$\text{var}(U - \hat{U}) = \sigma^2(1 + 1/n),$$

from which we can derive the correct prediction interval that accounts for estimation of μ. However, this approach works when $U|y$ follows the normal distribution. Now a question is how to perform a finite sample adjustment under a general model.

4.8 Finite sample adjustment

A full likelihood-based inference, including prediction (or confidence) intervals, for future (or realized values of) random parameters is a harder problem, particularly since in most applications the information available for a random parameter stays limited even if the total sample size increases, so we cannot rely on asymptotic normality. In the previous example, if U is not normal there is no asymptotically normal consistent estimator for $U = Y_{n+1}$.

A simplest way of prediction is to use the EB method

$$P^E(U|y) \equiv P_{\hat\theta}(U|y). \qquad (4.18)$$

We see that this MLE (plug-in method) is convenient and asymptotically optimal in estimating the predictive probability function $P_\theta(U|y)$, but ignores the added uncertainty due to the estimation of $\hat\theta$. When θ is unknown, the Hessian matrix of the estimated posterior $f_{\hat\theta}(U|y)$ for deriving an EB procedure underestimates $\text{var}(\hat U - U)$, because

$$\text{var}(\hat U - U) \geq E\{\text{var}(U|y)\},$$

where the right-hand side is in general the naive EB variance estimator obtainable from $P_{\hat\theta}(U|y)$. Various procedures have been suggested for the EB interval estimate (e.g., Carlin and Louis, 2000). Lee and Nelder (2004) showed that the proper variance estimator can be obtained from the Hessian matrix from the h-likelihood. However, the first two moment estimators are not sufficient for prediction intervals unless $U|y$ follows normal distribution. See Lee and Nelder (2009) for the h-likelihood approach when $U|y$ follows non-normal distributions.

Early non-Bayesian efforts to define a proper likelihood for random parameters — called predictive likelihood (predictive probability of this book) — can be traced to Lauritzen (1974) and Hinkley (1979). Suppose (y, U) has a joint density $f_\theta(y, U)$, and $R(y, U)$ is a non-trivial sufficient statistic for θ, so that the conditional distribution of (y, U) given $R = r$ is free of θ; thus the predictive probability of U alone is

$$L(U; (y, U)|r) = \frac{f_\theta(y, U)}{f_\theta(r(y, U))}.$$

Example 4.9: Suppose the observed y is binomial with parameters n and θ, and the unobserved U is binomial with parameters m and θ. The number of trials m and n are known. Intuitively, knowing y should tell us something about U. The statistic $r = y + U$ is sufficient for θ, and given r the conditional probability of $(Y = y, U = u)$ is given by the hypergeometric probability

$$P(Y = y, V = v|r) = \frac{\binom{n}{y}\binom{m}{v}}{\binom{m+n}{v+y}},$$

so the predictive likelihood of v is

$$L(v;(y,v)|r) = \frac{\binom{n}{y}\binom{m}{v}}{\binom{m+n}{v+y}},$$

which is non-normal and free of θ.□

When available, this predictive probability accounts for the finite sample size. However, the need to have a non-trivial sufficient statistic to remove the nuisance parameter θ restricts the application of this approach. We need a technique that works more generally.

Lee and Kim (2016) proposed using $c(\theta)$, the confidence density of θ described in Section 1.11, as a weight function to find a predictive probability free of θ:

$$P(U|y) \equiv \int f_\theta(U|y)c(\theta)d\theta, \qquad (4.19)$$

while the uncertainty in the unknown θ is propagated by $c(\theta)$. This marginalizing step is justified by the fact that $c(\theta)$ has a corresponding coverage probability interpretation. The result looks like a Bayesian posterior density, but since U has an objective distribution, e.g., it can be assessed from the data, and we never use any prior distribution for θ, this is fundamentally a non-Bayesian approach. To apply this method, we need the confidence density $c(\theta)$.

The simplest alternative is to assume $c(\theta)$ is concentrated at $\widehat{\theta}$, which we can obtain from the marginal likelihood $L(\theta;y)$ and so we get

$$P^E(U|y) \equiv f_{\widehat{\theta}}(U|y).$$

If the estimate of θ based on the marginal likelihood $L(\theta;y)$ is based on a large sample size, we may consider a normal approximation, i.e., $c(\theta)$ is the normal density function of

$$N(\widehat{\theta}, I^{-1}(\widehat{\theta})),$$

where $I^{-1}(\widehat{\theta})$ is the observed Fisher information from $L(\theta;y)$. When the normal approximation is in doubt, we might consider the normalized $L(\theta;y)$ as an approximate $c(\theta)$. In location problems, the approximation is in fact exact. In this case (4.19) becomes

$$\begin{aligned} P(U|y) &\equiv k\int f_\theta(U|y)L(\theta;y)d\theta \\ &= k\int L(\theta,U;y,U)d\theta, \qquad (4.20) \end{aligned}$$

where k is a normalizing constant, so $P(U|y)$ is a predictive probability.

This is equivalent to the Bayesian predictive posterior under the uniform prior on θ.

We now return to the prediction problem in the previous section, where $f_\mu(U|y) = f_\mu(U)$ is the density function of $N(\mu, \sigma^2)$ and $c(\mu)$ is that of $N(\bar{y}, \sigma^2/n)$. It can be shown that

$$P(U|y) = \int f_\mu(U|y)c(\mu)d\mu = \frac{1}{\sqrt{2\pi\sigma^2(1 + 1/n)}} \exp \frac{-(U - \bar{y})^2}{2\sigma^2(1 + 1/n)}$$

is the density of $N(\bar{y}, \sigma^2(1 + 1/n))$. Here approximated $c(\mu)$s based on a normal approximation and a normalized likelihood become exact $c(\mu)$ so that these probabilities give exact predictive intervals for U, accounting for information loss caused by estimating μ.

From (1.20) in Chapter 1, the Laplace approximation of the integral (4.20) is in fact given by the adjusted profile loglihood

$$\ell_a(U) \equiv \ell_e(\widehat{\theta}_U, U) - \frac{1}{2} \log |I(\widehat{\theta}_U)/(2\pi)|,$$

where $\widehat{\theta}_U$ is the MLE of θ for each value of U, and $I(\widehat{\theta}_U)$ is the observed Fisher information for fixed U. This shows clearly that (4.20) accounts for the estimation of the unknown θ in the same way as the adjusted profile likelihood.

Finally, because the bootstrap distribution can be used as an approximation to the confidence density $c(\theta)$ (Pawitan, 2001, Chapter 5), Lee and Kim (2016) propose to use the bootstrap method to calculate the predictive probability (4.19):

$$P^B(U|y) \equiv \frac{1}{B} \sum_{j=1}^{B} f_{\theta_j^*}(U|y),$$

where $\theta_1^*, \ldots, \theta_B^*$ are the bootstrap replicates of $\widehat{\theta}$. In complex models used in this book, it may not be straightforward to design the bootstrap scheme, so that it is convenient to generate the bootstrap replicates of $\widehat{\theta}$ from the asymptotic normal distribution of $\widehat{\theta}$ or the normalized likelihood. The latter works slightly better than the former because the latter procides a better approximation to confidence density. Via simulation studies, Lee and Kim (2016) demonstrate that these bootstrap methods provide excellent prediction intervals for future random effects. Bootstrap also gives good confidence intervals for realized values of random effects as in Example 4.8, but care is necessary to maintain the stated level at extreme realized values.

4.9 Is marginal likelihood enough for inference about fixed parameters?

In this chapter we demonstrated that the use of marginal likelihood for inferences about fixed parameters is in accordance with the classical likelihood principle. The question is whether the marginal likelihood provides all the useful likelihood procedures for inferences about fixed parameters. The answer is no when the assumed model is not correct. The h-likelihood can represent a new likelihood procedure that cannot be derived from the marginal likelihood. Error components in marginal models are often correlated, while those in conditional (random effect) models can be orthogonal, so that various residuals can be developed for model checking, as we shall see later. Lee (2002a) showed by an example that with h-likelihood we can define a robust sandwich variance estimator for models not currently covered. There are two unrelated procedures in extended likelihood:

(a) Marginalization by integrating out random effects

(b) Sandwich variance estimation

Starting with the extended likelihood, if we apply (a) and then (b), we get the current sandwich estimator used for GEE methods in Chapter 3, while if we apply (b) only we get a new sandwich estimator. The standard sandwich estimator is robust against broader model violation, while the new one is applicable to a wider class of models. Lee (2002a) showed that if we apply the two sandwich estimators to mixed linear models, the standard one is robust against misspecifications of correlations, while the new one is robust against heteroscedasticity only; the standard one cannot be applied to crossed designs, while the new one can. Thus, likelihood inferences can be enriched by use of the extended likelihood.

4.10 Summary: likelihoods in extended framework

Classical likelihood is for inferences about fixed parameters. For general models allowing unobserved random variables, the h-likelihood is the fundamental treatment from which the marginal (or classical) and restricted likelihoods can be derived as adjusted profile likelihoods for inferences of fixed effects. Furthermore, likelihood inferences are possible for latent random effects, missing data, and unobserved future observations based on the h-likelihood and the predictive probability . We now discuss some general issues concerning the use of extended likelihood inferences.

In dealing with random parameters, must we use the extended likelihood framework? There is no simple answer to this; we can go back one step and ask, in dealing with *fixed* parameters, must we use classical likelihood? From frequentist perspectives, we might justify the likelihood from large sample optimality, but in small samples there is no such guarantee. Here we might invoke the principle that the likelihood contains all the information about the parameter, although the process of estimation by maximization arises from convenience rather than following strictly from the principle. For many decades likelihood-based methods were not the dominant paths to estimation. The emergence of likelihood methods was a response to our needs in dealing with complex data, such as non-normal or censored data.

These reasonings seem to apply also to the extended likelihood, where the estimation or prediction of the random parameters typically relies on small samples, so that we cannot justify the likelihood from optimality consideration only. We have the extended likelihood principle to tell us why we should start with the extended likelihood, although it does not tell us what to do in practice. It is possible to devise non-likelihood-based methods, e.g., by minimizing the MSE, but they are not easily extendable to various non-normal models and censored data. We believe that the extended likelihood framework will fill our needs in the same way the classical likelihood helps us in modelling complex data.

The counter-examples associated with the extended likelihood can be explained as the result of a blind joint maximization of the likelihood. We show that such a maximization is meaningful only if the random parameter has a special scale, which in some sense is information-free for the fixed parameter. In this case the extended likelihood is called h-likelihood, and we show that joint inferences from the h-likelihood behave like inferences from an ordinary likelihood. The canonical scale definition did not appear in Lee and Nelder (1996), although they stated that the random effects must appear linearly in the linear predictor scale, which in the context of hierarchical GLMs amounts to a canonical-scale restriction (see Chapter 6).

Regarding the lack of invariance in the use of extended likelihood, it might be useful to draw an analogy: Wald tests or confidence intervals (Section 1.5) are well known to lack invariance in that trivial re-expression of the parameters can lead to different results. To use Wald statistics, we must be aware of what particular scale of the parameter is appropriate; once the scale is known, the Wald-based inference is very convenient and in common use.

If the canonical scale for a random effect exists, must we use it? Yes, if we want to use joint maximization of fixed and random effects from the extended likelihood. The use of canonical scale simplifies the inferences of all the parameters from the h-likelihood.

The canonical-scale requirement in using the extended likelihood seems restrictive. Maintaining invariance of inferences from the joint maximization of the extended loglihood for trivial re-expressions of the underlying model leads to a definition of the scale of random parameters for the h-likelihood (Chapter 6), which covers the broad class of GLM models.

We may regard this scale as a weak canonical and study models allowing such scale. However, there exist models which cannot be covered by such a condition. For those models we propose the adjusted profile likelihoods for inferences for fixed parameters, which often give satisfactory estimations. We see explicitly in Example 4.6 that this approach gives a correct estimation for fixed parameters even if the scale is incorrect. Thus, only the joint estimation is not possible. Adjusted profile likelihoods and predictive probabilities do not presume a canonical scale.

Even if we focus on inferences about fixed parameters from extended models, likelihood inferences are often hampered by analytically intractable integrals. Numerical integration is often not feasible when the dimensionality of the integral is high. This led Lee and Nelder (1996) to introduce the h-likelihood for the GLM with random effects. Another criticism concerns the statistical inefficiency of certain estimates derived from the h-likelihood caused by using the raw h-likelihood when the number of nuisance parameters increases with sample size.

We can avoid this problem by using the proper adjusted profile likelihood. The other problem related to the statistical efficiency of the h-likelihood method is an unawareness of the difference between h-likelihood and the severely biased penalized likelihood method of Breslow and Clayton (1993) (Chapter 6). We elaborate on this more in later chapters by explaining how the extended likelihood framework can give statistically efficient estimations.

With an extended likelihood framework the standard error estimates are straightforwardly obtained from the Hessian matrix. In other methods, such as the EM algorithm, a separate procedure may be necessary to obtain these estimates.

We show how to establish optimality of estimation of random parameters and their prediction. The predictive probability is useful but requires heavy computation. For many problems, we may only need to compute the ratios as we shall study in Chapter 13 because

$$\frac{P_\theta(u_1|y)}{P_\theta(u_2|y)} = \frac{f_\theta(u_1|y)f_\theta(y)}{f_\theta(u_2|y)f_\theta(y)} = \frac{f_\theta(u_1, y)}{f_\theta(u_1, y)}$$

allows us to obtain the ratio of predictive probability via the ratio of h-likelihoods. To maintain the invariance of likelihood ratio with respect to a transformation of random parameters, we should use the h-likelihood.

CHAPTER 5

Normal linear mixed models

In this chapter, linear models are extended to models with additional random components. We start with the general forms of the models and describe specific models as applications. Let y be an N-vector of responses, and X and Z be $N \times p$ and $N \times q$ model matrices for the *fixed effect* parameters β and *random effect* parameters v. The standard linear mixed model specifies

$$y = X\beta + Zv + e, \tag{5.1}$$

where $e \sim MVN(0, \Sigma)$, $v \sim MVN(0, D)$, and v and e are independent. The variance matrices Σ and D are parameterized by an unknown variance component parameter τ, so random effect models are also known as variance component models. The random effect term v is often assumed to be $MVN(0, \sigma_v^2 I_q)$, and the error term $MVN(0, \sigma_e^2 I_N)$, where I_k is a $k \times k$ identity matrix, so the variance component parameter is $\tau = (\sigma_e^2, \sigma_v^2)$.

If inferences are required about the fixed parameters only, they can be made from the implied multivariate normal model

$$y \sim MVN(X\beta, V),$$

where

$$V = ZDZ' + \Sigma.$$

For known variance components, the MLE

$$\hat{\beta} = (X^t V^{-1} X)^{-1} X^t V^{-1} y \tag{5.2}$$

is the BLUE and BUE. When the variance components are unknown, we can plug in the variance component estimators, resulting in a non-linear estimator for the mean parameters.

The simplest random effect model is the classical one-way layout

$$y_{ij} = \mu + v_i + e_{ij}, \ i = 1, \ldots, q, \ j = 1, \ldots, n_i \tag{5.3}$$

where μ is a fixed overall mean parameter. The index i typically refers to a cluster and the vector $y_i = (y_{i1}, \ldots, y_{in_i})$ to a set of measurements taken from the cluster. Thus, a cluster may define a person, a family or an arbitrary experimental unit on which we obtain multiple measurements. It is typically assumed that v_is are iid $N(0, \sigma_v^2)$, the e_{ij}s are iid $N(0, \sigma_e^2)$ and all these random quantities are independent. It is clear that the total variance of the observed y_{ij} is given by

$$\sigma_y^2 = \sigma_v^2 + \sigma_e^2,$$

so σ_v^2 and σ_e^2 are truly the components of the total variation. The most common question in one-way layout problems is whether there is a significant heterogeneity among clusters (cluster effect), i.e., whether $\sigma_v^2 > 0$ or $\sigma_v^2 = 0$. If the measurements include other predictors, we might consider a more complex model

$$y_{ij} = x_{ij}^t \beta + v_i + e_{ij},$$

where x_{ij} is the vector of covariates. In this model v_i functions as a random intercept term.

It is well known that Gauss and Legendre independently discovered the method of least squares to solve problems in astronomy. We may consider least squares as the original development of fixed effect models. It is less well known, however, that the first use of random effects was also for an astronomical problem. Airy (1861), as described in Scheffe (1956), essentially used the one-way layout (5.3) to model measurements on different nights. The ith night effect v_i, representing the *atmospheric and personal circumstances* peculiar to that night, was modelled as random. He then assumed that all the effects and the error terms were independent. To test the between-night variability $\sigma_v^2 = 0$, he used the mean absolute deviation statistic

$$d = \frac{1}{q} \sum_i |\bar{y}_{i.} - \bar{y}_{..}|,$$

where $\bar{y}_{i.}$ is the average from night i and $\bar{y}_{..}$ is the overall average. Fisher's (1918) paper on population genetics introduced the terms 'variance' and 'analysis of variance' and used a random effect model. The more general mixed model was implicit in Fisher's (1935) discussion of variety trials in many locations.

Example 5.1:   Figure 5.1(a) shows the foetal birth weights of 432 boys from 108 families of size 4. The data were plotted by families and these families were ordered by the family means $\bar{y}_{i.}$. It is clear from the plot that there is a strong familial effect in birth weight, presumably due to both genetic and environmental influences. The figure also indicates that the variability is largely constant across all the families. Subplot (b) shows that the within-family variation is normally distributed and (c) shows that the family means

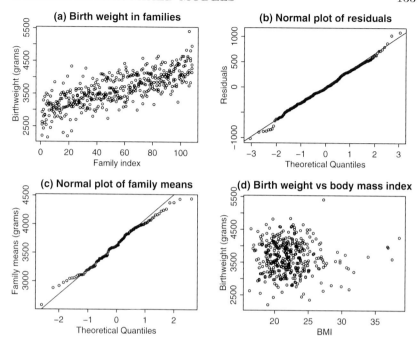

Figure 5.1 *Birth weights of 432 boys from 108 families of size 4. In (a) the data were plotted by families and these families were ordered by the family means $\bar{y}_{i\cdot}$.*

have slightly shorter tails than the normal distribution. Overall, the data follow the standard assumptions of the simple random effect model. The first question, whose answer seems obviously yes, is whether there is a significant between-family variation relative to within-family variation. Second, we might want to estimate the variance components and quantify the extent of the familial effect. Subplot (d) shows little association between foetal birth weight and the body-mass index of the mother, so the familial effect in birth weight is not simply due to the size of the mother. □

Example 5.2: Suppose from individual i we measure a response y_{ij} and corresponding covariate x_{ij}. We assume that each individual has his own regression line, i.e.,

$$
\begin{aligned}
y_{ij} &= (\beta_0 + v_{0i}) + (\beta_1 + v_{1i})x_{ij} + e_{ij}, \ i = 1, \ldots, q, \ j = 1, \ldots, n_i \\
&= \beta_0 + \beta_1 x_{ij} + v_{0i} + v_{1i} x_{ij} + e_{ij}.
\end{aligned}
$$

Assuming (v_{0i}, v_{1i}) are iid normal with mean zero and variance matrix D_i, we have a collection of regression lines with average intercept β_0 and average slopes β_1.

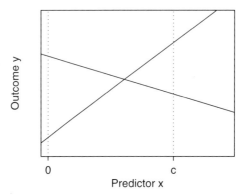

Figure 5.2 *In a random-slope regression model, the choice of predictor origin affects the correlation between the random intercept and the random slope. On the original scale x, the intercept and slope are negatively correlated, but if we shift the origin to x − c they are positively correlated.*

This small extension of the simple random effect model illustrates two important points when fitting mixed models. First, the model implies a particular structure of the covariance matrix of $(y_{i1}, \ldots, y_{in_i})$ which in turn depends on the observed covariates $(x_{i1}, \ldots, x_{in_i})$. Second, it is wise to allow some correlation between the random intercept and the random slope parameters. However, the correlation term may not be interpretable, since it is usually affected by the scale of x_{ij}, i.e., it can be changed if we shift the data to $x_{ij} - c$ for an arbitrary constant c. In Figure 5.2, on the original scale of predictor x, a high intercept is associated with a negative slope and vice versa, i.e., they are negatively correlated. But if we shift the origin by c, i.e., using $x - c$ as predictor, a high intercept is now associated with a positive slope and they are now positively correlated. □

5.1 Developments of normal mixed linear models

There are two distinct strands in the development of normal mixed models. The first occurred in experimental design, where the introduction of split-plot and split-split-plot designs led to models with several error components. Here the main interest is on inferences about means, namely treatment effects. The second strand arose in variance component models, for example, in the context of animal-breeding experiments, where the data are unbalanced, and the interest is not so much on the size of the variances of the random effects but rather on the estimation of the random effects themselves.

5.1.1 Experimental design

The first explicit formulation of a model for a balanced complete block design appears to be by Jackson (1939) in a paper on mental tests. In his model, the response y_{ij} of subject i on test j is assumed to be

$$y_{ij} = \mu + v_i + \beta_j + e_{ij}, \quad i = 1, \cdots, q; \ j = 1, \ldots, p \qquad (5.4)$$

where the subject effect $v_i \sim N(0, \sigma_v^2)$ is a random parameter and the test effect β_j is a fixed parameter. Here a contrast such as $\delta = \beta_1 - \beta_2$ can be estimated by $\hat{\delta} = \bar{y}_{\cdot 1} - \bar{y}_{\cdot 2}$, where $\bar{y}_{\cdot k} = \sum_i y_{ij}/q$. This linear estimator is BLUE and BUE under normality, and an exact F-test is available for significance testing. Even though $\hat{\delta}$ can be obtained from the general formula (5.2), the resulting estimator does not depend upon the variance components, i.e., $\hat{\delta}$ is the BLUE even when the variance components are unknown.

Furthermore, $\hat{\delta}$ can be obtained from the ordinary least squares (OLS) method by treating v_i as fixed; this we shall call the *intra-block estimator*. This helpful property holds in many balanced experimental designs, where we can then proceed with inferences using OLS methods.

Yates (1939) found that this happy story is not true in general. Consider the following balanced but incomplete block design, where three treatments are assigned to three blocks of the size two. Observations (treatments, A, B, C) are shown in the following table:

Block 1	Block 2	Block 3
y_{11} (A)	y_{21} (A)	y_{31} (B)
y_{12} (B)	y_{22} (C)	y_{32} (C)

We can consider the same model (5.4) with v_i for the block effect, but the design is incomplete because three treatments cannot be accommodated in blocks of size two. Here the intra-block estimator for $\delta = \beta_A - \beta_B$ (by treating v_i as fixed) is $\hat{\delta} = y_{11} - y_{12}$. However, assuming random block effects, another linear estimator is available from the block means, namely $2(\bar{y}_{2\cdot} - \bar{y}_{3\cdot})$. This means that the intra-block estimator does not use all the information in the data, i.e., there is information about δ in the inter-block contrasts and this should be recovered for efficient inferences.

Intra-block estimators can be obtained also from the conditional likelihood, by conditioning on block totals; see Section 6.3.1. However, as clearly seen in this simple example, conditioning loses information on inter-block contrasts. Thus the unconditional analysis of the random effect model in this case leads to the so-called recovery of inter-block information.

This result also holds when the design is unbalanced. Consider the one-way random effect model (5.3). We have $\bar{y}_{i.} \sim N(\mu, \sigma_i^2)$, where $\sigma_i^2 = \sigma_v^2\{1 + \sigma_e^2/(n_i\sigma_v^2)\}$, and the BLUE for μ is

$$\sum_i (\bar{y}_{i.}/\sigma_i^2) / \sum_i (1/\sigma_i^2).$$

As σ_v^2 approaches ∞, the unweighted sample mean

$$\sum_i \bar{y}_{i.}/q$$

becomes the BLUE, and has often been recommended for making confidence intervals with unbalanced models (Burdick and Graybill, 1992). In unbalanced models, Saleh et al. (1996) showed that such linear estimators do not use all the information in the data, so that they can be uniformly improved. The unweighted mean can be viewed as an extension of the intra-block estimator. Unweighted sums of squares are often used for inferences about random components, and this can be similarly extended (Eubank et al., 2003) as a limit as $\sigma_v^2 \to \infty$.

For general models (5.1) Zyskind (1967) showed that a linear estimator $a^t y$ is BLUE for $E(a^t y)$ if and only if

$$Va = Xc$$

for some c. It turns out, however, that linear estimators are generally not fully efficient, so that ML-type estimators should be used to exploit all the information in the data.

5.1.2 Generally balanced structures

Within the class of experimental designs, those exhibiting general balance have a particularly simple form for the estimates of both fixed effects and variance components (Nelder, 1965a, 1965b and 1968). Such designs can be defined by a block structure for the random effects, a treatment structure for the (fixed) treatment effects, and a model matrix showing which treatments are to be applied to each experimental unit (plot). The block structure is orthogonal, i.e., decomposes into mutually orthogonal subspaces called strata. Within each stratum the (fixed) parameters of each term in the treatment structure are estimated with equal information (which may be zero). The variance components are estimated by equating the error terms in the corresponding analysis of variance to their expectations. Finally, these estimates are used to form weights in combining information on the treatment effects over strata. For the most general account that introduces the idea of general combinability; see Payne and Tobias (1992). General balance enables one to recover inter-block information simply by combining information among strata.

5.2 Likelihood estimation of fixed parameters

If the interest is only about fixed parameters, marginal likelihood inferences can be made from the multivariate normal model

$$y \sim MVN(X\beta, V).$$

It is instructive to look closely at the theory of the simplest random effect model. Consider the one-way random effect model (5.3)

$$y_{ij} = \mu + v_i + e_{ij}, \ i = 1, \ldots, q, \ j = 1, \ldots, n_i \tag{5.5}$$

where for simplicity, we shall assume that the data are balanced in the sense that $n_i \equiv n$. Measurements within a cluster are correlated according to

$$\mathrm{Cov}(y_{ij}, y_{ik}) = \sigma_v^2, \ j \neq k,$$

and $\mathrm{var}(y_{ij}) = \sigma_v^2 + \sigma_e^2$. So, $y_i = (y_{i1}, \ldots, y_{in})^t$ is multivariate normal with mean $\mu 1$, and the variance matrix has the so-called compound-symmetric structure

$$V_i = \sigma_e^2 I_n + \sigma_v^2 J_n \tag{5.6}$$

where J_n is an $n \times n$ matrix of ones. Setting $\tau = (\sigma_e^2, \sigma_v^2)$, the loglihood of the fixed parameters is given by

$$\ell(\mu, \tau) = -\frac{q}{2} \log |2\pi V_i| - \frac{1}{2} \sum_i (y_i - \mu 1)^t V_i^{-1} (y_i - \mu 1),$$

where μ is subtracted element-by-element from y_i. To simplify the likelihood, we use the formulae (e.g., Rao 1973, page 67)

$$
\begin{aligned}
|V_i| &= \sigma_e^{2(n-1)}(\sigma_e^2 + n\sigma_v^2) \\
V_i^{-1} &= \frac{I_n}{\sigma_e^2} - \frac{\sigma_v^2}{\sigma_e^2(\sigma_e^2 + n\sigma_v^2)} J_n,
\end{aligned} \tag{5.7}
$$

where I_n is an $n \times n$ identity matrix and J_n is an $n \times n$ matrix of ones. Thus,

$$
\begin{aligned}
\ell(\mu, \tau) \;=\; & -\frac{q}{2}[(n-1)\log(2\pi\sigma_e^2) + \log\{2\pi(\sigma_e^2 + n\sigma_v^2)\}] \\
& -\frac{1}{2}\left\{\frac{SSE}{\sigma_e^2} + \frac{SSV + qn(\bar{y}_{..} - \mu)^2}{\sigma_e^2 + n\sigma_v^2}\right\}
\end{aligned}
\tag{5.8}
$$

where we have defined the error and cluster sums of squares respectively as

$$
SSE \;=\; \sum_i \sum_j (y_{ij} - \bar{y}_{i.})^2
$$

$$
SSV \;=\; n\sum_i (\bar{y}_{i.} - \bar{y}_{..})^2.
$$

It is clear that for any fixed (σ_e^2, σ_v^2), the MLE of μ is the overall mean $\bar{y}_{..}$, so the profile likelihood of the variance components is given by

$$
\begin{aligned}
\ell_p(\tau) \;=\; & -\frac{q}{2}[(n-1)\log(2\pi\sigma_e^2) + \log\{2\pi(\sigma_e^2 + n\sigma_v^2)\}] \\
& -\frac{1}{2}\left(\frac{SSE}{\sigma_e^2} + \frac{SSV}{\sigma_e^2 + n\sigma_v^2}\right).
\end{aligned}
\tag{5.9}
$$

This example illustrates that explicit formulation of the likelihood of the fixed parameters, even in this simplest case, is not trivial. In particular, it requires analysis of the marginal covariance matrix V. In general, V will be too complicated to allow an explicit determinant or inverse.

Example 5.3: For the birth weight data in Example 5.1, we first convert the weights into kilograms and obtain the following statistics

$$
\begin{aligned}
\bar{y}_{..} \;&=\; 3.6311 \\
SSV \;&=\; 65.9065 \\
SSE \;&=\; 44.9846.
\end{aligned}
$$

In this simple case it is possible to derive explicit formulae of the MLEs of (σ_e^2, σ_v^2) and their standard errors from (5.9). By setting the first derivatives to zero, provided all the solutions are non–negative, we find

$$
\begin{aligned}
\hat{\sigma}_e^2 \;&=\; SSE/\{q(n-1)\} \\
\hat{\sigma}_v^2 \;&=\; (SSV/q - \hat{\sigma}_e^2)/n,
\end{aligned}
\tag{5.10}
$$

but in general such explicit results in mixed models are rare. In practice we rely on various numerical algorithms to compute these quantities. In this example, it is more convenient to optimize (5.9) directly, including the numerical computation of the second derivative matrix and hence the standard errors; this gives

$$
\begin{aligned}
\hat{\sigma}_e^2 \;&=\; 0.1388 \;\pm\; 0.0077 \\
\hat{\sigma}_v^2 \;&=\; 0.1179 \;\pm\; 0.0148.
\end{aligned}
$$

The result confirms the significant variance component due to family effect. We might express the family effect in terms of the intra-class correlation

$$r = \frac{\hat{\sigma}_v^2}{\hat{\sigma}_v^2 + \hat{\sigma}_e^2} = 0.46,$$

or test the familial effect using the classical F statistic, which in this setting is given by

$$F = \frac{SSV/(q-1)}{SSE/\{q(n-1)\}} = 4.43$$

with $\{q-1, q(n-1)\} = \{107, 324\}$ degrees of freedom. This is highly significant with P-value < 0.000001; the 0.1% critical value of the null F distribution is 1.59. However, this exact normal-based test does not extend easily to unbalanced designs (Milliken and Johnson, 1984). Even in balanced designs, the results for normal responses are not easily extended to those for non-normal responses, as we shall see in the next chapter. □

5.2.1 Inferences about the fixed effects

From the multivariate normal model we have the marginal loglihood of the fixed parameters (β, τ) in the form

$$\ell(\beta, \tau) = -\frac{1}{2} \log |2\pi V| - \frac{1}{2}(y - X\beta)^t V^{-1}(y - X\beta), \qquad (5.11)$$

where the dispersion parameter τ enters through the marginal variance V.

First we show conceptually that multiple-component models are no more complex than single-component models. Extensions of (5.1) to include more random components take the form

$$y = X\beta + Z_1 v_1 + \cdots + Z_m v_m + e,$$

where the Z_i are $N \times q_i$ model matrices, and the v_i are independent $MVN_{q_i}(0, D_i)$. This extension can be written in the simple form (5.1) by combining the pieces appropriately

$$Z = [Z_1 \cdots Z_m]$$
$$v = (v_1 \cdots v_m).$$

The choice whether to have several random components is determined by the application, where separation of parameters may appear naturally.

In some applications, the random effects are iid, so the variance matrix is given by

$$D = \sigma_v^2 I_q.$$

It is also common to see a slightly more general variance matrix

$$D = \sigma_v^2 R,$$

with known matrix R. This can be reduced to the simple iid form by re-expressing the model in the form

$$
\begin{aligned}
y &= X\beta + ZR^{1/2}R^{-1/2}v + e \\
&= X\beta + Z^*v^* + e
\end{aligned}
$$

by defining $Z^* \equiv ZR^{1/2}$ and $v^* = R^{-1/2}v$, where $R^{1/2}$ is the square root matrix of R. Now v^* is $MVN(0, \sigma_v^2 I_q)$. This means that methods developed for the iid case can be applied more generally.

For fixed τ, taking the derivative of the loglihood with respect to β gives

$$
\frac{\partial \ell}{\partial \beta} = X^t V^{-1}(y - X\beta)
$$

so that the MLE of β is the solution of

$$
X^t V^{-1} X\hat{\beta} = X^t V^{-1}y,
$$

the well-known generalized least squares formula. Hence the profile likelihood of the variance parameter τ is given by

$$
\ell_p(\tau) = -\frac{1}{2}\log|2\pi V| - \frac{1}{2}(y - X\widehat{\beta}_\tau)^t V^{-1}(y - X\widehat{\beta}_\tau), \qquad (5.12)
$$

and the Fisher information of β is

$$
I(\widehat{\beta}_\tau) = X^t V^{-1} X.
$$

In practice, the estimated value of τ is plugged into the information formula from which we can find the standard error for the MLE $\widehat{\beta}$ in the form

$$
\begin{aligned}
\widehat{\beta} &= \widehat{\beta}_{\widehat{\tau}} \\
I(\widehat{\beta}) &= X^t V_{\widehat{\tau}}^{-1} X,
\end{aligned}
$$

where the dependence of V on the parameter estimate is made explicit. The standard errors computed from this plug-in formula do not take into account the uncertainty in the estimation of τ, but the formula is nevertheless commonly used. Because $E(\partial^2 \ell/\partial\beta\partial\tau) = 0$, i.e., the mean and dispersion parameters are orthogonal (Pawitan 2001, page 291), this variance inflation caused by the estimation of τ is fortunately asymptotically negligible. However, it could be non–negligible if the design is unbalanced in small samples. In such cases, numerical methods such as the jackknife are useful to estimate the variance inflation in finite samples (Lee, 1991). For finite sample adjustments to t- and F-tests see Kenward and Roger (1997).

In linear models, it is not necessary to have distributional assumptions about y, but only that

$$
E(Y) = X\beta \quad \text{and} \quad \text{var}(Y) = V,
$$

so that the MLE above is the BLUE for given dispersion parameters. The dispersion parameters are estimated by the method of moments using ANOVA. However, this simple technique is difficult to extend to more complex models.

5.2.2 Estimation of variance components

If we include the REML adjustment (Example 1.14) to account for the estimation of the fixed effect β, because $E(\partial^2 \ell / \partial\beta\partial\tau) = 0$, from (1.15) we get an adjusted profile likelihood

$$p_\beta(\ell|\tau) = \ell(\hat{\beta}_\tau, \tau) - \frac{1}{2}\log|X^t V^{-1} X/(2\pi)|.$$

In normal linear mixed models, this likelihood can be derived as an exact likelihood either by conditioning or marginalizing.

Conditional likelihood

Let $\hat{\beta} = Gy$ where $G = (X^t V^{-1} X)^{-1} X^t V^{-1}$. From

$$f(y) = |2\pi V|^{-1/2} \exp\left\{-\frac{1}{2}(y - X\beta)^t V^{-1}(y - X\beta)\right\}$$

and, for fixed τ, $\hat{\beta} \sim MVN(\beta, (X^t V^{-1} X)^{-1})$, so

$$f(\hat{\beta}) = |2\pi(X^t V^{-1} X)^{-1}|^{-1/2} \exp\left\{-\frac{1}{2}(\hat{\beta} - \beta)^t X^t V^{-1} X(\hat{\beta} - \beta)\right\},$$

giving the conditional likelihood

$$f(y|\hat{\beta}) = |2\pi V|^{-1/2}|X^t V^{-1} X/(2\pi)|^{-1/2} \exp\{-\frac{1}{2}(y - X\hat{\beta})^t V^{-1}(y - X\hat{\beta})\}.$$

The loglihood gives $p_\beta(\ell|\tau)$.

Marginal likelihood

The marginal likelihood is constructed from the residual vector. Let

$$P_X \equiv X(X^t X)^{-1} X^t$$

be the hat matrix with rank p. Let A be an $n \times (n-p)$ matrix satisfying $A^t A = I_{n-p}$ and $AA^t = I_n - P_X$. Now $R = A^t y$ spans the space of residuals and satisfies

$$E(R) = 0.$$

R and $\hat{\beta}$ are independent because

$$\mathrm{cov}(R, \hat{\beta}) = 0.$$

Let $T = (A, G)$. Matrix manipulation shows that

$$
\begin{aligned}
f(y) &= f(R, \hat{\beta})|T| \\
&= f(R, \hat{\beta})|T^t T|^{1/2} \\
&= f(R) f(\hat{\beta})|X^t X|^{-1/2}.
\end{aligned}
$$

This residual density $f(R)$ is proportional to the conditional density $f(y|\hat{\beta})$, and the corresponding loglihood is up to a constant term equal to the adjusted profile loglihood $p_\beta(\ell|\tau)$.

Example 5.4: In the simple random effect model (5.5), the model matrix X for the fixed effect is simply a vector of ones, and V^{-1} is a block diagonal matrix with each block given by V_i^{-1} in (5.7). The REML adjustment in the simple random effect model is given by

$$
\begin{aligned}
-\frac{1}{2} \log |X'V^{-1}X/(2\pi)| &= -\frac{1}{2} \log \sum_{i=1}^{q} (2\pi)^{-1} \left(\frac{n}{\sigma_e^2} - \frac{\sigma_v^2 n^2}{\sigma_e^2(\sigma_e^2 + n\sigma_v^2)} \right) \\
&= \frac{1}{2} \log \frac{2\pi(\sigma_e^2 + n\sigma_v^2)}{qn}.
\end{aligned}
$$

This term modifies the profile likelihood (5.9) only slightly: the term involving $\log\{2\pi(\sigma_e^2 + n\sigma_v^2)\}$ is modified by a factor $(q-1)/q$, so that when q is moderately large, the REML adjustment will have little effect on inferences. $\square$

The direct maximization of $p_\beta(\ell|\tau)$ gives the REML estimators for the dispersion parameters. However, we have found the resulting procedure to be very slow. In Section 5.4.4, we study an efficient REML procedure using the extended loglihood.

5.3 Classical estimation of random effects

For inferences about the random effects, it is obvious we cannot use the likelihood (5.11), so we need another criterion. Since we are dealing with random parameters, the classical approach is based on optimizing the mean square error (Section 4.6)

$$E||\hat{v} - v||^2.$$

The resulting $\hat{v}$ is the BLUE for $E(v|y)$ and the BLUP for v. In the general normal mixed model (5.1) we have

$$\hat{v} = E(v|y)|_{\beta=\hat{\beta}} = (Z^t \Sigma^{-1} Z + D^{-1})^{-1} Z^t \Sigma^{-1}(y - X\hat{\beta}). \qquad (5.13)$$

Even if the data are not normal, the formula is BLUE. If β is unknown, we can use its BLUE (5.2) and the resulting estimator is still BLUE for $E(v|y)$.

For the record, we should mention that Henderson *et al.* (1959) recognized that the estimates (5.2) and (5.13) derived for optimal estimation can be obtained by maximizing the *joint density function* [our emphasis] of y and v:

$$\log f(y, v) \propto -\frac{1}{2}(y - X\beta - Zv)^t \Sigma^{-1}(y - X\beta - Zv) - \frac{1}{2}v^t D^{-1}v, \quad (5.14)$$

with respect to β and v. In 1950 he called these the *joint maximum likelihood estimates*. We know from the previous chapter that such a joint optimization works only if the random effects v are the canonical scale for β and this is so here. However, the result is not invariant with respect to non-linear transformations of v; see, e.g., Example 4.3. Later in 1973 Henderson wrote that these estimates should not be called maximum likelihood estimates, since the function maximized is not a likelihood. It is thus clear that he used the joint maximization only as an algebraic device, and did not recognize the theoretical implications in terms of extended likelihood inference.

The derivative of $\log f(y, v)$ with respect to β is

$$\frac{\partial \log f}{\partial \beta} = X^t \Sigma^{-1}(y - X\beta - Zv).$$

Combining this with the derivative with respect to v and setting them to zero gives

$$\begin{pmatrix} X^t \Sigma^{-1} X & X^t \Sigma^{-1} Z \\ Z^t \Sigma^{-1} X & Z^t \Sigma^{-1} Z + D^{-1} \end{pmatrix} \begin{pmatrix} \widehat{\beta} \\ \widehat{v} \end{pmatrix} = \begin{pmatrix} X^t \Sigma^{-1} y \\ Z^t \Sigma^{-1} y \end{pmatrix}. \quad (5.15)$$

The estimates we get from these simultaneous equations are exactly those we get from (5.2) and (5.13). The joint equation that forms the basis for most algorithms in mixed models is often called Henderson's mixed model equation. When D^{-1} goes to zero, the resulting estimating equation is the same as that treating v as fixed. Thus, the so-called intra-block estimator can be obtained by taking $D^{-1} = 0$.

5.3.1 When should we use random effects?

The model (5.5) looks exactly the same whether we assume the v_i to be fixed or random. When should we take effects as random? A common rule which seems to date back to Eisenhart (1947) is that the effects are assumed fixed if the interest is on inferences about the specific values of effects. We believe this to be a misleading view, as it implies that it is not meaningful to estimate random effects. In fact there is a growing list of applications where the quantities of interest are the random effects. Examples are:

- Estimation of genetic merit or selection index in quantitative genetics. This is one of the largest applications of mixed model technology. In animal or plant breeding, the selection index is used to rank animals or plants for improvement of future progenies.
- Time series analysis and the Kalman filter. For tracking or control of a time series observed with noise, the underlying signal assumed is to be random.
- Image analysis and geostatistics. Problems in these large areas include noise reduction, image reconstruction and the so-called small area estimation, for example, in disease mapping. The underlying image or pattern is best modelled in terms of random effects.
- Non–parametric function estimation. This includes estimation of free shapes such as in regression and density functions.

There are also applications where we believe the responses depend on some factors, not all of which are known or measurable. Such unknown variables are usually modelled as random effects. When repeated measurements may be obtained for a subject, the random effect is an unobserved common variable for each subject and is thus responsible for creating the dependence between repeated measurements. These random effects may be regarded as a sample from some suitably defined population distribution.

One advantage of the use of the fixed effect model is that the resulting intra-block analysis does not depend upon distributional assumptions about random effects, even if the random effects were random samples. However, as a serious disadvantage, the use of a fixed effect model can result in a large number of parameters and loss of efficiency. For example, in the one-way random effect model, the full set includes

$$(\mu, \tau, v) \equiv (\mu, \sigma^2, \sigma_v^2, v_1, \ldots, v_q),$$

where $\tau \equiv (\sigma^2, \sigma_v^2)$, and $v \equiv (v_1, \ldots, v_q)$. Thus the number of parameters increases linearly with the number of clusters. For example, in the previous birth weight example, there are $3 + 108 = 111$ parameters. With a random effect specification, we gain significant parsimony as the number of parameters in (μ, τ) is fixed. In such situations, even if the true model is the fixed effect model, i.e., there is no random sampling involved, the use of random effect estimation has been advocated as shrinkage estimation (James and Stein, 1961); see below. Only when the number of random effects is small, for example three or four, will there be little gain from using the random effect model.

5.3.2 Estimation in one-way random effect models

Consider the one-way random effect model (5.3), where up to a constant term we have

$$\log f = -\frac{1}{2\sigma_e^2}\sum_{i=1}^{q}\sum_{j=1}^{n}(y_{ij} - \mu - v_i)^2 - \frac{1}{2\sigma_v^2}\sum_{i=1}^{q}v_i^2. \qquad (5.16)$$

For comparison, if we assume a fixed effect model, i.e., the v_i are fixed parameters, the classical loglihood of the unknown parameters is

$$\ell(\mu, \tau, v) = -\frac{1}{2\sigma_e^2}\sum_{i=1}^{q}\sum_{j=1}^{n}(y_{ij} - \mu - v_i)^2,$$

which does not involve the last term of (5.16). Using the constraint $\sum \hat{v}_i = 0$, we can verify that the MLE of fixed v_i is given by

$$\hat{v}_i^f = \overline{y}_{i.} - \overline{y}_{..}, \qquad (5.17)$$

where $\overline{y}_{i.}$ is the average of $y_{i1}, \ldots, y_{in}$, the MLE of μ is $\overline{y}_{..}$, and the MLE of $\mu_i = \mu + v_i$ is $\overline{y}_{i.}$ (regardless of what constraint is used on the v_i). The corresponding constraint in the random effect model is $E(v_i) = 0$.

For the moment assume that dispersion components τ are known. From the joint loglihood (5.16) we have a score equation

$$\frac{\partial \ell_e}{\partial v_i} = \frac{1}{\sigma_e^2}\sum_{j=1}^{n}(y_{ij} - \mu - v_i) - \frac{v_i}{\sigma_v^2} = 0,$$

which gives the BLUE for $E(v|y)$

$$\hat{v}_i = \alpha(\overline{y}_{i.} - \hat{\mu}) = E(v_i|y)|_{\mu=\hat{\mu}}, \qquad (5.18)$$

where $\alpha = (n/\sigma_e^2)/(n/\sigma_e^2 + 1/\sigma_v^2)$. There is a Bayesian interpretation of this estimate: if v_i is a fixed parameter with a prior $N(0, \sigma^2)$, then $\hat{v}_i$ is called a Bayesian estimate of v_i. Note, however, that there is nothing intrinsically Bayesian in the random effect model since the v_is have an objective distribution, so the coincidence is only mathematical.

In practice the unknown τ is replaced by its estimate. Thus

$$\hat{v}_i = \hat{\alpha}(\overline{y}_{i.} - \overline{y}_{..}), \qquad (5.19)$$

where $\hat{\alpha} = (n/\hat{\sigma}_e^2)/(n/\hat{\sigma}_e^2 + 1/\hat{\sigma}_v^2)$. Comparing this with (5.17) makes it clear that the effect of the random effect assumption is to shrink $\widehat{v}_i$ toward its zero mean. This is why the result is also called a 'shrinkage' estimate. The estimate of μ_i from the random effect model is given by

$$
\begin{aligned}
\widehat{\mu}_i &= \overline{y}_{..} + \widehat{v}_i \\
&= \overline{y}_{..} + \widehat{\alpha}(\overline{y}_{i.} - \overline{y}_{..}) \\
&= \widehat{\alpha}\overline{y}_{i.} + (1 - \widehat{\alpha})\overline{y}_{..}
\end{aligned}
$$

If n/σ_e^2 is large relative to $1/\sigma_v^2$ (i.e., there is a lot of information in the data about μ_i), α is close to one and the estimated mean is close to the i th family(or cluster) mean. On the other hand, if n/σ_e^2 is small relative to $1/\sigma_v^2$, the estimates are shrunk toward the overall mean. This is called an empirical Bayes estimate, as it can be thought of as implementing a Bayes estimation procedure on the mean parameter μ_i, with a normal prior that has mean μ and variance σ_v^2. It is *empirical* since the parameter of the prior is estimated from the data, but, as stated earlier, theoretically it is not a Bayesian procedure.

Example 5.5: Continuing Example 5.3, we use the estimated variance components $\widehat{\sigma}_e^2 = 0.1388$ and $\widehat{\sigma}_v^2 = 0.1179$ to compute $\widehat{\alpha} = 0.77$, so the random effect estimate is given by

$$
\widehat{\mu}_i = 0.77\overline{y}_{i.} + 0.23\overline{y}_{..},
$$

with $\overline{y}_{..} = 3.63$. For example, for the most extreme families (the two families with the smallest and the largest group means) we obtain

$$
\begin{aligned}
\overline{y}_{i.} &= 2.58 \quad \rightarrow \quad \widehat{\mu}_i = 2.82 \\
\overline{y}_{i.} &= 4.45 \quad \rightarrow \quad \widehat{\mu}_i = 4.26,
\end{aligned}
$$

so the random effect estimates are moderating the evidence provided by the extreme sample means. The overall mean has some impact here since the information on family means does not totally dominate the prior information (i.e., $\alpha = 0.77 < 1$).

To see the merit of random effect estimation, suppose we want to predict the weights of future children. Using the same data set, we first estimate the unknown parameters using the first three births, obtaining the random effect estimate

$$
\widehat{\mu}_i = 0.73\overline{y}_{i.} + 0.27\overline{y}_{..}.
$$

For prediction of the fourth child, we have the total prediction error

$$
\begin{aligned}
\sum_i (y_{i4} - \overline{y}_{i.})^2 &= 23.11 \\
\sum_i (y_{i4} - \widehat{\mu}_i)^2 &= 21.62,
\end{aligned}
$$

so the random effect estimates perform better than the family means. The improved prediction is greatest for the families with lowest averages:

$$\bar{y}_{i.} = 2.46, \quad \hat{\mu}_i = 2.77, \quad y_{i4} = 2.93$$
$$\bar{y}_{i.} = 2.74, \quad \hat{\mu}_i = 2.97, \quad y_{i4} = 3.29.$$

In this data set we observe that the fourth child is slightly larger than the previous three (by an average of 107 g), so the means of the largest families perform well as predictors. □

5.3.3 The James-Stein estimate

Estimation of a large number of *fixed* parameters cannot be done naively, even when these are the means of independent normals. Assuming a one-way layout with fixed effects, James and Stein (1961) showed that when $q \geq 3$, it is possible to beat the performance of the cluster means $\bar{y}_{i.}$ with a shrinkage estimate of the form

$$m_i = c + \left(1 - \frac{(q-2)\sigma_e^2/n}{\sum_i(\bar{y}_{i.} - c)^2}\right)(\bar{y}_{i.} - c) \qquad (5.20)$$

for some constant c, in effect shrinking the cluster means toward c. Specifically, they showed that

$$E\{\sum_i(m_i - \mu_i)^2\} \leq E\{\sum_i(\bar{y}_{i.} - \mu_i)^2\} = \frac{q\sigma_e^2}{n}, \qquad (5.21)$$

where the gain is a decreasing function of $\tau = \sum_i(\mu_i - c)^2$; it is largest when $\tau = 0$, i.e., all the means are in fact equal to c. If τ is large, the denominator $\sum_i(\bar{y}_{i.} - c)^2$ in the formula (5.20) tends to be large, so m_i should be close to the cluster mean. There are good reasons for choosing $c = \bar{y}_{..}$, although the James-Stein theory does not require it. If c is a fixed constant, the estimate is not invariant with respect to a simple translation of the data, i.e., if we add some constant a to the data, we do not get a new estimate $m_i + a$, as we should expect. With $c = \bar{y}_{..}$, the James-Stein estimate is translation invariant. Furthermore, it becomes very close to the random effect estimate:

$$m_i = \bar{y}_{..} + \left(1 - \frac{\sigma_e^2/n}{\sum_i(\bar{y}_{i.} - \bar{y}_{..})^2/(q-3)}\right)(\bar{y}_{i.} - \bar{y}_{..}),$$

(the term $(q-3)$ replaces $(q-2)$ since the dimension of the vector $\{\bar{y}_{i.} - \bar{y}_{..}\}$ is reduced by one), while the random effect estimate is

$$\hat{\mu}_{i.} = \bar{y}_{..} + \left(1 - \frac{1/\hat{\sigma}_v^2}{n/\hat{\sigma}_e^2 + 1/\hat{\sigma}_v^2}\right)(\bar{y}_{i.} - \bar{y}_{..})$$
$$= \bar{y}_{..} + \left(1 - \frac{\hat{\sigma}_e^2/n}{\hat{\sigma}_v^2 + \hat{\sigma}_e^2/n}\right)(\bar{y}_{i.} - \bar{y}_{..}).$$

Recall from Section 6.3.1 that we set $SSV = n \sum_i (\overline{y}_{i.} - \overline{y}_{..})^2$. In effect, the James-Stein shrinkage formula is estimating the parameter $1/(\sigma_v^2 + \sigma_e^2/n)$ by

$$\frac{1}{SSV/\{n(q-3)\}},$$

while from (5.10), the normal random effect approach uses the MLE

$$\frac{1}{SSV/(nq)}.$$

This similarity means that the random effect estimate can be justified for general use, even when we think of the parameters as fixed. The only condition for the moment is that there should be a large number of parameters involved. For further analogy between shrinkage estimation and random effect estimation, see Lee and Birkes (1994).

The theoretical result (5.21) shows that the improvement over the cluster means occurs if population means are spread over a common mean. If not so, they might not outperform the sample mean. In fact, the estimates of certain means might be poor. For example (Cox and Hinkley, 1974, p. 449), suppose for large q and ρ of order q with n fixed,

$$\mu_1 = \sqrt{\rho}, \ \mu_2 = \cdots = \mu_q = 0$$

and we use the James-Stein estimate (5.20) with $c = 0$, so

$$m_1 \approx \left(1 - \frac{q\sigma_e^2/n}{\rho}\right) \overline{y}_{1.}$$

and the mean squared error is approximately

$$E(m_1 - \mu_1)^2 \approx \left(\frac{q\sigma_e^2/n}{\rho}\right)^2 \rho,$$

which is of order q, while the mean squared error of the sample mean $\overline{y}_{1.}$ is σ_e^2/n, so μ_1 is badly estimated by the shrinkage estimate.

If μ_i follow some distribution such as normal (i.e., a random effect model is plausible) we should always use the shrinkage estimators because they combine information about the random effects from data and from the fact that it has been sampled from a distribution which we can check. However, the previous example highlights the advantage of the modelling approach, where it is understood that the normal distribution assumption of random effects can be wrong, and it is an important duty of the analyst to check whether the assumption is reasonable. In this example, the model checking plot will immediately show an outlier, so that we can see that the normal assumption is violated. In such cases a remedy would be the use of structured dispersion (Chapter 7) or the use of a model with a heavy tail (Chapter 10). Such models give improved estimates, but also can perform automatic variable selection (Chapter 11).

5.3.4 Inferences about random effects

Since the TOE of v is $E(v|y)$, the inference is also naturally based on the conditional variance $\text{var}(v|y)$. We can justify this theoretically: because v is random, a proper variance of the estimate is $\text{var}(\hat{v} - v)$ rather than $\text{var}(\hat{v})$. Assuming all the fixed parameters are known, in the general normal mixed model (5.1), we have

$$
\begin{aligned}
\text{var}(\hat{v} - v) &= E(\hat{v} - v)^2 \\
&= E\{E((v - \hat{v})^2|y)\} \\
&= E\{\text{var}(v|y)\},
\end{aligned}
$$

where we have used the fact that $E((\hat{v} - v)|y)$ is zero. In this case,

$$
\text{var}(v|y) = (Z^t \Sigma^{-1} Z + D^{-1})^{-1}
$$

is constant and equal to $\text{var}(\hat{v} - v)$. For example, the standard errors of $\hat{v} - v$ can be computed as the square root of the diagonal elements of the conditional variance matrix. Confidence intervals at level $1 - \alpha$ for v are usually computed element by element using

$$
\hat{v}_i \pm z_{\alpha/2} \; \text{se}(\hat{v}_i - v_i)
$$

where $z_{\alpha/2}$ is an appropriate value from the normal table. Note that

$$
\text{var}(\hat{v}) \geq \text{var}(\hat{v} - v)
$$

and for confidence intervals for random v we should use $\text{var}(\hat{v} - v)$.

In the simple random effect model, if the fixed parameters are known, the conditional variance is given by

$$
\text{var}(v_i|y) = \left(\frac{n}{\sigma_e^2} + \frac{1}{\sigma_v^2} \right)^{-1}
$$

compared with σ_e^2/n if v_i is assumed fixed. Consequently the standard error of $\hat{v}_i - v_i$ under the random effect model is smaller than the standard error under the fixed effect model.

Considering $f(v)$ as the prior density of v, the posterior distribution of v is normal with mean $\hat{v}$ and variance $(Z^t \Sigma^{-1} Z + D^{-1})^{-1}$. This is the empirical Bayes interpretation of the general result.

Since the fixed parameters are usually not known, in practice we simply plug their estimates into the above procedures. Note, however, that this method does not take into account the uncertainty in the estimation of those parameters.

5.3.5 Augmented linear model

It is interesting to see that the previous joint estimation of β and v can be derived also via a classical linear model with *both* β and v appearing as fixed parameters. First consider an augmented linear model

$$\begin{pmatrix} y \\ \psi_M \end{pmatrix} = \begin{pmatrix} X & Z \\ 0 & I \end{pmatrix} \begin{pmatrix} \beta \\ v \end{pmatrix} + e^*,$$

where the error term e^* is normal with mean zero and variance matrix

$$\Sigma_a \equiv \begin{pmatrix} \Sigma & 0 \\ 0 & D \end{pmatrix},$$

and the augmented quasi–data $\psi_M = 0$ are assumed to be normal with mean $E\psi_M = v$ and variance D and independent of y. Here the subscript M is a label referring to the mean model. Results of this chapter are extended to the dispersion models later.

By defining the quantities appropriately:

$$y_a \equiv \begin{pmatrix} y \\ \psi_M \end{pmatrix}, \quad T \equiv \begin{pmatrix} X & Z \\ 0 & I \end{pmatrix}, \quad \delta = \begin{pmatrix} \beta \\ v \end{pmatrix}$$

we have a classical linear model

$$y_a = T\delta + e^*.$$

Now, by taking $\psi_M \equiv Ev = 0$, the weighted least squares equation

$$(T^t \Sigma_a^{-1} T)^{-1} \widehat{\delta} = T^t \Sigma_a^{-1} y_a \tag{5.22}$$

is exactly the mixed model equation (5.15). The idea of an augmented linear model does not seem to add anything new to the analysis of normal mixed models, but it turns out to be a very useful device in the extension to non-normal mixed models.

5.3.6 Fitting algorithm

In the classical approach the variance components can be estimated using, for example, the marginal or restricted likelihood. In summary, the estimation of (β, τ, v) in the linear mixed model can be done by an iterative algorithm as follows, where for clarity we show all the required equations.

1. Start with an estimate of the variance parameter τ.

2. Given the current estimate of τ, update $\widehat{\beta}$ and $\widehat{v}$ by solving the mixed model equation:

$$\left(\begin{array}{cc} X^t\Sigma^{-1}X & X^t\Sigma^{-1}Z \\ Z^t\Sigma^{-1}X & Z^t\Sigma^{-1}Z + D^{-1} \end{array} \right) \left(\begin{array}{c} \widehat{\beta} \\ \widehat{v} \end{array} \right) = \left(\begin{array}{c} X^t\Sigma^{-1}y \\ Z^t\Sigma^{-1}y \end{array} \right). \quad (5.23)$$

However, in practice, computing this jointly is rarely the most efficient way. Instead, it is often simpler to solve the following two equations

$$\begin{aligned} (X^t\Sigma^{-1}X)\widehat{\beta} &= X^t\Sigma^{-1}(y - Z\widehat{v}) \\ (Z^t\Sigma^{-1}Z + D^{-1})\widehat{v} &= Z^t\Sigma^{-1}(y - X\widehat{\beta}). \end{aligned}$$

The updating equation for β is easier to solve than (5.2) since there is no term involving V^{-1}.

3. Given the current values of $\widehat{\beta}$ and $\widehat{v}$, update τ by maximizing either the marginal loglihood $\ell(\beta, \tau)$ or the adjusted profile likelihood $p_\beta(\ell|\tau)$. The former gives the ML estimators and the latter the REML estimators for the dispersion parameters. The ML estimation is fast but biased in small samples or when the number of βs increases with the sample size. The REML procedure gets rid of the bias but is slower computationally.

4. Iterate between 2 and 3 until convergence.

5.4 H-likelihood approach

For normal linear mixed models, the classical approach provides sensible inferences about β and the random parameters v; for further discussion, see Robinson (1991). However, its extension to non-normal models is not straightforward. To prepare for the necessary extensions later, we study h-likelihood inference for linear mixed models.

The general model (5.1) can be stated equivalently as follows: conditional on v, the outcome y is normal with mean

$$E(y|v) = X\beta + Zv$$

and variance Σ, and v is normal with mean zero and variance D. From Section 4.1, the extended loglihood of all the unknown parameters is given by

$$\begin{aligned} \ell_e(\beta, \tau, v) &= \log f(y, v) = \log f(y|v) + \log f(v) \\ &= -\frac{1}{2}\log|2\pi\Sigma| - \frac{1}{2}(y - X\beta - Zv)^t\Sigma^{-1}(y - X\beta - Zv) \\ &\quad -\frac{1}{2}\log|2\pi D| - \frac{1}{2}v^t D^{-1}v, \end{aligned} \quad (5.24)$$

where the dispersion parameter τ enters via Σ and D.

To use the h-likelihood framework, from Section 4.5, first we need to establish the canonical scale for the random effects. Given the dispersion parameters, by maximizing the extended likelihood, we obtain

$$\hat{v} = (Z^t \Sigma^{-1} Z + D^{-1})^{-1} Z^t \Sigma^{-1} (y - X\hat{\beta})$$

and from the second derivative ℓ_e with respect to v, we get the Fisher information

$$I(\hat{v}) = (Z^t \Sigma^{-1} Z + D^{-1}).$$

Since the Fisher information depends on the dispersion parameter τ, but not on β, the scale v is not canonical for τ, but it can be for β. In fact it is the canonical scale. This means that the extended likelihood is an h-likelihood, allowing us to make joint inferences about β and v, but estimation of τ requires a marginal likelihood. Note that $\hat{v}$ is a function of fixed parameters, so that we use notations $\hat{v}$, $\hat{v}(\beta, \tau)$ and $\hat{v}_{\beta,\tau}$ for convenience. This is important when we need to maximize adjusted profile likelihoods.

From Section 4.5, the canonical scale v is unique up to linear transformations. For non-linear transformations of the random effects, the h-likelihood must be derived following the invariance principle given in Section 4.5, i.e.,

$$H(\beta, \tau, u(v)) \equiv H(\beta, \tau, v).$$

With this, joint inferences of β and v from the h-likelihood are invariant with respect to any monotone transformation (or re-expression) of v.

We compare the h-likelihood inference with the classical approach:

- All inferences including those for the random effects are made within the (extended) likelihood framework.
- Joint estimation of β and v is possible because v is canonical for β.
- Estimation of the dispersion parameter requires an adjusted profile likelihood.
- Extensions to non-normal models are immediate as we shall see in later chapters.

5.4.1 Inference for mean parameters

From the optimization of the log density in Section 5.3, given D and Σ, the h-likelihood estimates of β and v satisfy the mixed model equation (5.15). Let H be the square matrix of the left-hand side of the equation,

$V = ZDZ^t + \Sigma$ and $\Lambda = Z^t\Sigma^{-1}Z + D^{-1}$. The solution for β gives the MLE, satisfying

$$X^tV^{-1}X\hat{\beta} = X^tV^{-1}y.$$

and the solution for v gives the empirical BUE

$$
\begin{aligned}
\hat{v} &= \widehat{E(v|y)} = E(v|y)|_{\beta=\hat{\beta}} \\
&= DZ^tV^{-1}(y - X\hat{\beta}) \\
&= \Lambda^{-1}Z^t\Sigma^{-1}(y - X\hat{\beta}).
\end{aligned}
$$

Furthermore, H^{-1} gives the estimate of

$$E\left\{ \left(\begin{array}{c} \hat{\beta} - \beta \\ \hat{v} - v \end{array} \right) \left(\begin{array}{c} \hat{\beta} - \beta \\ \hat{v} - v \end{array} \right)^t \right\}.$$

This yields $(X^tV^{-1}X)^{-1}$ as a variance estimate for $\hat{\beta}$, which coincides with that for the ML estimate. We now show that H^{-1} also gives the correct estimate for $E\left\{(\hat{v}-v)(\hat{v}-v)^t\right\}$, one that accounts for the uncertainty in $\hat{\beta}$.

The TOE of the random effect estimate is given by

$$\delta = E(v|y),$$

so we have

$$\mathrm{var}(\delta - v) = E\left\{(\delta - v)(\delta - v)^t\right\} = E\left\{\mathrm{var}(v|y)\right\},$$

where

$$\mathrm{var}(v|y) = D - DZ^tV^{-1}ZD = \Lambda^{-1}.$$

When β is known, Λ^{-1} gives a proper estimate of the variance of $\delta - v$. However, when β is unknown, the plugged-in empirical Bayes estimate $\Lambda^{-1}|_{\beta=\hat{\beta}}$ for $\mathrm{var}(\hat{v} - v)$ does not properly account for the extra uncertainty due to estimating β.

By contrast, the h-likelihood computation gives a straightforward correction. Now we have

$$\mathrm{var}(\hat{v} - v) = E\left\{\mathrm{var}(v|y)\right\} + E\left\{(\hat{v} - \delta)(\hat{v} - \delta)^t\right\},$$

where the second term shows the variance inflation caused by estimating the unknown β. As an estimate for $\mathrm{var}(\hat{v}-v)$, the appropriate component of H^{-1} gives

$$\{\Lambda^{-1} + \Lambda^{-1}Z^t\Sigma^{-1}X(X^tV^{-1}X)^{-1}X^t\Sigma^{-1}Z\Lambda^{-1}\}|_{\beta=\hat{\beta}}.$$

Because $\hat{v} - \delta = -DZ^tV^{-1}X(\hat{\beta} - \beta)$ we can show that

$$E\left\{(\hat{v} - \delta)(\hat{v} - \delta)^t\right\} = \Lambda^{-1}Z^t\Sigma^{-1}X(X^tV^{-1}X)^{-1}X^t\Sigma^{-1}Z\Lambda^{-1}.$$

Thus, the h-likelihood approach correctly handles the variance inflation caused by estimating the fixed effects. From this we can construct confidence bounds for unknown v.

5.4.2 Estimation of variance components

We previously derived the profile likelihood for the variance component parameter τ, but the resulting formula (5.12) is complicated by terms involving $|V|$ or V^{-1}. In practice these matrices are usually too unstructured to deal with directly. Instead we can use formulae derived from the h-likelihood. First, the marginal likelihood of (β, τ) is

$$
\begin{aligned}
L(\beta, \tau) \;=\; & |2\pi\Sigma|^{-1/2} \int \exp\{-\frac{1}{2}(y - X\beta - Zv)^t \Sigma^{-1}(y - X\beta - Zv)\} \\
& \times |2\pi D|^{-1/2} \exp\{-\frac{1}{2} v^t D^{-1} v\} dv \\
\;=\; & |2\pi\Sigma|^{-1/2} \exp\{-\frac{1}{2}(y - X\beta - Z\widehat{v}_{\beta,\tau})^t \Sigma^{-1}(y - X\beta - Z\widehat{v}_{\beta,\tau})\} \\
& \times |2\pi D|^{-1/2} \exp\{-\frac{1}{2}\widehat{v}_{\beta,\tau}^t D^{-1}\widehat{v}_{\beta,\tau}\} \\
& \times \int \exp\{-\frac{1}{2}(v - \widehat{v}_{\beta,\tau})^t I(\widehat{v}_{\beta,\tau})(v - \widehat{v}_{\beta,\tau})\} dv \\
\;=\; & |2\pi\Sigma|^{-1/2} \exp\{-\frac{1}{2}(y - X\beta - Z\widehat{v}_{\beta,\tau})^t \Sigma^{-1}(y - X\beta - Z\widehat{v}_{\beta,\tau})\} \\
& \times |2\pi D|^{-1/2} \exp\{-\frac{1}{2}\widehat{v}_{\beta,\tau}^t D^{-1}\widehat{v}_{\beta,\tau}\} \\
& \times |I(\widehat{v}_{\beta,\tau})/(2\pi)|^{-1/2}.
\end{aligned}
$$

(Going from the first to the second formula involves tedious matrix algebra.) We can obtain the marginal loglihood in terms of the adjusted profile likelihood:

$$
\begin{aligned}
\ell(\beta, \tau) \;=\; & h(\beta, \tau, \widehat{v}_{\beta,\tau}) - \frac{1}{2}\log|I(\widehat{v}_{\beta,\tau})/(2\pi)|, \qquad (5.25) \\
\;=\; & p_v(h|\beta, \tau)
\end{aligned}
$$

where, from before,

$$
I(\widehat{v}_{\beta,\tau}) = -\left.\frac{\partial^2 h}{\partial v \partial v^t}\right|_{v=\widehat{v}_{\beta,\tau}} = Z^t \Sigma^{-1} Z + D^{-1}.
$$

The constant (2π) is kept in the adjustment term to make the loglihood an exact log density; this facilitates comparisons between models as in the example below. Thus the marginal likelihood in the mixed effects models is equivalent to an adjusted profile likelihood obtained by *profiling* out the random effects. For general non-normal models, this result will be only approximately true, up to the Laplace approximation (1.19) of the integral.

Example 5.6: In the one-way random effect model

$$y_{ij} = \mu + v_i + e_{ij}, \ i = 1, \ldots, q, \ j = 1, \ldots, n \qquad (5.26)$$

from our previous derivations, given the fixed parameters (μ, τ),

$$\hat{v}_i = \left(\frac{n}{\sigma_e^2} + \frac{1}{\sigma_v^2}\right)^{-1} \frac{n}{\sigma_e^2}(\bar{y}_{i.} - \mu)$$

$$= \frac{n\sigma_v^2}{\sigma_e^2 + n\sigma_v^2}(\bar{y}_{i.} - \mu)$$

$$I(\hat{v}_i) = \frac{n}{\sigma_e^2} + \frac{1}{\sigma_v^2},$$

so the adjusted profile loglihood becomes

$$\begin{aligned}
p_v(h|\mu, \tau) &= -\frac{qn}{2}\log(2\pi\sigma_e^2) - \frac{1}{2\sigma_e^2}\sum_{i=1}^{q}\sum_{j=1}^{n}(y_{ij} - \mu - \hat{v}_i)^2 \\
&\quad -\frac{q}{2}\log(2\pi\sigma_v^2) - \frac{1}{2\sigma_v^2}\sum_{i=1}^{q}\hat{v}_i^2 - \frac{q}{2}\log\frac{\sigma_e^2 + n\sigma_v^2}{2\pi\sigma_v^2\sigma_e^2} \\
&= -\frac{q}{2}[(n-1)\log(2\pi\sigma_e^2) + \log\{2\pi(\sigma_e^2 + n\sigma_v^2)\}] \\
&\quad -\frac{1}{2\sigma_e^2}\sum_{i=1}^{q}\sum_{j=1}^{n}(y_{ij} - \mu - \hat{v}_i)^2 - \frac{1}{2\sigma_v^2}\sum_{i=1}^{q}\hat{v}_i^2 \\
&= -\frac{q}{2}[(n-1)\log(2\pi\sigma_e^2) + \log\{2\pi(\sigma_e^2 + n\sigma_v^2)\}] \\
&\quad -\frac{1}{2}\left\{\frac{SSE}{\sigma_e^2} + \frac{SSV + qn(\bar{y}_{..} - \mu)^2}{\sigma_e^2 + n\sigma_v^2}\right\}, \qquad (5.27)
\end{aligned}$$

as we have shown earlier in (5.9), but now derived much more simply since we do not have to deal with the variance matrix V directly.

Note that the h-loglihood $h(\mu, \tau, v)$ and information matrix $I(\hat{v}_i)$ are unbounded as σ_v^2 goes to zero, even though the marginal loglihood $\ell(\mu, \sigma_e^2, \sigma_v^2 = 0)$ exists. The theoretical derivation here shows that the offending terms cancel out. Numerically, this means that we cannot use $p_v(h)$ at $(\mu, \sigma_e^2, \sigma_v^2 = 0)$. This problem occurs more generally when we have several variance components. In these cases we should instead compute $p_v(h)$ based on the h-likelihood of the reduced model when one or more random components is absent. For this reason the constant 2π should be kept in the adjusted profile loglihood.(Lee and Nelder, 1996) □

5.4.3 REML estimation of variance components

In terms of the h-likelihood, the profile likelihood of the variance components (5.12) can be rewritten as

$$
\begin{aligned}
\ell_p(\tau) &= \ell(\widehat{\beta}_\tau, \tau) \\
&= h(\widehat{\beta}_\tau, \tau, \widehat{v}_\tau) - \frac{1}{2}\log|I(\widehat{v}_\tau)/(2\pi)|, \qquad (5.28)
\end{aligned}
$$

where τ enters the function through Σ, D, $\widehat{\beta}_\tau$ and $\widehat{v}_\tau$, and as before $I(\widehat{v}_\tau) = Z^t\Sigma^{-1}Z + D^{-1} = \Lambda$, since $I(\widehat{v}_{\beta,\tau})$ is not a function of β. The joint estimation of $\widehat{\beta}$ and $\widehat{v}$ as a function of τ was given previously by (5.15).

If we include the REML adjustment for the estimation of the fixed effect β from Section 5.2.2, we have

$$
\begin{aligned}
p_\beta(\ell|\tau) &= \ell(\widehat{\beta}_\tau, \tau) - \frac{1}{2}\log|X^t V^{-1}X/(2\pi)| \\
&= h(\widehat{\beta}_\tau, \tau, \widehat{v}_\tau) - \frac{1}{2}\log|I(\widehat{v}_\tau)/(2\pi)| - \frac{1}{2}\log|X^t V^{-1}X/(2\pi)| \\
&= p_{\beta,v}(h|\tau), \qquad (5.29)
\end{aligned}
$$

where the $p()$ notation allows the representation of the adjusted profiling of *both* fixed and random effects simultaneously. Hence, in the normal case, the different forms of likelihood of the fixed parameters match exactly the adjusted profile likelihood derived from the h-likelihood. Since $\ell(\beta, \tau) = p_v(h)$, we also have

$$
p_{\beta,v}(h) = p_\beta\{p_v(h)\},
$$

a useful result that will be only approximately true in non-normal cases.

5.4.4 Fitting algorithm

The h-likelihood approach provides an insightful fitting algorithm, particularly with regard to the estimation of the dispersion parameters. The normal case is a useful prototype for the general case dealt with in the next chapter. Consider an augmented classical linear model as in Section 5.3.5:

$$
y_a = T\delta + e_a,
$$

where $e_a \sim MVN(0, \Sigma_a)$, and

$$
y_a \equiv \begin{pmatrix} y \\ \psi_M \end{pmatrix}, \quad T \equiv \begin{pmatrix} X & Z \\ 0 & I \end{pmatrix}, \quad \delta = \begin{pmatrix} \beta \\ v \end{pmatrix}
$$

$$e_a = \begin{pmatrix} e \\ e_M \end{pmatrix}, \quad \Sigma_a \equiv \begin{pmatrix} \Sigma & 0 \\ 0 & D \end{pmatrix}.$$

In this chapter, $\Sigma = \sigma_e^2 I$ and $D = \sigma_v^2 I$. Because the augmented linear model is a GLM with a constant variance function, we can apply the REML methods for the joint GLM in Chapter 3 to fit the linear mixed models. Here the deviance components corresponding to e are the squared residuals

$$d_i = (y_i - X_i\hat{\beta} - Z_i\hat{v})^2$$

and those corresponding to e_M are

$$d_{Mi} = (\psi_M - \hat{v}_i)^2 = \hat{v}_i^2,$$

and the corresponding leverages are diagonal elements of

$$T(T^t\Sigma_a^{-1}T)^{-1}T^t\Sigma_a^{-1}.$$

Table 5.1 *Inter-connected GLMs for parameter estimation in linear mixed models.*

Component	β (fixed)	σ_e^2 (fixed)
Response	y	d^*
Mean	μ	σ_e^2
Variance	σ_e^2	σ_e^4
Link	$\eta = \mu$	$\xi = h\left(\sigma_e^2\right)$
Linear predictor	$X\beta + Zv$	γ
Deviance	d	$\Gamma(d^*, \sigma_e^2)$
Prior weight	$1/\sigma_e^2$	$(1 - q)/2$

Component	v (random)	σ_v^2 (fixed)
Response	ψ_M	d_M^*
Mean	v	σ_v^2
Variance	σ_v^2	σ_v^4
Link	$\eta_M = v$	$\xi_M = h_M\left(\sigma_v^2\right)$
Linear predictor	v	γ_m
Deviance	d_M	$\Gamma(d_M^*, \sigma_v^2)$
Prior weight	$1/\sigma_v^2$	$(1 - q_M)/2$

$d = (y - X\hat{\beta} - Z\hat{v})^2$.
$d_M = \hat{v}^2$.
$\Gamma(d^*, \phi) = 2\{-\log(d^*/\phi) + (d^* - \phi)/\phi\}$.
(q, q_M) are leverages given by diagonal elements of $T(T^t\Sigma_a^{-1}T)^{-1}T^t\Sigma_a^{-1}$.

The estimation of (β, τ, v) in the linear mixed model can be done by IWLS for the augmented linear model as follows, where for clarity we show all the required equations:

1. Start with an estimate of the variance parameter τ.

2. Given the current estimate of τ, update $\hat{\delta}$ by solving the augmented generalized least squares equation:

$$T^t \Sigma_a^{-1} T \hat{\delta} = T^t \Sigma_a^{-1} y_a.$$

3. Given the current values of $\hat{\delta}$, get an update of τ; the REML estimators can be obtained by fitting a gamma GLM as follows. The estimator for σ_e^2 is obtained from the GLM, characterized by a response $d^* = d/(1-q)$, a gamma error, a link $h()$, a linear predictor γ (intercept–only model), and a prior weight $(1-q)/2$. The estimator for σ_v^2 is obtained by the GLM, characterized by a response $d_M^* = d_M/(1-q_M)$, a gamma error, a link $h_M()$, a linear predictor γ_M (intercept only model), and a prior weight $(1-q_M)/2$. Note here that

$$E(d_i^*) = \sigma_e^2 \text{ and } \mathrm{var}(d_i^*) = 2\sigma_e^4/(1-q_i),$$

and

$$E(d_{Mi}^*) = \sigma_v^2 \text{ and } \mathrm{var}(d_{Mi}^*) = 2\sigma_v^4/(1-q_{Mi}).$$

This algorithm is often much faster than the ordinary REML procedure of the previous section. The MLE can be obtained by taking the leverages to be zero.

4. Iterate between 2 and 3 until convergence. At convergence, the standard error of $\hat{\beta}$ and $\hat{v} - v$ can be computed from the inverse of the information matrix H^{-1} from the h-likelihood and the standard errors of $\hat{\tau}$ are computed from the Hessian of $p_{\beta,v}(h|\tau)$ at $\hat{\tau}$. Typically there is no explicit formula for this quantity.

This is an extension of the REML procedure for joint GLMs to linear mixed models. Fitting involves inter-connected component GLMs. Connections are marked by lines in Table 5.1. Each connected GLM can be viewed as a joint GLM. Then the joint GLMs are connected by an augmented linear model for β and v components.

5.4.5 Residuals in linear mixed models

If we use the marginal model with a multivariate normal distribution the natural residuals would be

$$\hat{r}_i = y_i - X_i\hat{\beta} = Z_i\hat{v} + \hat{e}_i.$$

EXAMPLE   159

With these residuals, model checking about assumptions on either v or e is difficult, because they cannot be separated. This difficulty becomes worse as model assumptions about these components become more complicated.

For two random components v and e, our ML procedure gives the two sets of residuals $\hat{v}_i$ and $\hat{e}_i = y_i - X_i\hat{\beta} - Z_i\hat{v}$, while our REML procedure provides the two sets of standardized residuals $\hat{v}_i/\sqrt{1 - q_{Mi}}$ and $\hat{e}_i/\sqrt{1 - q_i}$. Thus, assumptions about these two random components can be checked separately. Moreover, the fitting algorithm for the variance components implies that another two sets of (deviance) residuals from gamma GLMs are available for checking the dispersion models.

Table 5.1 shows that a linear mixed model can be decomposed into four GLMs. The components β and v have linear models, while components σ_e^2 and σ_v^2 have gamma GLMs. Thus, any of the four separate GLMs can be used to check model assumptions about their components. If the number of random components increases by one, it produces two additional GLM components: a normal model for v and a gamma model for σ_v^2. From this it is possible to develop regression models with covariates for the components σ_e^2 and σ_v^2, as we shall show in later chapters. This is a great advantage of using h-likelihood.

Because
$$T^t\Sigma_a^{-1}\hat{e}_a = 0,$$
we immediately have $\sum_i \hat{e}_i = 0$ and $\sum_i \hat{v}_i = 0$; this is an extension of $\sum_i \hat{e}_i = 0$ in classical linear models with an intercept. In classical linear models the $\hat{e}_i$ are plotted against $X_i\hat{\beta}$ to check systematic departures from model assumptions. In the corresponding linear mixed models, a plot of $\hat{e}_i = y_i - X_i\hat{\beta} - Z_i\hat{v}$ against $\hat{\mu}_i = X_i\hat{\beta} + Z_i\hat{v}$ yields an unwanted trend caused by correlation between $\hat{e}_i$ and $\hat{\mu}_i$, so that Lee and Nelder (2001a) recommend a plot of $\hat{e}_i$ against $X_i\hat{\beta}$. This successfully removes the unwanted trend and we use these plots throughout the book.

5.5 Example

In an experiment on the preparation of chocolate cakes conducted at Iowa State College, three recipes for preparing the batter were compared (Cochran and Cox, 1957). Recipes I and II differed in that the chocolate was added at $40°C$ and $60°C$, respectively, while recipe III contained extra sugar. In addition, six baking temperatures were tested: these ranged in $10°C$ steps from $175°C$ to $225°C$. For each mix, enough batter was prepared for six cakes, each of which was baked at a different temperature. Thus the recipes are the whole-unit treatments, while

Table 5.2 *Breaking angles (degrees) of chocolate cakes.*

Recipe	Replication	175°	185°	195°	205°	215°	225°
I	1	42	46	47	39	53	42
	2	47	29	35	47	57	45
	3	32	32	37	43	45	45
	4	26	32	35	24	39	26
	5	28	30	31	37	41	47
	6	24	22	22	29	35	26
	7	26	23	25	27	33	35
	8	24	33	23	32	31	34
	9	24	27	28	33	34	23
	10	24	33	27	31	30	33
	11	33	39	33	28	33	30
	12	28	31	27	39	35	43
	13	29	28	31	29	37	33
	14	24	40	29	40	40	31
	15	26	28	32	25	37	33
II	1	39	46	51	49	55	42
	2	35	46	47	39	52	61
	3	34	30	42	35	42	35
	4	25	26	28	46	37	37
	5	31	30	29	35	40	36
	6	24	29	29	29	24	35
	7	22	25	26	26	29	36
	8	26	23	24	31	27	37
	9	27	26	32	28	32	33
	10	21	24	24	27	37	30
	11	20	27	33	31	28	33
	12	23	28	31	34	31	29
	13	32	35	30	27	35	30
	14	23	25	22	19	21	35
	15	21	21	28	26	27	20
III	1	46	44	45	46	48	63
	2	43	43	43	46	47	58
	3	33	24	40	37	41	38
	4	38	41	38	30	36	35
	5	21	25	31	35	33	23
	6	24	33	30	30	37	35
	7	20	21	31	24	30	33
	8	24	23	21	24	21	35
	9	24	18	21	26	28	28
	10	26	28	27	27	35	35
	11	28	25	26	25	38	28
	12	24	30	28	35	33	28
	13	28	29	43	28	33	37
	14	19	22	27	25	25	35
	15	21	28	25	25	31	25

the baking temperatures are the sub-unit treatments. There were 15 replications, and it will be assumed that these were conducted serially according to a randomized block scheme: one replication was completed before starting the next, so that differences among replicates represent time differences. A number of measurements were made on the cakes.

EXAMPLE   161

The measurement presented here was the breaking angle. One half of a slab of cake was held fixed, while the other half was pivoted about the middle until breakage occured. The angle through which the moving half revolved was read on a circular scale. Since breakage was gradual, the reading tended to have a subjective element. The data are shown in Table 5.2.

We consider the following linear mixed model: for $i = 1, ..., 3$ recipes, $j = 1, ..., 6$ temperatures and $k = 1, ..., 15$ replicates,

$$y_{ijk} = \mu + \gamma_i + \tau_j + (\gamma\tau)_{ij} + v_k + v_{ik} + e_{ijk},$$

where γ_i are main effects for recipe, τ_j are main effects for temperature, $(\gamma\tau)_{ij}$ are recipe-temperature interactions, v_k are random replicates, v_{ik} are whole-plot error components and $e_{ijk,}$ are white noise. We assume all error components are independent and identically distributed. Cochran and Cox (1957) treated v_k as fixed, but because of the balanced design structure the analyses are identical. Residual plots for the $e_{ijk,}$ component are in Figure 5.3.

The second upper plot indicates a slight increasing tendency. The deviance is

$$-2p_v(h|y_{ijk}; \beta, \tau) = \log L(\beta, \tau) = 1639.07.$$

We found that the same model with responses $\log y_{ijk}$ gave a better fit. The corresponding deviance is

$$-2p_v(h| \log y_{ijk}; \beta, \tau) + 2 \sum \log y_{ijk} = 1617.24,$$

where the second term is the Jacobian for the data transformation. Because the assumed models are the same, with only the scale of the response differing, we can use the AIC to select the log-normal linear mixed model. However, gamma linear mixed model in the next chapter has

$$-2p_v(h|y_{ijk}; \beta, \tau) = 1616.12$$

and therefore it is better than the log-normal linear mixed model under the AIC rule. Residual plots of the log-normal model for the $e_{ijk,}$ component are in Figure 5.4. Normal probability plots for v_k, v_{ik}, and the error component for the dispersion model are in Figure 5.5. We found that replication effects seem to not follow a normal distribution, so that it may be appropriate to take them as fixed. We see here how the extended likelihood framework gives us sets of residuals to check model assumptions.

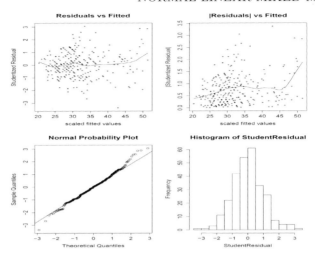

Figure 5.3 *Residual plots of the e_{ijk} component in the normal linear mixed model for the cake data.*

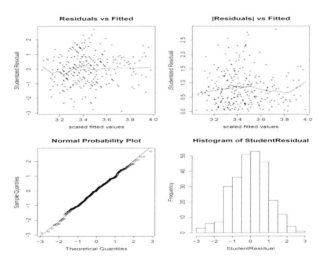

Figure 5.4 *Residual plots of the e_{ijk} component in the log-normal linear mixed model for the cake data.*

5.6 Invariance and likelihood inference

In random effect models we often see several alternative representations of the same model. Suppose we have two alternative random effect models

$$Y = X\beta + Z_1 v_1 + e_1 \tag{5.30}$$

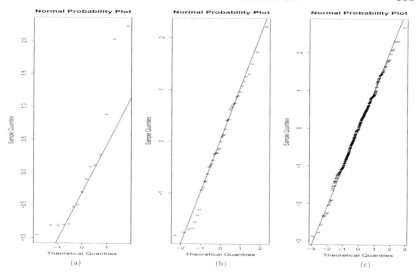

Figure 5.5 *Normal probability plots of (a) v_k, (b) v_{ik}, and (c) error component for the dispersion model in the log-normal linear mixed model for the cake data.*

and

$$Y = X\beta + Z_2 v_2 + e_2, \tag{5.31}$$

where $v_i \sim N(0, D_i)$ and $e_i \sim N(0, \Sigma_i)$. Marginally, these two random effect models lead to multivariate-normal (MVN) models, for $i = 1, 2$,

$$Y \sim MVN(X\beta, V_i)$$

where $V_i = Z_i D_i Z_i^t + \Sigma_i$. If the two models are the same, i.e., $V_1 = V_2$, likelihood inferences from these alternative random effect models should be identical. It is clear that the extended likelihood framework leads to identical inferences for fixed parameters because the two models lead to identical marginal likelihoods based upon the multivariate normal model. The question is whether inferences for random effects are also identical.

Let Θ_i $i = 1, 2$ be parameter spaces which span V_i. When $\Theta_1 = \Theta_2$, both random effect models should have the same likelihood inferences. Now suppose that $\Theta_1 \subset \Theta_2$, which means that given V_1 there exists V_2 such that

$$V_1 = V_2 \in \Theta_1.$$

If $\hat{V}_1$ is the ML or REML estimator for model (5.30) satisfying $\hat{V}_1 = \hat{V}_2$, then $\hat{V}_2$ is the ML or REML estimator for model (5.31) under the parameter space $V_2 \in \Theta_1$. When $\Theta_1 \subset \Theta_2$, we call (5.30) the reduced model and (5.31) the full model. Furthermore, given the equivalent dispersion estimates $\hat{V}_1 = \hat{V}_2$, the ML and REML estimators for the parameters and the h-likelihood estimators for random effects are preserved for equivalent elements(Lee and Nelder, 2006b). To get the same prediction from the two models $X\hat{\beta}+Z_2\hat{v}_2 = X\hat{\beta}+Z_1\hat{v}_1$, first we assume that the residual variances of two models are the same, $\Sigma_1 = \Sigma_2$. Then $V_1 = V_2$ implies $Z_1D_1Z_1^t = Z_2D_2Z_2^t$.

Result. Suppose $\Theta_1 \subset \Theta_2$. Provided that the estimator for V_2 of the full model lies in Θ_1, likelihood-type estimators are identical for equivalent elements.

Proof: The proof of this result for fixed parameters is obvious, so that we prove it only for the random parameter estimation. Consider the two estimating equations from the two models for $i = 1, 2$

$$\left(\begin{array}{cc} X^t\Sigma^{-1}X & X^t\Sigma^{-1}Z_1 \\ Z_i^t\Sigma^{-1}X & Z_i^t\Sigma^{-1}Z_i + D_i^{-1} \end{array} \right) \left(\begin{array}{c} \hat{\beta} \\ \hat{v}_i \end{array} \right) = \left(\begin{array}{c} X^t\Sigma^{-1}Y \\ Z_i^t\Sigma^{-1}Y \end{array} \right).$$

Because BLUEs can be written as

$$\hat{v}_i = D_i Z_i^t P_i y,$$

where $P_i = V_i^{-1} - V_i^{-1}X(X^tV_i^{-1}X)^{-1}V_i^{-1}$, we have

$$Z_1\hat{v}_1 = Z_2\hat{v}_2$$

as $V_1 = V_2$ and $P_1 = P_2$. Thus, given the dispersion parameters $Z_1D_1Z_1^t = Z_2D_2Z_2^t$, the BLUEs from the two mixed model equations satisfy $Z_1\hat{v}_1 = Z_2\hat{v}_2$. This completes the proof. $\square$

This apparently obvious result has not been well recognized and exploited in statistical literature. Various correlation structures are allowed in the bottom-level error e (SPLUS) or in random effects v (SAS, GenStat, etc.). Thus, there may exist several representations of the same model. For example we may consider a model with

$$Z_1 = Z_2 = I, \quad v_1 \sim AR(1), \quad e_1 \sim N(0, \phi I),$$

$$v_2 \sim N(0, \phi I), \quad e_2 \sim AR(1),$$

where $AR(1)$ stands for the autoregressive model of order 1. Here the first model assumes $AR(1)$ for random effects, while the second model assumes it for the bottom-level errors. Here, $Z_1\hat{v}_1 \neq Z_2\hat{v}_2$. However, because $\Theta_1 = \Theta_2$, i.e., $V_1 = D + \phi I = V_2$ with D being the covariance induced by $AR(1)$, both models are equivalent and must lead to identical inferences about error components.This may not be immediately apparent when there are additional random effects. The result shows that ML and/or REML inferences for the parameters (β, ϕ, D) and inferences for the errors are identical; e.g., $\hat{e}_1 = Y - X\hat{\beta} - Z_1\hat{v}_1 = \hat{v}_2$ and

$\hat{v}_1 = \hat{e}_2 = Y - X\hat{\beta} - Z_2\hat{v}_2$. Another example is that $AR(1)$ and the compound symmetric model become identical when there are only two time points.

If $\hat{V}_1$ is the ML or REML estimator for the reduced model and there exists $\hat{V}_2$ satisfying $\hat{V}_1 = \hat{V}_2$, then $\hat{V}_2$ is for the full model under the parameter space $V_2 \in \Theta_1$, which is not necessarily the likelihood estimator under $V_2 \in \Theta_2$. Care is necessary if the likelihood estimator $\hat{V}_2$ for the full model does not lie in Θ_1, because likelihood inferences from the reduced model are no longer the same as those from the full model.

Example 5.7: Rasbash *et al.* (2000) analyzed some educational data with the aim of establishing whether some schools were more effective than others in promoting student learning and development, taking account of variations in the characteristics of students when they started secondary school. The response was the exam score obtained by each student at age 16 and the covariate was the score for each student at age 11 on the London Reading Test, both normalized. They fitted a random coefficient regression model

$$y_{ij} = \beta_0 + x_{ij}\beta_1 + a_i + x_{ij}b_i + e_{ij},$$

where $e_{ij} \sim N(0, \sigma^2)$, (a_i, b_i) are bivariate normal with mean zeros, and $\mathrm{var}(a_i) = \lambda_{11}$, $\mathrm{var}(b_i) = \lambda_{22}$, and $\mathrm{cov}(a_i, b_i) = \lambda_{12}$. Here subscript j refers to student and i refers to school so that y_{ij} is the score of student j from school i at age 16.

Lee and Nelder (2006b) showed that this model can be fitted with a random effect model with independent random components. Using the SAS MIXED procedure, they first fitted an independent random component model

$$y_{ij} = \beta_0 + x_{ij}\beta_1 + w_{i1} + x_{ij}w_{i2} + (1 + x_{ij})w_{i3} + (1 - x_{ij})w_{i4} + e_{ij},$$

where $w_{ik} \sim N(0, \varpi_k)$. Because

$$\mathrm{var}(a_i + x_{ij}b_i) = \lambda_{11} + 2\lambda_{12}x_{ij} + \lambda_{22}x_{ij}^2$$

and

$$\mathrm{var}(w_{i1} + x_{ij}w_{i2} + (1 + x_{ij})w_{i3} + (1 - x_{ij})w_{i4}) = \gamma_{11} + 2\gamma_{12}x_{ij} + \gamma_{22}x_{ij}^2,$$

where $\gamma_{11} = \varpi_1 + \varpi_3 + \varpi_4$, $\gamma_{22} = \varpi_2 + \varpi_3 + \varpi_4$ and $\gamma_{12} = \varpi_3 - \varpi_4$. Even though one parameter is redundant, the use of four parameters is useful when the sign of $\hat{\gamma}_{12}$ is not known. For example, $\hat{\varpi}_3 = 0$ implies $\hat{\gamma}_{12} < 0$, while $\hat{\varpi}_4 = 0$ implies $\hat{\gamma}_{12} > 0$. Because $|\gamma_{12}| \leq \min\{\gamma_{11}, \gamma_{22}\}$ while the λ_{ij} have no such restriction, the independent random component model is the submodel. From this model we get the REML estimates

$$\hat{\varpi}_1 = 0,0729, \quad \hat{\varpi}_2 = 0, \quad \hat{\varpi}_3 = 0.0157, \quad \hat{\varpi}_4 = 0.$$

Because $\hat{\varpi}_4 = 0$, $\hat{\gamma}_{12} > 0$. From this we have the REML estimates

$$\hat{\gamma}_{11} = \hat{\varpi}_1 + \hat{\varpi} = 0.0886, \quad \hat{\gamma}_{22} = 0.0157, \quad \hat{\gamma}_{12} = 0.0157.$$

This gives estimators that lie on the boundary of the allowed parameter space

$|\hat{\gamma}_{12}| \leq \hat{\gamma}_{22}$. However, the true REML estimator would satisfy $\hat{\lambda}_{12} \geq 0$ and $|\hat{\lambda}_{12}| > \hat{\lambda}_{22}$. Thus we now fit a model

$$y_{ij} = \beta_0 + x_{ij}\beta_1 + w_{i1} + x_{ij}^* w_{i2} + (1 + x_{ij}^*)w_{i3} + e_{ij},$$

where $x_{ij}^* = cx_{ij}$, $c = (\hat{\gamma}_{22}/\hat{\gamma}_{11})^{1/2} = 0.4204$ and $w_{ik} \sim N(0, \varpi_k^*)$ to make $\hat{\varpi}_1^* \simeq \hat{\varpi}_2^*$ giving $\hat{\gamma}_{11}^* = \hat{\varpi}_1^* + \hat{\varpi}_3^* \simeq \hat{\gamma}_{22}^* = \hat{\varpi}_2^* + \hat{\varpi}_3^*$. When $\hat{\gamma}_{11}^* \simeq \hat{\gamma}_{22}^*$, the constraint $|\hat{\gamma}_{12}^*| \leq \min\{\hat{\gamma}_{11}^*, \hat{\gamma}_{22}^*\}$ no longer restricts the parameter estimates. From the SAS MIXED procedure we have

$$\hat{\varpi}_1^* = 0.048, \ \hat{\varpi}_2^* = 0.041, \ \hat{\varpi}_3^* = 0.044,$$

$$\text{var}(\hat{\varpi}^*) = \begin{pmatrix} 0.00034 & & \\ -0.00003 & 0.00055 & \\ -0.00013 & -0.00006 & 0.00028 \end{pmatrix}.$$

Thus, we get the REML estimates for λ, with their standard error estimates in parentheses

$$\begin{aligned} \hat{\lambda}_{11} &= \hat{\varpi}_1^* + \hat{\varpi}_3^* = 0.092 \ (0.019), \\ \hat{\lambda}_{22} &= c^2(\hat{\varpi}_2^* + \hat{\varpi}_3^*) = 0.015 \ (0.005), \text{ and} \\ \hat{\lambda}_{12} &= c\hat{\varpi}_3^* = 0.018 \ (0.007) \end{aligned}$$

which are the same as Rasbash $et \ al.$'s (2000) REML estimates. Now $|\hat{\lambda}_{12}|$ can satisfy $|\hat{\lambda}_{12}| > \hat{\lambda}_{22}$. Because $\hat{\lambda}_{ij}$ are linear combinations of $\hat{\varpi}_i^*$ we can compute their variance estimates. For example

$$\text{var}(\hat{\lambda}_{22}) = c^4\{\text{var}(\hat{\varpi}_2^*) + 2\text{Cov}(\hat{\varpi}_2^*, \hat{\varpi}_3^*) + \text{var}(\hat{\varpi}_3^*)\}.$$

Finally, we find the ML estimate of $E(y_{ij}) = \beta_0 + \beta_1 x_{ij}$ to be $-0.012 + 0.557x_{ij}$, with standard errors 0.040 and 0.020. □

We have seen that in normal mixed models the h-likelihood gives

- The BLUE for $E(v|y)$ and BLUP for v
- The marginal ML estimator for β because v is canonical to β
- $p_{v,\beta}(h|\tau)$ to provide restricted MLEs for dispersion parameters
- Equivalent inferences for alternative random effect models, leading to the same marginal model

All these results are exact. Furthermore, many test statistics based upon sum of squares follow the exact χ^2 or F-distributions, and these distributions uniquely determine the test statistics (Seely and El-Bassiouni, 1983; Seely $et \ al.$, 1997), leading to numerically efficient confidence intervals for variance ratios (Lee and Seely, 1996). In non–normal random effect models from Chapter 6, most results hold only asymptotically.

Hierarchical GLMS

In this chapter, we introduce HGLMs as a synthesis of two widely-used existing model classes: GLMs (Chapter 2) and normal linear mixed models (Chapter 5). In an unpublished technical report, Pierce and Sands (Oregon State University, 1975) introduced generalized linear mixed models (GLMMs), where the linear predictor of a GLM is allowed to have, in addition to the usual fixed effects, one or more random components with assumed normal distributions. Although the normal distribution is convenient for specifying correlations among the random effects, the use of other distributions for the random effects greatly enriches the class of models. Lee and Nelder (1996) extended GLMMs to hierarchical GLMs (HGLMs), in which the distribution of random components is extended to conjugates of arbitrary distributions from the GLM family.

6.1 HGLMs

Lee and Nelder (1996) originally defined HGLMs as follows:

(a) Conditional on random effects u, the responses y follow a GLM family, satisfying

$$E\left(y|u\right) = \mu \quad \text{and} \quad \text{var}\left(y|u\right) = \phi V\left(\mu\right),$$

for which the kernel of the loglihood is given by

$$\sum \{y\theta - b(\theta)\}/\phi,$$

where $\theta = \theta(\mu)$ is the canonical parameter. The linear predictor takes the form

$$\eta = g\left(\mu\right) = X\beta + Zv, \tag{6.1}$$

where $v = v\left(u\right)$ for some monotone function $v()$ are the random effects and β are the fixed effects.

(b) The random component u follows a distribution conjugate to a GLM family of distributions with parameters λ.

For simplicity of argument, we first consider models with just one random vector u.

Example 6.1: The normal linear mixed model in Chapter 5 is an HGLM because

(a) $y|u$ follows a GLM distribution with

$$\text{var}\,(y|u) = \phi, \quad \text{with} \ \ \phi = \sigma^2 \ \ \text{and} \ \ V\,(\mu) = 1,$$

$$\eta = \mu = X\beta + Zv,$$

where $v = u$.

(b) $u \sim N(0, \lambda)$ with $\lambda = \sigma_v^2$.

We call this the normal-normal HGLM, where the first adjective refers to the distribution of the $y|u$ component and the second to the u component.

Example 6.2: Suppose $y|u$ is Poisson with mean

$$\mu = E(y|u) = \exp(X\beta)u.$$

With the log link we have

$$\eta = \log \mu = X\beta + v,$$

where $v = \log u$. If the distribution of u is gamma, v has a log-gamma distribution and we call the model the Poisson-gamma HGLM. The GLMM assumes a normal distribution for v, so the distribution of u is log-normal. The corresponding Poisson GLMM could be called the Poisson-log-normal HGLM under the $v = \log u$ parameterization. Note that a normal distribution for u is the conjugate of the normal for $y|u$, and the gamma distribution is that for the Poisson. It is not necessary for the distribution of u to be the conjugate of that for $y|u$. If it is, we call the resulting model a conjugate HGLM. Both the Poisson-gamma model and Poisson GLMM belong to the class of HGLMs, the former being a conjugate HGLM while the latter is not.

6.1.1 Constraints in models

In an additive model such as $X\beta + v$, the location of v is unidentifiable since

$$X\beta + v = (X\beta + a) + (v - a),$$

while in a multiplicative model such as $\exp(X\beta)u$, the scale of u is unidentifiable since

$$(\exp X\beta)u = (a \exp X\beta)(u/a)$$

for $a > 0$. Thus, in defining random effect models we may impose constraints either on the fixed effects or on the random effects. In linear models and linear mixed models, constraints have been put on random effects such as $E(e) = 0$ and $E(v) = 0$. In GLMMs, it is standard to assume that $E(v) = 0$.

Lee and Nelder (1996) noted that imposing constraints on random effects is more convenient when we move to models with more than one random component. In the Poisson-gamma model we set constraints $E(u) = 1$. These constraints affect the estimates of the parameters on which constraints are put. For random effects with $E(v_i) = 0$, we have $\sum_{i=1}^{q} \hat{v}_i/q = 0$, while for those with $E(u_i) = 1$, we have $\sum_{i=1}^{q} \hat{u}_i/q = 1$. Thus, care is necessary in comparing results from two different HGLMs. Note that in the Poisson-gamma model we have $E(y) = \exp(X\beta)$, while in the Poisson GLMM $E(y) = \exp(X\beta + \lambda/2)$ with $\text{var}(v) = \lambda$. This means that fixed effects in Poisson-gamma HGLMs and Poisson GLMMs cannot be compared directly because they assume different parameterizations by having different constraints on their estimates (Lee and Nelder, 2004).

6.2 H-likelihood

In last two chapters we have seen that for inferences from HGLMs we should define the h-loglihood of the form

$$h \equiv \log f_{\beta,\phi}(y|v) + \log f_{\lambda}(v), \tag{6.2}$$

where (ϕ, λ) are dispersion parameters and in normal mixed linear models v is the canonical scale for β. However, this definition is too restrictive because, for example, there may not exist a canonical scale for non-normal GLMMs.

6.2.1 Weak canonical scale

Given that some extended-likelihood should serve as the basis for statistical inferences of a general nature, we want to find one whose maximization gives meaningful estimators of the random parameters. Lee and Nelder (2005) showed that maintaining invariance of inferences from the extended likelihood for trivial re-expressions of the underlying model leads to a definition of the h-likelihood. For further explanation, we need the following property of extended likelihoods.

Property. The extended likelihoods $L(\theta, u; y, u)$ and $L(\theta, u; y, k(u))$ give identical inferences about the random effects if $k(u)$ is a linear function of u.

This property of extended likelihoods has an analogue in the BUE property, which is preserved only under linear transformation: $E\{k(u)|y\} = k(E(u|y))$ only if $k()$ is linear.

Consider a simple normal-normal HGLM of the form: for $i = 1, ..., m$ and $j = 1, ..., n$ with $N = mn$

$$y_{ij} = \beta + v_i + e_{ij}, \qquad (6.3)$$

where $v_i \sim i.i.d.\ N(0, \lambda)$ and $e_{ij} \sim i.i.d.\ N(0, 1)$. Consider a linear transformation $v_i = \sigma v_i^*$ where $\sigma = \lambda^{1/2}$ and $v_i^* \sim i.i.d.\ N(0, 1)$. The joint loglihoods $\ell(\theta, v; y, v)$ and $\ell(\theta, v^*; y, v^*)$ give the same inference for v_i and v_i^*. In (6.2) the first term $\log f_{\beta,\phi}(y|v)$ is invariant with respect to reparameterizations; in fact $f_{\beta,\phi}(y|v) = f_{\beta,\phi}(y|u)$ functionally for one-to-one parameterization $v = v(u)$. Let $\hat{v}_i$ and $\hat{v}_i^*$ maximize $\ell(\theta, v; y, v)$ and $\ell(\theta, v^*; y, v^*)$, respectively. We then have invariant estimates $\hat{v}_i = \sigma \hat{v}_i^*$ because

$$-2\log f_\lambda(v) = m\log(2\pi\sigma^2) + \sum v_i^2/\sigma^2 = -2\log f_\lambda(v^*) + m\log(\sigma^2),$$

and these loglihoods differ only by a constant.

Consider model (6.3), with a different parameterization

$$y_{ij} = \beta + \log u_i + e_{ij}, \qquad (6.4)$$

where $\log(u_i) \sim i.i.d.\ N(0, \lambda)$. Let $\log(u_i) = \sigma \log u_i^*$ and $\log(u_i^*) \sim i.i.d.$ $N(0, 1)$. Here we have

$$
\begin{aligned}
-2\log f_\lambda(u) &= m\log(2\pi\lambda) + \sum(\log u_i)^2/\lambda + 2\sum \log u_i \\
&= -2\log f_\lambda(u^*) + m\log(\lambda) + 2\sum \log(u_i/u_i^*).
\end{aligned}
$$

Let $\hat{u}_i$ and $\hat{u}_i^*$ maximize $\ell(\theta, u; y, u)$ and $\ell(\theta, u^*; y, u^*)$, respectively. Then, $\log \hat{u}_i \neq \sigma \log \hat{u}_i^*$ because $\log u_i = \sigma \log u_i^*$, i.e., $u_i = u_i^{*\sigma}$, is no longer a linear transformation.

Clearly the models (6.3) and (6.4) are equivalent, so if h-likelihood is to be a useful notion we need their corresponding h-loglihoods to be equivalent as well. This implies that to maintain invariance of inference with respect to equivalent modellings, we must define the h-likelihood on the specific scale $v(u)$ on which the random effects combine additively with the fixed effects β in the linear predictor. We call this the *weak canonical scale* and for model (6.4) the scale is $v = \log u$. This weak canonical scale can be always defined if we can define the linear predictor.

The ML estimator of β is invariant with respect to equivalent models, and can be obtained by joint maximization if v is canonical to β. Thus, if a canonical scale for β exists, it also satisfies the weak canonical property in that the resulting estimator of β is invariant with respect to equivalent models.

With this definition the h-likelihood for model (6.4) is

$$L(\theta, v; y, v) \equiv f_{\beta,\phi}(y|\log u) f_{\lambda}(\log u),$$

giving

$$\eta_{ij} = \mu_{ij} = \beta + v_i \quad \text{with} \quad \mu_{ij} = E(y_{ij}|v_i).$$

For simplicity of argument, let $\lambda = 1$, so that there is no dispersion parameter, but only a location parameter β. The h-loglihood $\ell(\theta, v; y, v)$ is given by

$$-2h = -2\ell(\theta, v; y, v) \equiv \{N \log(2\pi) + \sum_{ij}(y_{ij} - \beta - v_i)^2\}$$

$$+\{m \log(2\pi) + \sum_i v_i^2\}.$$

This has its maximum at the TOE:

$$\hat{v}_i = E(v_i|y) = \frac{n}{n+1}(\bar{y}_{i.} - \beta).$$

Suppose that we estimate β and v by joint maximization of h. The solution is

$$\hat{\beta} = \bar{y}_{..} = \sum_{ij} y_{ij}/N \quad \text{and} \quad \hat{v}_i = \frac{n}{n+1}(\bar{y}_{i.} - \bar{y}_{..}) = \sum_j (y_{ij} - \bar{y}_{..})/(n+1).$$

Now $\hat{\beta}$ is the ML estimator and $\hat{v}_i$ is the BLUP of v_i defined by

$$\hat{v}_i = \widehat{E(v_i|y)},$$

and can also be justified as the BLUE of the TOE (Chapter 4).

The extended loglihood $\ell(\beta, u; y, u)$ gives

$$-2\ell(\beta, u; y, u) \equiv \{N \log(2\pi) + \sum(y_{ij} - \beta - \log u_i)^2\}$$

$$+\{m \log(2\pi) + \sum(\log u_i)^2 + 2\sum(\log u_i)\}$$

with an estimate

$$\hat{v}_i = \log \hat{u}_i = \frac{n}{n+1}(\bar{y}_{i.} - \beta) - 1/(n+1).$$

The joint maximization of $L(\beta, u; y, u)$ leads to

$$\hat{\beta} = \bar{y}_{..} + 1 \quad \text{and} \quad \hat{v}_i = \frac{n}{n+1}(\bar{y}_{i.} - \bar{y}_{..}) - 1.$$

Thus, in this example, joint maximization of the h-loglihood provides satisfactory estimates of both the location and random parameters for either parameterization, while that of an extended loglihood may not.

6.2.2 GLM family for the random components

A key aspect of HGLMs is the flexible specification of the distribution of the random effects u, which can come from an exponential family with log density proportional to

$$\sum\{k_1 c_1(u) + k_2 c_2(u)\}$$

for some functions $c_1()$ and $c_2()$, and parameters k_1 and k_2. The weak canonical scale gives a clear representation of loglihood for random effects, which can be written as

$$\sum\{\psi_M \theta_M(u) - b_M(\theta_M(u))\}/\lambda, \qquad (6.5)$$

for some known functions $\theta_M()$ and $b_M()$, so that it looks conveniently like the kernel of the GLM family, and choosing a random effect distribution becomes similar to choosing a GLM model. Examples of these functions based on common distributions are given in Table 6.1. (We use the label M to refer to the mean structure.) In the next section, we exploit this structure in our algorithm. Allowing for the constraint on $E(u)$ as discussed above, the constant ψ_M takes a certain value, so the family (6.5) is actually indexed by a single parameter λ. Table 6.1 also provides the corresponding values for ψ_M in the various families. As we shall show later, in conjugate distributions we have

$$E(u) = \psi_M \quad \text{and} \quad \text{var}(u) = \rho V_M(\psi_M).$$

Recall that the loglihood based on $y|v$ is

$$\sum\{y\theta(\mu) - b(\theta(\mu))\}/\phi.$$

Now, by choosing the specific functions $\theta_M(u) = \theta(u)$ and $b_M(\theta_M) = b(\theta)$, we obtain the conjugate loglihood

$$\sum\{\psi_M \theta(u) - b(\theta(u))\}/\lambda \qquad (6.6)$$

for the random effects. Cox and Hinkley (1974, p. 370) defined the so-called conjugate distribution. We call (6.6) the conjugate loglihood to highlight that it is not a log density for ψ_M. The corresponding HGLM is called a conjugate HGLM , but there is no need to restrict ourselves to such models. It is worth noting that the weak canonical scale of v leads to this good representation of conjugacy. In conjugate distributions, the scale of random effects is not important when they are to be integrated out, while in conjugate likelihood the scale is important, leading to effective inferential procedures.

In principle, various combinations of GLM distribution and link for $y|v$

Table 6.1 *GLM families for response $y|v$ and conjugates of GLM families for random effects u.*

$y|v$ distribution	$V(\mu)$	$\theta(\mu)$	$b(\theta)$
Normal	1	μ	$\theta^2/2$
Poisson	μ	$\log \mu$	$\exp \theta$
Binomial	$\mu(m-\mu)/m$	$\log\{\mu/(m-\mu)\}$	$\log\{1+\exp\theta\}$
Gamma	μ^2	$-1/\mu$	$-\log\{-\theta\}$

u distribution	$V_M(u)$	$\theta_M(u)$	$b_M(\theta_M)$	ψ_M	ρ
Normal	1	u	$\theta_M^2/2$	0	λ
Gamma	u	$\log u$	$\exp\theta_M$	1	λ
Beta	$u(1-u)$	$\log\{u/(1-u)\}$	$\log\{1+\exp\theta_M\}$	1/2	$\lambda/(1+\lambda)$
Inverse gamma	u^2	$-1/u$	$-\log\{-\theta_M\}$	1	$\lambda/(1-\lambda)$

and a conjugate to any GLM distribution and link for v can be used to construct HGLMs. Examples of some useful HGLMs are shown in Table 6.2. Note that the idea allows a quasi–likelihood extension to the specification of the random effects distribution, via specification of the mean and variance functions.

Table 6.2 *Examples of HGLMs.*

$y|u$ dist.	$g(\mu)^*$	u dist.	$v(u)$	Model
Normal	id	Normal	id	Conjugate HGLM
				Linear mixed
Binomial	logit	Beta	logit	Conjugate HGLM
				Beta-binomial
Binomial	logit	Normal	id	Binomial GLMM
Binomial	comp	Gamma	log	HGLM
Gamma	recip	Inverse gamma	recip	Conjugate HGLM
Gamma	log	Inverse gamma	recip	Conjugate HGLM
				with non-canonical link
Gamma	log	Gamma	log	HGLM
Poisson	log	Normal	id	Poisson GLMM**
Poisson	log	Gamma	log	Conjugate HGLM

* id = identity, recip = reciprocal, comp = complementary log-log.
** In GLMMs, we take $v = v(u) = u$.

Example 6.3: Consider a Poisson-gamma model having
$$\mu_{ij} = E(y_{ij}|u_i) = (\exp x_{ij}\beta)u_i$$
and random effects u_i being iid with the gamma distribution
$$f_\lambda(u_i) = (1/\lambda)^{1/\lambda}(1/\Gamma(1/\lambda))u_i^{1/\lambda-1}\exp(-u_i/\lambda),$$
so that
$$\psi_M = E(u_i) = 1 \quad \text{and} \quad \text{var}(u_i) = \lambda\psi_M = \lambda.$$
The log link leads to a linear predictor
$$\eta_{ij} = \log \mu_{ij} = x_{ij}\beta + v_i,$$
where $v_i = \log u_i$.

The loglihood contribution of the $y|v$ part comes from the Poisson density:
$$\sum(y_{ij}\log \mu_{ij} - \mu_{ij}),$$
and the loglihood contribution of v is
$$\ell(\lambda; v) = \log f_\lambda(v) = \sum_i \{(\psi_M \log u_i - u_i)/\lambda - \log(\lambda)/\lambda - \log \Gamma(1/\lambda)\},$$

with $\psi_M = Eu_i = 1$. We can recognize the kernel (6.5), which in this case is a conjugate version (6.6).

Note here that the standard gamma GLM for the responses y has $V(\mu) = \mu^2$ with $\mu = E(y)$ and reciprocal canonical links, but that for the random effect distribution has $V_M(u) = u$ and log canonical links . $\square$

Example 6.4: In the beta-binomial model we assume

(a) $Y_{ij}|u_i \sim \text{Bernoulli}(u_i)$ and

(b) $u_i \sim \text{beta}(\alpha_1, \alpha_2)$, having

$$\psi_M = E(u) = \alpha_1/(\alpha_1 + \alpha_2), \quad var(u) = \rho\psi_M(1 - \psi_M),$$

where $\rho = 1/(1 + \alpha_1 + \alpha_2)$. Since $v_i = \theta(u_i) = \log\{u_i/(1 - u_i)\}$ we have

$$\ell(\lambda; v) = \sum_i ([\{\psi_M v_i - \log\{1/(1 - u_i)\}]/\lambda - \log B(\alpha_1, \alpha_2)),$$

where $\lambda = 1/(\alpha_1 + \alpha_2)$ and $B(,)$ is the beta function. Thus, $\rho = \lambda/(1 + \lambda)$. Here the parameters α_1 and α_2 represent asymmetry in the distribution of u_i. Because the likelihood surface is often flat with respect to α_1 and α_2, Lee and Nelder (2001a) proposed an alternative model as follows:

(a) $Y_{ij}|u_i \sim \text{Bernoulli}(p_{ij})$, giving

$$\eta_{ij} = \log p_{ij}/(1 - p_{ij}) = x_{ij}\beta + v_i$$

and

(b) $u_i \sim \text{beta}(1/\alpha, 1/\alpha)$, giving $\psi_M = 1/2$ and $\lambda = \alpha/2$. They put a constraint on the random effects

$$E(u_i) = 1/2.$$

With this model we can accommodate arbitrary covariates x_{ij} for fixed effects and it has better convergence. $\square$

Example 6.5: For the gamma-inverse gamma HGLM, suppose that $u_i \sim$ inverse gamma$(1 + \alpha, \alpha)$ with $\alpha = 1/\lambda$,

$$\psi_M = E(u_i) = 1, var(u_i) = 1/(\alpha - 1) = \rho\psi_M^2$$

and $\rho = \lambda/(1 - \lambda)$. Since $v_i = \theta(u_i) = -1/u_i$ we have

$$l(\lambda; v) = \sum_i [\{\psi_M v_i - \log(u_i)\}/\lambda + (1 + 1/\lambda)\log(\psi_M/\lambda) - \log\{\Gamma(1/\lambda)/\lambda\}],$$

where $\Gamma(1/\lambda)/\lambda = \Gamma(1/\lambda + 1)$. This is the loglihood for the conjugate pair of the gamma GLM for $Y|u_i$. $\square$

6.2.3 Augmented GLMs and IWLS

In the joint estimation of the fixed and random effect parameters in the normal linear mixed model (Section 5.3.5), we show that the model can

be written as an augmented classical linear model involving fixed effect
parameters only. As a natural extension to HGLM, we should expect an
augmented classical GLM also with fixed effect parameters only. This is
not strange; during the estimation, the random effects u were treated
as fixed unknown values. Now the model (6.5) can be immediately interpreted as a GLM with *fixed canonical parameter* $\theta_M(u)$ and *response*
ψ_M, satisfying

$$
\begin{aligned}
E(\psi_M) &= u = b'_M(\theta_M(u))\\
\mathrm{var}(\psi_M) &= \lambda V_M(u) = \lambda b''_M(\theta_M(u)).
\end{aligned}
$$

As is obvious in the linear model case and the examples above, during the estimation, the response ψ_M takes the value determined by the
constraint on $E(u)$.

Thus the h-likelihood estimation for an HGLM can be viewed as that
for an augmented GLM with the response variables $(y^t, \psi_M^t)^t$, where

$$
\begin{aligned}
E(y) = \mu, &\qquad E(\psi_M) = u,\\
\mathrm{var}(y) = \phi V(\mu), &\quad \mathrm{var}(\psi_M) = \lambda V_M(u),
\end{aligned}
$$

and the augmented linear predictor

$$
\eta_{Ma} = (\eta^t, \eta_M^t)^t = T_M \omega,
$$

where $\eta = g(\mu) = X\beta + Zv$; $\eta_M = g_M(u) = v$, $\omega = (\beta^t, v^t)^t$ are fixed
unknown parameters and quasi–parameters, and the augmented model
matrix is

$$
T_M = \begin{pmatrix} X & Z \\ 0 & I \end{pmatrix}.
$$

For conjugate HGLMs, $V_M() = V()$. For example, in the Poisson-gamma
models $V_M(u) = V(u) = u$, while in Poisson GLMM $V_M(u) = 1 \neq
V(u) = u$. Note also that the gamma distribution for the u component
has $V_M(u) = u$ (not its square) because it is the conjugate pair for the
Poisson distribution.

As an immediate consequence, given (ϕ, λ), the estimate of two components $\omega = (\beta^t, v^t)^t$ can be computed by iterative weighted least squares
(IWLS) from the augmented GLM

$$
T_M^t \Sigma_M^{-1} T_M \omega = T_M^t \Sigma_M^{-1} z_{Ma}, \tag{6.7}
$$

where $z_{Ma} = (z^t, z_M^t)^t$ and $\Sigma_M = \Gamma_M W_{Ma}^{-1}$ with $\Gamma_M = \mathrm{diag}(\Phi, \Lambda)$, $\Phi =
\mathrm{diag}(\phi_i)$ and $\Lambda = \mathrm{diag}(\lambda_i)$. The adjusted dependent variables $z_{Mai} =
(z_i, z_{Mi})$ are defined by

$$
z_i = \eta_i + (y_i - \mu_i)(\partial \eta_i / \partial \mu_i)
$$

for the data y_i, and

$$z_{Mi} = v_i + (\psi_M - u_i)(\partial v_i / \partial u_i)$$

for the augmented data ψ_M. The iterative weight matrix

$$W_{Ma} = \text{diag}(W_{M0}, W_{M1})$$

contains elements

$$W_{M0i} = (\partial \mu_i / \partial \eta_i)^2 V(\mu_i)^{-1}$$

for the data y_i, and

$$W_{M1i} = (\partial u_i / \partial v_i)^2 V_M(u_i)^{-1}$$

for the augmented data ψ_M.

6.3 Inferential procedures using h-likelihood

In normal linear mixed models, the scale v is canonical to the fixed mean parameter β, so that the marginal ML estimator for β can be obtained from the joint maximization, and the h-loglihood gives the BLUP for the particular scale v_i. These properties also hold in some non-normal HGLMs. Consider a Poisson-gamma model having

$$\mu_{ij} = E(Y_{ij}|u) = (\exp x_{ij}^t \beta) u_i$$

where $x_{ij} = (x_{1ij}, ..., x_{pij})^t$.

The kernel of the marginal loglihood is

$$\ell = \sum_{ij} y_{ij} x_{ij}^t \beta - \sum_i (y_{i+} + 1/\lambda) \log(\mu_{i+} + 1/\lambda),$$

where $y_{i+} = \sum_j y_{ij}$ and $\mu_{i+} = \sum_j \exp x_{ij}^t \beta$, which gives

$$\partial \ell / \partial \beta_k = \sum_{ij} (y_{ij} - \frac{y_{i+} + 1/\lambda}{\mu_{i+} + 1/\lambda} \exp x_{ij}^t \beta) x_{kij}.$$

The kernel of h-loglihood is

$$
\begin{aligned}
h &= \log f_\beta(y|v) + \log f_\lambda(v) \\
&= \sum_{ij} (y_{ij} \log \mu_{ij} - \mu_{ij}) + \sum_i \{(v_i - u_i)/\lambda - \log(\lambda)/\lambda - \log \Gamma(1/\lambda)\},
\end{aligned}
$$

giving

$$\partial h / \partial v_i = (y_{i+} + 1/\lambda) - (\mu_{i+} + 1/\lambda) u_i, \tag{6.8}$$

$$\partial h / \partial \beta_k = \sum_{ij} (y_{ij} - \mu_{ij}) x_{kij}. \tag{6.9}$$

This shows that the scale v is canonical for β, so that the marginal ML estimator for β can be obtained from the joint maximization. Furthermore, the h-loglihood gives the BLUP for u_i as

$$E(u_i|y)|_{\theta=\hat{\theta}} = (y_{i+} + 1/\hat{\lambda})/(\hat{\mu}_{i+} + 1/\hat{\lambda}).$$

Thus, some properties of linear mixed models continue to hold here. However, they no longer hold in Poisson GLMMs and Poisson-gamma HGLMs with more than one random component.

6.3.1 Analysis of paired data

In this section we illustrate why care is necessary in developing likelihood inferences for binary HGLMs. Consider the normal linear mixed model for the paired data: for $i = 1, \ldots, m$ and $j = 1, 2$,

$$y_{ij} = \beta_0 + \beta_j + v_i + e_{ij},$$

where β_0 is the intercept, β_1 and β_2 are treatment effects for two groups, random effects $v_i \sim N(0, \lambda)$ and the white noise $e_{ij} \sim N(0, \phi)$. Suppose we are interested in inferences about $\beta_2 - \beta_1$. Here the marginal MLE is

$$\widehat{\beta_2 - \beta_1} = \bar{y}_{\cdot 2} - \bar{y}_{\cdot 1}$$

where $\bar{y}_{\cdot j} = \sum_i y_{ij}/m$. This estimator can be also obtained as the intra-block estimator, the OLS estimator treating v_i as fixed.

Now consider the conditional approach. Let $y_{i+} = y_{i1} + y_{i2}$. Because

$$y_{i1}|y_{i+} \sim N(\{y_{i+} + (\beta_1 - \beta_2)\}/2, \phi/2),$$

the use of this conditional likelihood is equivalent to using the distribution

$$y_{i2} - y_{i1} \sim N(\beta_2 - \beta_1, 2\phi).$$

Thus all three estimators from the ML, intra-block and conditional approaches lead to the same estimates.

Now consider the Poisson-gamma HGLM for paired data, where

$$\eta_{ij} = \log \mu_{ij} = \beta_0 + \beta_j + v_i.$$

Suppose we are interested in inferences about $\theta = \mu_{i2}/\mu_{i1} = \exp(\beta_2 - \beta_1)$. Then, from (6.9) we have the estimating equations

$$\bar{y}_{\cdot 1} = \exp(\beta_0 + \beta_1) \sum_i \hat{u}_i/m,$$

$$\bar{y}_{\cdot 2} = \exp(\beta_0 + \beta_2) \sum_i \hat{u}_i/m$$

giving

$$\exp(\widehat{\beta_2} - \widehat{\beta_1}) = \bar{y}_{\cdot 2}/\bar{y}_{\cdot 1}.$$

This proof works even when the v_i are fixed, so that the intra-block estimator is the same as the marginal MLE. From Example 1.9 we see that the conditional estimator is also the same. This result holds for Poisson GLMMs too. This means that the results for normal linear mixed models for paired data also hold for Poisson HGLMs.

Now consider the models for binary data: suppose $y_{ij}|v_i$ follows the Bernoulli distribution with p_{ij} such that

$$\eta_{ij} = \log\{p_{ij}/(1 - p_{ij})\} = \beta_0 + \beta_j + v_i.$$

Here the three approaches give different inferences. Andersen (1970) showed that the intra-block estimator, obtained by treating v_i as fixed, is severely biased. With binary data this is true in general. Patefield (2000) showed the bias of intra-block estimators for cross–over trials and Lee (2002b) for therapeutic trials. Now consider the conditional estimator, conditioned upon the block totals y_{i+} that have possible values of $0, 1, 2$. The concordant pairs (when $y_{i+} = 0$ or 2) carry no information, because $y_{i+} = 0$ implies $y_{i1} = y_{i2} = 0$, and $y_{i+} = 2$ implies $y_{i1} = y_{i2} = 1$. Thus, in the conditional likelihood, only the discordant pairs $y_{i+} = 1$ carry information. The conditional distribution of $y_{i1}|(y_{i+} = 1)$ follows the Bernoulli distribution with

$$
\begin{aligned}
p &= \frac{P(y_{i1} = 1, y_{i2} = 0)}{P(y_{i1} = 1, y_{i2} = 0) + P(y_{i1} = 0, y_{i2} = 1)} \\
&= \frac{\exp(\beta_1 - \beta_2)}{1 + \exp(\beta_1 - \beta_2)},
\end{aligned}
$$

which is equivalent to

$$\log\{p/(1 - p)\} = \beta_1 - \beta_2.$$

Thus, $\beta_1 - \beta_2$ can be obtained from the GLM of discordant data, giving what might be called the McNemar (1947) estimator. This conditional estimator is consistent. In binary matched pairs, the conditional likelihood estimator of the treatment effect is asymptotically fully efficient (Lindsay, 1983). However, if there are other covariates, the conditional estimator is not always efficient; Kang et al. (2005) showed that the loss of information could be substantial. In the general case, the MLE from the HGLM should be used to exploit all the information in the data.

6.3.2 Fixed versus random effects

Fixed effects can describe systematic mean patterns such as trend, while random effects may describe either correlation patterns between repeated measures within subjects or heterogeneities between subjects or both. The correlation can be represented by random effects for subjects, and heterogeneities by saturated random effects. In practice, it is often necessary to have both types of random components. However, sometimes it may not be obvious whether effects are to be treated as fixed or random. For example, there has been much debate among econometricians about two alternative specifications of fixed and random effects in mixed linear models: see Baltagi (1995) and Hsiao (1995). When v_i are random, the ordinary least square estimator for β, treating v_i as fixed is in general not fully efficient, but is consistent under wide conditions. By contrast, estimators for β, treating v_i as random can be biased if random effects and covariates are correlated (Hausman, 1978).

Thus, even if random effects are an appropriate description for v_i we may still prefer to treat the v_i as fixed unless the assumptions about the random effects can be confirmed. Without sufficient random effects to check their assumed distribution, it may be better to treat them as fixed. This produces what is known as the intra-block analysis, and it is robust against assumptions about the random effects in normal linear mixed models. Econometrics models are mainly based upon the normality assumption. However, with binary data, the robustness property of intra-block estimators no longer holds. In general there is no guarantee that the intra-block analysis will be robust.

6.3.3 Inferential procedures

From the h-loglihood we have two useful adjusted profile loglihoods: the marginal loglihood ℓ and the restricted loglihood of Section 5.2.2,

$$\log f_{\phi,\lambda}(y|\tilde{\beta}),$$

where $\tilde{\beta}$ is the marginal ML estimator given $\tau = (\phi, \lambda)$. Following Cox and Reid (1987) (see Section 1.9), the restricted loglihood can be approximated by

$$p_\beta(\ell|\phi, \lambda).$$

In principle we should use the h-loglihood h for inferences about v, the marginal loglihood ℓ for β and the restricted loglihood $\log f_{\phi,\lambda}(y|\tilde{\beta})$ for the dispersion parameters. If the restricted loglihood is hard to obtain, we may use the adjusted profile likelihood $p_\beta(\ell)$. When ℓ is numerically hard to obtain, Lee and Nelder (1996, 2001) proposed to use $p_v(h)$ as an approximation to ℓ and $p_{\beta,v}(h)$ as an approximation to $p_\beta(\ell)$, and therefore to $\log f_{\phi,\lambda}(y|\tilde{\beta})$; $p_{\beta,v}(h)$ gives approximate restricted ML (REML) estimators for the dispersion parameters and $p_v(h)$ approximate ML estimators for the location parameters. Because $\log f_{\phi,\lambda}(y|\tilde{\beta})$ has no explicit form except in normal mixed models, in this book we call dispersion estimators that maximize $p_{\beta,v}(h)$ the REML estimators.

In Poisson-gamma models, v is canonical for β, but not for τ, in the sense that

$$\frac{L(\beta_1, \tau, \hat{v}_{\beta_1,\tau}; y, v)}{L(\beta_2, \tau, \hat{v}_{\beta_2,\tau}; y, v)} = \frac{L(\beta_1, \tau; y)}{L(\beta_2, \tau; y)}. \tag{6.10}$$

Given τ, joint maximization of h gives the marginal ML estimators for β. This property may hold approximately under a weak canonical scale in HGLMs, e.g., the deviance based upon h is often close to ℓ, so that Lee and Nelder (1996) proposed joint maximization for β and v. We have often found that the MLE of β from the marginal likelihood is close numerically to the joint maximizer $\hat{\beta}$ from the h-likelihood. To establish this, careful numerical studies are necessary on a model-by-model basis. With binary data we found that the joint maximization results in non-negligible bias and that $p_v(h)$ must be used for estimating β.

In binary cases with small cluster size, this method gives non-ignorable biases in dispersion parameters, which causes biases in β. For the estimation of dispersion parameters, Noh and Lee (2007a) use the second-order Laplace approximation

$$p^s_{v,\beta}(h) = p_{v,\beta}(h) - F/24,$$

where

$$F = \mathrm{tr}[-\{3(\partial^4 h/\partial v^4) + 5(\partial^3 h/\partial v^3)D(h,v)^{-1}(\partial^3 h/\partial v^3)\}D(h,v)^{-2}]|_{v=\tilde{v}}.$$

Let $B^{jkl}_{3i} = (\partial^3 h_i/(\partial v_j \partial v_k \partial v_l))|_{v=\tilde{v}}$, $B^{jklm}_{4i} = (\partial^4 h_i/(\partial v_j \partial v_k \partial v_l \partial v_m))|_{v=\tilde{v}}$ and b_{ij} be the (i,j)th element of $D(h,\boldsymbol{v})^{-1}$. Then,

$$\mathrm{tr}(F)/24 = \sum_{i=1}^n C_{1i} + \sum_{i=1}^n \sum_{i'=1}^n C_{2(i,i')},$$

where

$$C_{1i} = \sum_{m=1}^N \sum_{l=1}^N \sum_{k=1}^N \sum_{j=1}^N (B^{jklm}_{4i} b_{jk} b_{lm})/8,$$

$$C_{2(i,i')} = \sum_{t=1}^N \sum_{s=1}^N \sum_{m=1}^N \sum_{l=1}^N \sum_{k=1}^N \sum_{j=1}^N \{B^{jkl}_{3i} B^{rst}_{3i'} (b_{jk} b_{lr} b_{st}/4 + b_{jr} b_{ks} b_{lt}/6)\}/2.$$

Thus, the computation of $tr(F)$ is the order of the $O\{m \times (N^4 + N^6)\} = O(mN^6)$ with $N = n \times m$, which is generally computationally too intensive. For models with crossed random effects for salamander data (McCullagh and Nelder, 1989), Shun (1997) showed that the computation of $tr(F)$ can be reduced by the order of $O(N^2)$. Noh and Lee (2007a) showed that it can be further reduced by the order of $O(N)$. However, their reduction method is restricted to crossed two random effect models. For factor models in Chapter 14, Noh et al. (2016) showed that the computation of $tr(F)$ can be generally in the order of $O(N)$.

Example 6.6: Consider the Poisson-gamma model in Example 6.2, having the marginal loglihood

$$\ell = \sum_{ij}\{y_{ij}x_{ij}^t\beta - \log\Gamma(y_{ij}+1)\} + \sum_i[-(y_{i+}+1/\lambda)\log(\mu_{i+}+1/\lambda)$$
$$- \log(\lambda)/\lambda + \log\{\Gamma(y_{i+}+1/\lambda)/\Gamma(1/\lambda)\}].$$

Here the h-loglihood is given by

$$h = \sum_{ij}\{y_{ij}x_{ij}^t\beta - \log\Gamma(y_{ij}+1)\} + \sum_i[(y_{i+}+1/\lambda)v_i - (\mu_{i+}+1/\lambda)u_i$$
$$- \log(\lambda)/\lambda - \log\{\Gamma(1/\lambda)\}].$$

Now, v is canonical for β, but not for λ. Note that the adjustment term for $p_v(h)$

$$D(h,v)|_{u_i=\hat{u}_i} = -\partial^2h/\partial v^2|_{u_i=\hat{u}_i} = (\mu_{i+}+1/\lambda)\hat{u}_i = y_{i+}+1/\lambda,$$

is independent of β but depends upon λ. Note also that

$$p_v(h) = [h - \frac{1}{2}\log\det\{D(h,v)/(2\pi)\}]|_{u=\hat{u}}$$
$$= \sum_{ij}\{y_{ij}x_{ij}^t\beta - \log\Gamma(y_{ij}+1)\} + \sum_i[-(y_{i+}+1/\lambda)\log(\mu_{i+}+1/\lambda)$$
$$+ (y_{i+}+1/\lambda)\log(y_{i+}+1/\lambda) - (y_{i+}+1/\lambda) - \log(\lambda)/\lambda$$
$$- \log\{\Gamma(1/\lambda)\} - \log(y_{i+}+1/\lambda)/2 + \log(2\pi)/2],$$

which is equivalent to approximating ℓ by the first-order Stirling approximation

$$\log\Gamma(x) \doteq (x-1/2)\log(x) + \log(2\pi)/2 - x \qquad (6.11)$$

for $\Gamma(y_{i+}+1/\lambda)$. Thus, the marginal MLE for β (maximizing $p_v(h)$) can be obtained by maximization of h. Furthermore, a good approximation to the ML estimator for λ can be obtained by using $p_v(h)$ if the first-order Stirling approximation works well. It can be further shown that the second-order Laplace approximation $p_v^s(h)$ is equivalent to approximating ℓ by the second-order Stirling approximation

$$\log\Gamma(x) \doteq (x-1/2)\log(x) + \log(2\pi)/2 - x + 1/(12x). \qquad (6.12)$$

Example 6.7: Consider the binomial-beta model in Example 6.4:

(a) $Y_{ij}|u_i \sim$ Bernoulli(u_i) for $j = 1, \cdots, m_i$

(b) $u_i \sim$ beta(α_1, α_2)

In this model, the marginal loglihood can be written explicitly as

$$\ell = \sum_i \{A_i - B_i + C_i\}$$

where $A_i = \log \text{Beta}(y_{i+} + \psi_M/\lambda, m_i - y_{i+} + (1 - \psi_M)/\lambda)$, $B_i = \log \text{Beta}(\psi_M/\lambda, (1 - \psi_M)/\lambda)$, $\log \text{Beta}(\alpha_1, \alpha_2) = \log\{\Gamma(\alpha_1)\} + \log\{\Gamma(\alpha_2)\} - \log\{\Gamma(\alpha_1 + \alpha_2)\}$ and $C_i = \log[m_i!/\{y_{i+}!(m_i - y_{i+})!\}]$. Note that

$$p_v(h) = \sum_i \{A_i^f - B_i + C_i\},$$

where A_i^f, which is $(y_{i+} + \psi_M/\lambda - 1/2)\log(y_{i+} + \psi_M/\lambda) + \{m_i - y_{i+} + (1 - \psi_M)/\lambda - 1/2\}\log\{m_i - y_{i+} + (1 - \psi_M)/\lambda\} - (m_i + 1/\lambda - 1/2)\log(m_i + 1/\lambda) + \log(2\pi)/2$, can be shown to be the first-order Stirling approximation (6.11) to A_i. We can further show that

$$p_{v,\beta}^s(h) = p_v(h) - F/24 = \sum_i \{A_i^s - B_i + C_i\},$$

where A_i^s, which is $A_i^f + 1/\{12(y_i + \psi_M/\lambda)\} + 1/\{12(m_i - y_i + (1 - \psi_M)/\lambda)\} - 1/\{12(m_i + 1/\lambda)\}$, can be shown to be the second-order Stirling approximation (6.12) to A_i. This gives very good approximation to ℓ, yielding estimators close to the MLE when $m_i = 1$ (Lee *et al.*, 2007).

In summary, for estimation of dispersion parameters we should use the adjusted profile likelihood $p_{v,\beta}(h)$. However, for the mean parameter β, the h-likelihood is often satisfactory as an estimation criterion. For sparse binary data, the first-order adjustment $p_v(h)$ should be used for β and the second-order approximation $p_{v,\beta}^s(h)$ can be advantageous in estimating the dispersion parameters (Noh and Lee, 2007b and Lee *et al.*, 2007).

6.4 Penalized quasi–likelihood

Schall (1991) and Breslow and Clayton (1993) proposed the penalized quasi–likelihood (PQL) method for GLMMs. Consider the approximate linear mixed model for the GLM adjusted dependent variable in Section 2.2.1:

$$z = X\beta + Zv + e,$$

where $e = (y - \mu)(\partial\eta/\partial\mu) \sim MVN(0, W^{-1})$, $v \sim MVN(0, D)$, and $W = (\partial\mu/\partial\eta)^2\{\phi V(\mu_i)\}^{-1}$. Breslow and Clayton assumed that the GLM iterative weight W varied slowly (or not at all) as a function of μ. This means that they can use Henderson's method in Section 5.3 directly for estimation in GLMMs.

For estimating the dispersion parameters, they maximize the approximate likelihood

$$|2\pi V|^{-1/2}|X^t V^{-1} X/(2\pi)|^{-1/2}\exp\{-\frac{1}{2}(z - X\hat{\beta})^t V^{-1}(z - X\hat{\beta})\},$$

where $V = ZDZ' + W^{-1}$. Intuitively, as in the normal case, the approximate marginal distribution of z carries the information about the dispersion parameters. Unfortunately, this is in general a poor approximation to the adjusted profile likelihood $\log f_{\phi,\lambda}(y|\tilde{\beta})$. Pawitan (2001) identified two sources of bias in the estimation:

- In contrast to the linear normal case, the adjusted dependent variate z in the IWLS is a function of the dispersion parameters, and when the latter are far from the true value, z carries biased information.
- The marginal variance of z should have used $E(W^{-1})$, but in practice W^{-1} is used, which adds another bias in the estimation of the dispersion parameters.

Pawitan (2001) suggested a computation-based bias correction for PQL.

Given dispersion parameters, the PQL estimators for (v, β) are the same as the h-likelihood estimators, which jointly maximize h. In Poisson and binomial GLMMs, given (v, β), the PQL dispersion estimators are different from the first-order h-likelihood estimator maximizing $p_{v,\beta}(h)$, because they ignore the terms $\partial \hat{v}/\partial \phi$ and $\partial \hat{v}/\partial \lambda$. The omission of these terms results in severe biases (Lee and Nelder, 2001a); see the example in Section 6.7. Breslow and Lin (1995) and Lin and Breslow (1996) proposed corrected PQL (CPQL) estimators, but, as we shall see in the next section, they still suffer non-ignorable biases. In HGLMs we usually ignore terms $\partial \hat{\beta}/\partial \phi$ and $\partial \hat{\beta}/\partial \lambda$, but we should not ignore them when the number of β grows with sample size (Ha and Lee, 2005a) .

6.4.1 Asymptotic bias

Suppose that we have a binary GLMM: for $j = 1, ..., 6$ and $i = 1, ..., n$,

$$\log\{p_{ij}/(1 - p_{ij})\} = \beta_0 + \beta_1 j + \beta_2 x_i + b_i, \qquad (6.13)$$

where $\beta_0 = \beta_1 = 1$, x_i is 0 for the first half of individuals and 1 otherwise and $b_i \sim N(0, \sigma_s^2)$. First consider the model without x_i, i.e., $\beta_2 = 0$. Following Noh and Lee (2007b), we show in Figure 6.1 the asymptotic biases (as n goes to infinity) of the h-likelihood, PQL and CPQL estimators. For the first- and second-order h-likelihood estimators for β_1 at $\sigma_s = 3$, these are 0.3% and 1%, for the CPQL estimator 5.5% and for the PQL estimator 11.1%.

Lee (2001) noted that biases in the h-likelihood method could arise from the fact that regardless of how many distinct b_i are realized in the model, there are only a few distinct values for $\hat{b}_i$. For the example in (6.13) we have

$$\partial h/\partial b_i = \sum_j (y_{ij} - p_{ij}) - b_i/\sigma_s^2 = 0,$$

and in this model $\sum_j p_{ij}$ is constant for all i. Thus, there are only seven distinct values for $\hat{b}_i$ because they depend only upon $\sum_j y_{ij}$, which can be 0, ..., 6. Now consider the model (6.13) with $\beta_2 = 1$. This model has 14 distinct values for $\hat{b}_i$ because $\sum_j p_{ij}$ has 2 distinct values depending upon the value of x_i. Figure 6.1(b) shows that the asymptotic bias of the first-order h-likelihood estimator for β_1 at $\sigma_s = 3$ is 0.3%, that of CPQL estimator is 3.1% and that of PQL estimator is 6.5%. Figure 6.1(c) shows that the corresponding figures for β_2 at $\sigma_s = 3$ are 0.2%, 2.2% and 4.6%, respectively. The second-order h-likelihood method has essentially no practical bias, but it can be demanding to compute in multi-component models.

It is interesting to note that biases may be reduced with more covariates. Thus, the first-order h-likelihood method is often satisfactory for many covariates giving many distinct values for the random effect estimates. Noh *et al.* (2005) showed that it is indeed satisfactory for the analysis of large family data because of the many additional covariates. There seems no indication of any practical bias at all. For paired binary data Kang *et al.* (2005) showed that the conditional estimator is also greatly improved with additional covariates. In binary data when the cluster size is small with only one covariate we should use the second-order method. Noh and Lee (2007a) proposed a method which entirely eliminates the bias in such binary data.

6.4.2 Discussion

Many methods have been developed for the computation of the ML and/or REML estimators for GLMMs. The marginal likelihood approach often requires analytically intractable integration. A numerical method such as Gauss-Hermite quadrature (GHQ) is not feasible when high-dimensional integrals are required, for example, when the random effects have a crossed design. To overcome this difficulty, various simulation methods such as the Monte Carlo EM method (McCulloch, 1994; Vaida and Meng, 2004), Monte Carlo Newton–Raphson method, simulated maximum likelihood method (McCulloch, 1997) and the Gibbs sampling method (Karim and Zeger, 1992) can be considered. However,

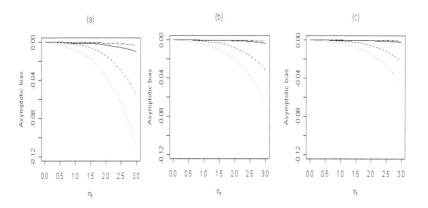

Figure 6.1 *Asymptotic bias of the regression parameter estimates in model (6.13) for $0 \leq \sigma_s \leq 3$. (a) Asymptotic bias of $\widehat{\beta}_1$ when $\beta_2 = 0$ for first-order (—) and second-order h-likelihood estimator ($- \cdot - \cdot -$), PQL estimator ($\cdots$), CPQL estimator (- - -). (b) and (c) Asymptotic bias of $\widehat{\beta}_1$ and $\widehat{\beta}_2$, respectively, when $\beta_2 = 1$.*

these simulation-based methods are all computationally intensive and can result in incorrect estimates, which may not be detected (Hobert and Casella, 1996).

Approximate-likelihood approaches for the analysis of clustered binary outcomes include PQL, CPQL and the methods of Shun and McCullagh (1995) and Shun (1997). All of them, except for the h-likelihood method, are limited (i.e., restricted to GLMMs and/or to some particular design structures) and miss some terms (Noh and Lee, 2007a). There is no evidence that simulation-based methods such as Markov chain Monte Carlo (MCMC) work better than h-likelihood methods. Noh and Lee (2007a) showed that simulation-based methods suffer biases in small samples. Furthermore the h-likelihood method works well in general: see the simulation studies with binary HGLMs with beta random effects (Lee *et al.*, 2005), Poisson and binomial HGLMs (Lee and Nelder, 2001a), frailty models (Ha *et al.*, 2001, Ha and Lee, 2003) and mixed linear models with censoring (Ha *et al.*, 2002).

Because of an apparent similarity of the h-likelihood method to the PQL method, the former has been wrongly criticized. PQL and h-likelihood methods differ and result in large differences in performance with binary data.

6.5 Deviances in HGLMs

Lee and Nelder (1996) proposed three deviances based upon $f_\theta(y, v)$, $f_\theta(y)$ and $f_\theta(y|\hat{\beta})$ for testing various components of HGLMs. For testing random effects they proposed the deviance $-2h$, for fixed effects -2ℓ and for dispersion parameters $-2\log f_\theta(y|\hat{\beta})$. When ℓ is numerically hard to obtain, they used $p_v(h)$ and $p_{\beta,v}(h)$ as approximations to ℓ and $\log f_\theta(y|\hat{\beta})$.

When testing hypotheses on the boundary of the parameter space, for example for $\lambda = 0$, the critical value is $\chi^2_{2\alpha}$ for a size-α test. This results from the fact that the asymptotic distribution of likelihood-ratio test is a 50:50 mixture of χ^2_0 and χ^2_1 distributions (Chernoff, 1954; Self and Liang, 1987): for application to random effect models see Stram and Lee (1994), Vu $et\ al.$, (2001), Vu and Knuiman (2002), Verbeke and Molenberghs (2003) and Ha $et\ al.$ (2007).

Based upon $\log f_\theta(y|v)$, Lee and Nelder (1996) proposed the scaled deviance for the goodness-of-fit test, defined by

$$D = D(y, \widehat{\mu}\ ) = -2\{\ell(\ \widehat{\mu}\ ; y|v) - \ell(y; y|v)\},$$

where $\ell(\ \widehat{\mu}\ ; y|v) = \log\{f(y|v; \widehat{\beta})\}$ and $\mu = E(y|v)$, having the estimated degrees of freedom d.f. $= n - p_D$, where

$$p_D = \text{trace}\{(T_m^t \Sigma_m^{-1} T_m)^{-1} T_m^t \Sigma_0^{-1} T_m\}$$

and $\Sigma_0^{-1} = W_{ma}\{\text{diag}(\Phi^{-1}, 0)\}$: see Equation (6.7). Lee and Nelder (1996) showed that $E(D)$ can be estimated by the estimated degrees of freedom $E(D) \approx n - p_D$ under the assumed model. Spiegelhalter $et\ al.$ (2002) viewed p_D as a measure of model complexity. This is an extension of the scaled deviance test for GLMs to HGLMs.

If ϕ is estimated by the REML method based upon $p_{\beta,v}(h)$, the scaled deviances $D/\hat{\phi}$ become the degrees of freedom $n - p_D$ again as in Chapter 2, so that the scaled deviance test for lack of fit is not useful when ϕ is estimated, but it can indicate that a proper convergence has been reached in estimating ϕ.

For model selection for fixed effects β, the information criterion based upon the deviance ℓ, and therefore $p_v(h)$, can be used, while for model selection for dispersion parameters, the information criterion based upon the deviance $p_\beta(\ell)$, and therefore $p_{v,\beta}(h)$, can be used. However, these information criteria cannot be used for models involving random parameters. For those Spiegelhalter $et\ al.$ (2002) proposed use in their Bayesian framework of an information criterion based upon D.

We claim that we should use the information criterion based upon the conditional loglihood $\log f_\theta(y|v)$ instead of D. Suppose $y \sim N(X\beta, \phi I)$ where the model matrix X is $n \times p$ matrix with rank p. There are two ways of constructing the information criterion; one is based upon the deviance and the other is based upon the conditional loglihood. First suppose that ϕ is known; then the conditional AIC based upon the conditional loglihood is

$$\text{cAIC} = n \log \phi + \sum (y_i - x_i^t \hat{\beta})^2/\phi + 2p_D,$$

while the information criterion based upon the deviance D is

$$\text{DIC} = \sum (y_i - x_i^t \hat{\beta})^2/\phi + 2p_D.$$

Here the two criteria differ by a constant and both try to balance the sum of the residual sum of squares, $\sum (y_i - x_i^t \hat{\beta})^2$ and the model complexity p_D. Spiegelhalter *et al.* (2002) proposed the use of DIC for a model selection in a Bayesian prespective, while Vaida and Blanchard (2005) proposed the use of cAIC in a frequentist perspective.

Now suppose ϕ is unknown. Then,

$$\text{DIC} = \sum (y_i - x_i^t \hat{\beta})^2/\hat{\phi} + 2p_D,$$

which becomes $n + 2p_D$ if the ML estimator is used for ϕ and $n + p_D$ if the REML estimator is used. So DIC always chooses the simplest model of which the extreme is the null model, having $p_D = 0$. Here

$$\text{cAIC} = n \log \hat{\phi} + \sum (y_i - x_i^t \hat{\beta})^2/\hat{\phi} + 2p_D,$$

which becomes $n \log \hat{\phi} + n + 2p_D$ if the ML estimator is used for ϕ and $n \log \hat{\phi} + n + p_D$ if the REML estimator is used. Thus, the cAIC still tries to balance the residual sum of squares $\sum (y_i - x_i^t \hat{\beta})^2$ and the model complexity p_D. This means that we should always use the conditional likelihood rather than the deviance. Thus, we use $-2 \log f_\theta(y|v) + 2p_D$ for model selection involving random parameters. In this book, four deviances, based upon h, $p_v(h)$, $p_{\beta,v}(h)$ and $\log f_\theta(y|v)$, are used for model selection and for testing different aspects of models.

6.6 Examples

6.6.1 Salamander mating data

McCullagh and Nelder (1989) presented a data set on salamander mating. Three experiments were conducted: two were done with the same salamanders in the summer and autumn and another one in the autumn of the same year using different salamanders. The response variable is binary, indicating success of mating. In each experiment, 20 females and 20 males from two populations called whiteside, denoted by W, and rough butt, denoted by R, were paired six times for mating with individuals from their own and the other population, resulting in 120 observations in each experiment. Covariates are, Trtf = 0, 1 for female R and W and Trtm = 0, 1 for male R and W. For $i, j = 1, ..., 20$ and $k = 1, 2, 3$, let y_{ijk} be the outcome for the mating of the ith female with the jth male in the kth experiment. The model can be written as

$$\log\{p_{ijk}/(1 - p_{ijk})\} = x_{ijk}^t \beta + v_{ik}^f + v_{jk}^m,$$

where $p_{ijk} = P(y_{ijk} = 1 | v_{ik}^f, v_{jk}^m)$, $v_{ik}^f \sim N(0, \sigma_f^2)$ and $v_{jk}^m \sim N(0, \sigma_m^2)$ are independent female and male random effects assumed independent of each other. The covariates x_{ijk} comprise an intercept, indicators Trtf and Trtm, and their interaction Trtf·Trtm.

In this model the random effects are crossed, so that numerical integration, using Gauss-Hermite quadrature, is not feasible, since high-dimensional integrals are required. Various estimation methods that have been developed are shown in Table 6.3. HL(i) for $i \geq 1$ is the ith-order h-likelihood method and H(0) is the use of joint maximization for β in HL(1). CPQL(i) is the ith-order CPQL, with CPQL(1) as the standard. For comparison, we include the Monte Carlo EM method (Vaida and Meng, 2004) and the Gibbs sampling method (Karim and Zeger, 1992).

The Gibbs sampling approach tends to give larger estimates than the Monte-Carlo EM method, which itself is the most similar to HL(2). Lin and Breslow (1996) reported that CPQL(i) has large biases when the variance components have large values. The results for CPQL(2) are from Lin and Breslow (1996) and show that CPQL(2) should not be used. Noh and Lee (2007a) showed that HL(2) has the smallest bias among HL(i), PQL and CPQL(i). While HL(2) has the smallest bias, HL(1) is computationally very efficient and satisfactory in practice. Table 6.3 shows that MCEM works well in that it gives similar estimates to HL(2). This example shows that statistically efficient estimation is possible without a computationally extensive method.

6.6.2 Data from a cross–over study comparing drugs

Koch *et al.* (1977) described a two-period crossover study for comparing three drugs. The data are shown in Table 6.4. Patients were divided into

Table 6.3 *Estimates of fixed effects for Salamander mating data.*

Method	Intercept	Trtf	Trtm	Trtf·Trtm	σ_f	σ_m
PQL	0.79	−2.29	−0.54	2.82	0.85	0.79
CPQL(1)	1.19	−3.39	−0.82	4.19	0.99	0.95
CPQL(2)	0.68	−2.16	−0.49	2.65	—	—
HL(0)	0.83	−2.42	−0.57	2.98	1.04	0.98
HL(1)	1.04	−2.98	−0.74	3.71	1.17	1.10
HL(2)	1.02	−2.97	−0.72	3.66	1.18	1.10
Gibbs	1.03	−3.01	−0.69	3.74	1.22	1.17
MCEM	1.02	−2.96	−0.70	3.63	1.18	1.11

two age groups and 50 patients were assigned to each of three treatment sequences in each age group, i.e., there were six distinct sequences, so that the total assigned patients were 300. The response was binary; $y = 1$ if the drug was favourable and $y = 0$ otherwise. Period and Drug were within-patient covariates and Age was a between-patient covariate.

Table 6.4 *Two-period crossover data (Koch et al., 1977).*

Age	Sequence[a]	Response profile at Period I vs. Period II[b]				Total
		FF	FU	UF	UU	
Older	A:B	12	12	6	20	50
Older	B:P	8	5	6	31	50
Older	P:A	5	3	22	20	50
Younger	B:A	19	3	25	3	50
Younger	A:P	25	6	6	13	50
Younger	P:B	13	5	21	11	50

[a]Sequence: a pair of drugs which were administered at Period I(left) and Period II(right).
[b]Response profile: a pair of whether F(favourable) or U(unfavourable) at Period I(left) and Period II(right).

Suppose v_i is the effect of the ith patient and $y_{ijkl}|v_i \sim \text{Bernoulli}(p_{ijkl})$ with $p_{ijkl} = P(y_{ijkl} = 1|v_i)$. We consider the following HGLM: for $i = 1, \ldots, 300$ and $j = 1, 2$

$$\log\left(\frac{p_{ijkl}}{1 - p_{ijkl}}\right) = \mu + \alpha_j + \beta_k + \tau_{f(j,l)} + \gamma_{jk} + v_i$$

where α_j is the period effect with null Period II ($j = 2$) effect, β_k is the age effect with null Younger ($k = 2$) effect, $k = 1, 2$, l indexes the treatment sequences, $l = 1, \cdots 6$, $\tau_{f(j,l)}$ represents the drug effect with null Drug P effect (for example $\tau_{f(1,1)}$ is Drug A effect and $\tau_{f(1,2)}$ is Drug B effect, etc.), γ_{jk} is the period-age interaction with $\gamma_{12} = \gamma_{21} = \gamma_{22} = 0$ and $v_i \sim N(0, \sigma_v^2)$ are independent. We impose these constraints to obtain conditional likelihood (CL) estimators identical to those from Stokes *et al.* (1995, pp.256–261).

The results are reported in Table 6.5. Here we use the first-order h-likelihood method. For each parameter, it has larger absolute t-values than the conditional likelihood (CL) method. In the HGLM analysis, the period-age interaction is clearly significant, while in the CL approach it is only marginally significant. Also in the HGLM analysis, Drug B effect is marginally significant, while in the CL approach it is not significant. With the CL method, inferences about age, the between-patient covariate, cannot be made. Furthermore, the conditional likelihood based upon the discordant pairs does not carry information about the variance component σ_v^2. The h-likelihood method allows recovery of inter-patient information. The magnitude of estimates tends to be larger, but with smaller standard errors: see more detailed in discussion in Kang *et al.* (2005).

Thus, h-likelihood (and therefore ML) estimation should be used to extract all the information in the data. The use of the CL estimator has been proposed because it is insensitive to the distributional assumption about the random effects. However, we shall show in Chapter 11 how such robustness can be achieved by allowing heavy-tailed distributions for the random effects.

Table 6.5 *Estimation results for crossover data from HGLM and CL methods.*

Covariate	H-likelihood			CL		
	Estimate	SE	t-value	Estimate	SE	t-value
Intercept	0.642	0.258	2.49			
Drug						
Drug A	1.588	0.249	6.37	1.346	0.329	4.09
Drug B	0.438	0.244	1.79	0.266	0.323	0.82
Period	−1.458	0.274	−5.32	−1.191	0.331	−3.60
Age	−1.902	0.306	−6.21			
d Period·age	0.900	0.384	2.34	0.710	0.458	1.55
$\log(\sigma_v^2)$	0.223	0.246	0.91			

SE = standard error.

6.6.3 Fabric data

In Bissell's (1972) fabric study, the response variable is the number of faults in a bolt of fabric of length l. Fitting the Poisson model

$$\log \mu = \alpha + x\beta,$$

where $x = \log l$, gives a deviance of 64.5 with 30 d.f., clearly indicating overdispersion. However, it may have arisen from assuming an incorrect Poisson regression model. Azzalini *et al.* (1989) and Firth *et al.* (1991) introduced non-parametric tests for the goodness of fit of the Poisson regression model, and found that the overdispersion is necessary. One way of allowing overdispersion is by using the quasi–likelihood approach of Chapter 3. Alternatively, an exact likelihood approach is available for the analysis of such overdispersed count data. Bissell (1972) proposed the use of the negativebinomial model

$$\mu_c = E(y|u) = \exp(\alpha + x\beta)u,$$

$$\mathrm{var}(y|u) = \mu_c,$$

where u follows the gamma distribution. This is a Poisson-gamma HGLM with saturated random effects. These two approaches lead to two different forms of variance function (Lee and Nelder, 2000b):

$$\text{QL model: } \mathrm{var}(y) = \phi\mu$$

$$\text{Negativebinomial model: } \mathrm{var}(y) = \mu + \lambda\mu^2.$$

The deviance for the QL model based upon (4.8) is 178.86, while that based upon the approximate marginal loglihood $p_v(h)$ is 175.76; AIC prefers the negativebinomial model. From Table 6.6, we see that the two models give similar analyses.

Table 6.6 *Estimates from models for fabric data.*

Covariate	QL model			Negativebinomial model		
	Estimate	SE	t-value	Estimate	SE	t-value
α	−4.17	1.67	−2.51	−3.78	1.44	−2.63
β	1.00	0.26	3.86	0.94	0.23	4.16
$\log(\phi)$	0.77	0.26	2.97			
$\log(\lambda)$				−2.08	0.43	−4.86

SE = standard error.

6.6.4 Cake data

We analyzed the cake data using the log-normal linear mixed model in Section 5.5, and it gave a deviance of 1617.24. Here we consider the gamma GLMM

$$\log \mu_{ijk} = \mu + \gamma_i + \tau_j + (\gamma\tau)_{ij} + v_k + v_{ik}.$$

This model has a deviance $-2p_v(h|y_{ijk}; \beta, \theta) = 1616.12$ and therefore it is slightly better than the log-normal linear mixed model under the AIC. Normal probability plots for v_k, v_{ik}, and the error component for the dispersion model are shown in Figure 6.2. We again found that the effects for replication (v_k) do not follow a normal distribution, and it may be appropriate to take them as fixed. In the log-normal linear mixed model in Section 5.5, the normal probability plot for the v_{ik} component in Figure 5.5 (b) has a discrepancy in the lower left-hand corner. It vanishes with a gamma assumption for the $y|v$ component (Figure 6.2 (b)).

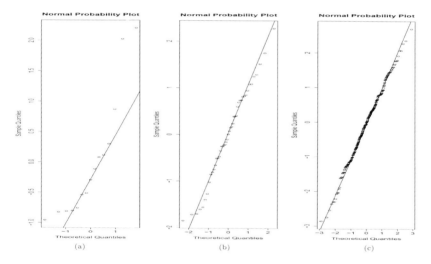

Figure 6.2 *Normal probability plots of (a) v_k, (b) v_{ik} and (c) error component for the dispersion model in the gamma GLMM for the cake data.*

6.7 Choice of random effect scale

The weak canonical scale is always defined if we can define the linear predictor. However, for some models the predictor may not be uniquely determined. In this section we discuss how to handle such situations.

Exponential-exponential model

Let us return to Examples 4.1 and 4.4. An alternative representation of the model

$$y|u \sim \exp(u) \quad \text{and} \quad u \sim \exp(\theta) \qquad (6.14)$$

is given by

$$y|w \sim \exp(w/\theta) \quad \text{and} \quad w \sim \exp(1) \qquad (6.15)$$

where $E(w) = 1$ and $E(u) = 1/\theta$. Here we have the marginal loglihood

$$\ell = \log L(\theta; y) = \log \theta - 2\log(\theta + y).$$

The marginal MLE is given by $\hat{\theta} = y$ and its variance estimator by $\widehat{\text{var}(\hat{\theta})} = 2y^2$.

Here $v = \log w$ is canonical for θ. In the model (6.15), because

$$\mu = E(y|w) = \theta/w,$$

the log link achieves additivity

$$\eta = \log \mu = \beta + v,$$

where $\beta = \log \theta$ and $v = -\log w$. This leads to the h-loglihood $h = \ell(\theta, v; y, v)$ and we saw that the joint maximization of h is a convenient tool to compute an exact ML estimator and its variance estimates.

In the model (6.14) $E(y|u) = u$, so that there is only one random effect u and no fixed effect. Thus, it is not clear which link function would yield a linear additivity of effects. Furthermore, suppose that we do not know the canonical scale v, and we choose an incorrect scale u to define the h-loglihood

$$h = l(\theta, u; y, u) = \log f(y|u) + \log f_\theta(u) = \log u + \log \theta - u(\theta + y),$$

where θ is a dispersion parameter appearing in $f_\theta(u)$. The equation $\partial h/\partial u = 1/u - (\theta + y) = 0$ gives $\tilde{u} = 1/(\theta + y)$. From this we get

$$
\begin{aligned}
p_u(h) &= \log \tilde{u} + \log \theta - \tilde{u}(\theta + y) - \frac{1}{2}\log\{1/(2\pi\tilde{u}^2)\} \\
&= \log \theta - 2\log(\theta + y) - 1 + \frac{1}{2}\log 2\pi,
\end{aligned}
$$

which is proportional to the marginal loglihood ℓ, yielding the same inference for θ, but $-\partial^2 h/\partial u^2|_{u=\tilde{u}} = 1/\tilde{u}^2 = (\theta + y)^2$ and thus h and $p_v(h)$ are no longer proportional.

The PQL method for GLMMs is analogous to maximizing $p_u(h)$, but ignores $\partial\tilde{u}/\partial\theta$ in the dispersion estimation. Now suppose that the $\partial\tilde{u}/\partial\theta$

term is ignored in maximizing $p_u(h)$. We then have the estimating equation

$$1 = \theta\tilde{u} = \theta/(\theta + y), \quad \text{for } y > 0$$

which gives an estimator $\hat{\theta} = \infty$. Thus, the term $\partial\tilde{u}/\partial\theta$ cannot be ignored; if it is, it can result in a severe bias in estimation and a distortion of the standard error estimate, for example,

$$\widehat{\text{var}(\hat{\theta})} = \hat{\theta}^2 = \infty.$$

This highlights the consequence of ignoring $\partial\tilde{u}/\partial\theta$.

Poisson exponential model 1

Consider the following two equivalent Poisson exponential models: for $i = 1, ..., m$,

$$y_i|u_i \sim \text{Poisson}(\delta u_i) \quad \text{and} \quad u_i \sim \exp(1) \tag{6.16}$$

and

$$y_i|w_i \sim \text{Poisson}(w_i) \quad \text{and} \quad w_i \sim \exp(1/\delta), \tag{6.17}$$

where $w_i = \delta u_i$; so we have $E(u_i) = 1$ and $E(w_i) = \delta$. and the marginal loglihood

$$\ell = \sum_i \{y_i \log \delta - (y_i + 1) \log(1 + \delta)\}$$

with marginal ML estimator $\hat{\delta} = \bar{y}$.

In model (6.16), the use of the log link on which the fixed and random effects are additive leads to

$$\log \mu_i = \beta + v_i,$$

where $\mu_i = E(y_i|u_i) = E(y_i|w_i)$, $\beta = \log \delta$, and $v_i = \log u_i$. Here v is the canonical scale for β.

Suppose we choose a wrong scale w to construct the h-loglihood in model (6.17), giving

$$h = \sum_i \{y_i \log w_i - w_i - \log y_i! - \log \delta - w_i/\delta\},$$

for which $\tilde{w}_i = y_i\delta/(\delta + 1)$. Because

$$-\partial^2 h/\partial w_i^2|_{w_i=\tilde{w}_i} = y_i/\tilde{w}_i^2 = (1 + \delta)^2/(y_i\delta^2),$$

it can be shown that

$$\begin{aligned} p_w(h) &= \sum_i \{y_i \log \delta - (y_i + 1) \log(1 + \delta) + (y_i + 1/2) \log y_i \\ &\quad - \log(y_i - 1)! + \log(2\pi)/2, \end{aligned}$$

so that $p_w(h)$ gives the MLE for the parameter δ.

Poisson exponential model 2

We have shown that *additivity on the linear predictor* (weak canonical scale) may not uniquely determine the scale to define the h-likelihood if the model has only one random effect without any fixed effect. This can also happen for models with fixed effects. Consider the following two equivalent models

$$y_{ij}|u_i \sim \text{Poisson}(\beta u_i), \quad u_i - 1 \sim \exp(\alpha)$$

and

$$y_{ij}|u_i^* \sim \text{Poisson}(\beta + u_i^*), \quad u_i^* \sim \exp(\delta),$$

where $\delta = \alpha/\beta$. In the first model, the weak canonical condition leads to $v = \log u$, and in the second model to $v = u^*$. Here, we define the h-likelihood with $v = \log u$ and find that $p_v(h)$ gives numerically satisfactory statistical inference.

These three examples show that as long as the adjusted profile loglihood is used, a wrong choice of scale to define the h-loglihood may not matter for inferences about fixed parameters. When the choice is not obvious, we recommend the scale that makes the range of v the whole real line because the adjusted profile likelihood is the Laplace approximation to the integrated likelihood (Chapter 1). If $\hat{v}$ is outside the required region (for example $\hat{v} < 0$ when v should be positive), another form of Laplace approximation is needed. For the prediction based on the predictive probability, such a scale is not required. Thus, inferences of random effects can be made using the predictive probability described in Chapter 4.

CHAPTER 7

HGLMs with structured dispersion

In the previous chapter, HGLMs were developed as a synthesis of two model classes, GLMs (Chapter 2) and normal models with additional random effects (Chapter 5). Further extensions can be made by adding additional features to HGLMs. In this chapter, we allow the dispersion parameters to have structures defined by their own sets of covariates. This brings together the HGLM class and the joint modelling of mean and dispersion (Chapter 3). We also discuss how the QL approach can be adapted to correlated errors.

In Chapter 3, we showed that the structured dispersion model can be viewed as a GLM with responses defined by deviances derived from the mean model, so that the fitting of the two interconnected component GLMs for the mean and dispersion suffices for models with structured dispersion. In Chapter 6, we showed that the h-likelihood estimation of HGLMs leads to augmented GLMs. In this chapter, these two methods are combined to give inferences from HGLMs as a synthesis of GLMs, random effect models and structured dispersions. In our framework GLMs play the part of building blocks and the extended class of models is composed of component GLMs.

It is very useful to build a complex model by combining component GLMs. In our framework adding an additional feature implies adding more component GLMs. The complete model is then decomposed into several components, and this decomposition provides insights into the development, extension, analysis and checking of new models. Statistical inferences from a complicated model can then be made from decomposing it into diverse components. This avoids the necessity of developing complex statistical methods on a case-by-case basis.

7.1 Description of model

Heterogeneity is common in many data and arises from various sources. It is often associated with unequal variances where, if not properly modelled, it can cause inefficiency or even an invalid analysis. In statistical literature, compared with modelling of the mean, modelling of the dispersion has often been neglected. In quality improvement engineering applications, achieving high precision is as important as determining the target values. Thus, the variance can be as important as the mean. To find a way of describing factors affecting the variance, we need a regression model for the dispersion.

To describe the model in its generality, consider a HGLM composed of two components:

(a) Conditional on random effects u, the responses y follow a GLM family, characterized by

$$E\left(y|u\right) = \mu \quad \text{and} \quad \text{var}\left(y|u\right) = \phi V\left(\mu\right)$$

with linear predictor

$$\eta = g\left(\mu\right) = X\beta + Zv,$$

where $v = v(u)$ for some known strictly monotonic function $v()$.

(b) The random component u follows a conjugate distribution of some GLM family whose loglihood is characterized by the quasi–relationship

$$
\begin{aligned}
E(\psi_M) &= u \\
\text{var}(\psi_M) &= \lambda V_M(u),
\end{aligned}
$$

where λ is the dispersion parameter for random effects u and ψ_M is the quasi–data described in Section 6.2.2. As before, the subscript M indicates that the random effect appears in the predictor for the mean. In Chapter 11, we allow random effects in the linear predictor of the dispersion.

We allow structured dispersions such that (ϕ, λ) are assumed to follow the models

$$\xi = h(\phi) = G\gamma, \tag{7.1}$$

and

$$\xi_M = h_M\left(\lambda\right) = G_M\gamma_M, \tag{7.2}$$

where $h()$ and $h_M()$ are link functions, and γ and γ_M are fixed effects for the ϕ and λ components, respectively.

With structured dispersion the extension of results from one-component to multi-component models is straightforward. Suppose we have a multi-component model for random effects in the form

$$Z_1 v^{(1)} + Z_2 v^{(2)} + \cdots + Z_k v^{(k)},$$

where Z_r $(r = 1, 2, \ldots, k)$ are the model matrices corresponding to the random effects $v^{(r)}$. This model can be written as a one-component model,

$$Zv,$$

where $Z = (Z_1, Z_2, \ldots, Z_k)$, and $v = (v^{(1)t}, v^{(2)t}, \ldots, v^{(k)t})^t$, but with a structured dispersion $\lambda = (\lambda_1, \ldots, \lambda_k)^T$. We allow that the random components $v^{(r)}$ can be from a different family of distributions. Thus, the method for single random component models with structured dispersions can be applied directly to multi-component models with structured dispersions.

7.2 Quasi–HGLMs

From the form of the h-likelihood, we can see that the maximum h-loglihood estimator of $\omega = (\beta^t, v^t)^t$ can be obtained from the IWLS equations for GLMs as shown in Section 2.2. In this section we study the estimation of dispersion parameters for HGLMs by extending HGLMs that structured dispersions to quasi–HGLMs that do not require exact likelihoods for the components of the model. By extending the EQLs for JGLMs (Chapter 3) to HGLMs, we derive a uniform algorithm for the estimation of dispersion parameters of quasi–HGLMs. Use of the EQL can result in biases in estimation because the result is not a true likelihood. For models allowing true likelihoods for component GLMs, we can modify the algorithm to avoid bias by using exact likelihoods for the corresponding component GLMs.

Consider a quasi–HGLM which can be viewed as an *augmented quasi–GLM* with the response variables $(y^t, \psi_M^t)^t$, having

$$\mu = E(y|v), \qquad u = E(\psi_M)$$
$$\text{var}(y|v) = \phi V(\mu), \quad \text{var}(\psi_M) = \lambda V_M(u).$$

The GLM distributions of the components $y|v$ and v are characterized by their variance functions. For example, if $V(\mu) = \mu$ with $\phi = 1$, the component $y|v$ has the Poisson distribution and if $V_M(u) = 1$, the component $v = v(u) = u$ in Table 6.2 has the normal distribution and the resulting model is a Poisson GLMM. If $V(\mu) = \mu$ with $\phi > 1$, the component $y|v$ has the overdispersed Poisson distribution and if $V_M(u) = u$, the component u has the gamma distribution. The resulting model is a quasi–Poisson-gamma HGLM. We can also allow models with multi-components having different distributions.

7.2.1 Double EQL

For inferences from quasi–likelihood models, Lee and Nelder (2001) proposed to use the double EQL, which uses EQLs to approximate both $\log f_{\beta,\phi}(y|u)$ and $\log f_\lambda(v)$ in the h-likelihood as follows. Let

$$q^+ = q(\theta(\mu),\phi;y|u) + q_M(u;\psi_M),$$

where

$$q(\theta(\mu),\phi;y|u) = -\sum[d_i/\phi + \log\{2\pi\phi V(y_i)\}]/2$$

$$q_M(u;\psi_M) = -\sum[\{d_{Mi}/\lambda + \log(2\pi\lambda V_M(\psi_M))\}/2]$$

with

$$d_i = 2\int_{\mu_i}^{y_i}(y_i - s)/V(s)ds$$

$$d_{Mi} = 2\int_{u_i}^{\psi_M}(\psi - s)/V_M(s)ds$$

being the deviance components of $y|u$ and u respectively. The function $q_M(u;\psi_M)$ has the form of an EQL for the quasi–data ψ_M.

The forms of deviance components for the GLM distributions and their conjugate distributions are set out in Table 7.1. The two deviances have the same form for conjugate pairs replacing y_i and $\hat{\mu}_i$ in d_i by ψ_i and $\hat{u}_i$ in d_{Mi}. The beta conjugate distribution assumes the binomial denominator to be 1.

Example 7.1: In the Poisson-gamma HGLM, the deviance component for the u component is given by

$$d_{Mi} = 2(u_i - \log u_i - 1) = (u_i - 1)^2 + o_p(\lambda),$$

where $\text{var}(u_i) = \lambda$. The Pearson-type (or method of moments) estimator can be extended by using the assumption

$$E(u_i - 1)^2 = \lambda.$$

This method gives a robust estimate for λ with the family of the model satisfying the moment assumption above. However, this orthodox BLUP method in Section 4.4. (Ma *et al.* (2003)) is not efficient for large λ because it differs from the h-likelihood method (and therefore the marginal likelihood method). Another disadvantage is that it is difficult to develop a REML adjustment, so that the method suffers serious bias when the number of β grows with the sample size, while with the h-likelihood approach it does not because it uses the adjusted profile likelihood in Chapter 6 (Ha and Lee, 2005a). □

Example 7.2: Common correlation model for binary data: suppose there

Table 7.1 Variance functions and corresponding deviances.

$y|u$ distribution	$V(\mu_i)$	d_i
Normal	1	$(y_i - \hat{\mu}_i)^2$
Poisson	μ_i	$2\{y_i \log(y_i/\hat{\mu}_i) - (y_i - \hat{\mu}_i)\}$
Binomial	$\mu_i(m_i - \mu_i)/m_i$	$2\{y_i \log(y_i/\hat{\mu}_i) + (m_i - y_i) \log\{(m_i - y_i)/(m_i - \hat{\mu}_i)\}]$
Gamma	μ_i^2	$2\{-\log(y_i/\hat{\mu}_i) + (y_i - \hat{\mu}_i)/\hat{\mu}_i\}$

u distribution	$V_M(u_i)$	d_{Mi}	ψ_i
Normal	1	$(\psi_i - \hat{u}_i)^2 = \hat{u}_i^2$	0
Gamma	u_i	$2\{\psi_i \log(\psi_i/\hat{u}_i) - (\psi_i - \hat{u}_i)\} = 2(-\log \hat{u}_i - (1 - \hat{u}_i))$	1
Beta*	$u_i(1 - u_i)$	$2\{\psi_i \log(\psi_i/\hat{u}_i) + (m_i - \psi_i) \log\{(m_i - \psi_i)/(m_i - \hat{u}_i)\}\}$	$1/2$
Inverse gamma	u_i^2	$2\{-\log(\psi_i/\hat{u}_i) + (\psi_i - \hat{u}_i)/\hat{u}_i\} = 2\{\log \hat{u}_i + (1 - \hat{u}_i)/\hat{u}_i\}$	1

$*m_i = 1$.

are k groups of individuals, and that the i–th group contains m_i individuals, each having a binary response y_{ij} $(i = 1, \cdots, k; j = 1, \cdots, m_i)$. Suppose that the two possible values of y_{ij} can be regarded as success and failure, coded as one and zero, respectively. The probability of success is assumed the same for all individuals, irrespective of group, i.e.,

$$\Pr(y_{ij} = 1) = E(y_{ij}) = \psi$$

for all i, j. The responses of individuals from different groups are assumed to be independent, while within each group, the correlation between any pair (the intra-class correlation) of responses (y_{ij}, y_{il}) for $j \neq l$ is ρ. This is sometimes called the common-correlation model. Let $y_i = \sum_j y_{ij}$ denote the total number of successes in the i–th group. Then, the y_i satisfy

$$E(y_i) = \mu_i = m_i\psi \text{ and } \text{var}(y_i) = \phi_i V(\mu_i),$$

where $\phi_i(= m_i\rho + (1-\rho) \geq 1)$ are dispersion parameters and $V(\mu_i) = \mu_i(m_i - \mu_i)/m_i$. Ridout et $al.$ (1999) showed that the use of EQL for this model could give an inefficient estimate of ρ. They proposed the pseudo-likelihood (PL) estimator. $\square$

One well known example of a common-correlation model occurs when y_{ij} follows the beta-binomial model of Example 6.4:

(a) $y_{ij}|u_i \sim$Bernoulli(u_i)

(b) $u_i \sim$beta(α_1, α_2),

where $E(u_i) = \psi = \alpha_1/(\alpha_1 + \alpha_2)$ and $\text{var}(u_i) = \rho\psi(1 - \psi)$ with $\rho = 1/(\alpha_1 + \alpha_2 + 1)$.

The assumption (a) is equivalent to

(a') $y_i|u_i \sim$binomial(m_i, u_i), where $\mu_i = E(Y_i|u_i) = m_i u_i$ and $\text{var}(Y_i|u_i) = m_i u_i(1 - u_i)$

Lee (2004) proposed that instead of approximating the marginal likelihood for y_i by using the EQL, we should approximate the components of the h-likelihood by using the double EQL(DEQL) for a quasi–HGLM:

(a) $y_i|u_i$ follows the quasi–GLM characterized by the variance function $V(\mu_i) = \mu_i(m_i - \mu_i)/m_i$

(b) ψ follows the quasi–GLM characterized by the variance function $V_M(u_i) = u_i(1 - u_i)$

This model gives a highly efficient estimator for a wide range of parameters as judged by its MSE, sometimes even better than results from the ML estimator in finite samples. However, it loses efficiency for large ρ, but this can be avoided by using the h-likelihood.

Here α_1 and α_2 are the dispersion parameters appearing in $f_{\alpha_1,\alpha_2}(v)$. In Example 6.7 we show that

$$\ell = \sum_i \{A_i - B_i + C_i\}$$

and

$$p_v(h) = \sum_i \{A_i^f - B_i + C_i\},$$

where A_i^f is the first-order Stirling approximation (6.11) to A_i. Lee (2004) showed that

$$p_v(q^+) = \sum_i \{A_i^f - B_i^f + C_i^f\},$$

where B_i^f and C_i^f are respectively the first-order Stirling approximations (6.11) to B_i and C_i. The second-order Laplace approximation gives a better approximation, which is the same as using the second-order Stirling approximations in corresponding terms.

Example 7.3: Agresti (2002, p. 152) studied data from a teratology experiment (Shepard *et al.*, 1980) in which female rats on iron-deficient diets were assigned to four groups. Rats in group 1 were given placebo injections, and rats in other groups were given injections of an iron supplement: treatments were given weekly in group 4, on days 7 and 10 in group 2, and on days 0 and 7 in group 3. The 58 rats were made pregnant, sacrificed after 3 weeks, and then the total number of dead foetuses was counted in each litter. In teratology experiments, overdispersion often occurs, due to unmeasured covariates and genetic variability. Note that $\phi = 1$ only when $\rho = 0$. In Table 7.2, all four methods indicate that the overdispersion is significant. Note that EQL and PL values are similar, while those for ML and EQL are not. DEQL gives a very good approximation to ML. This shows that the approximation of the likelihood is better made by applying EQLs to components instead of applying them to the marginal likelihood of y_i. □

7.2.2 REML estimation

For REML estimation of dispersion parameters for quasi–HGLMs, we use the adjusted profile loglihood

$$p_{v,\beta}(q^+),$$

which gives score equations for γ_k in (7.1),

$$2\{\partial p_{v,\beta}(q)/\partial\gamma_k\} = \sum_{i=1}^{n} g_{ik}(\partial\phi_i/\partial\xi_i)(1-q_i)\{(d_i^* - \phi_i)/\phi_i^2\} = 0$$

Table 7.2 *Parameter estimates with standard errors from fitting four methods to teratology data (Shepard* et al., *1980).*

Parameter	Method			
	ML	DEQL	EQL	PL
Intercept	1.346 (0.244)	1.344 (0.238)	1.215 (0.231)	1.212 (0.223)
Group 2	−3.114 (0.477)	−3.110 (0.458)	−3.371 (0.582)	−3.370 (0.562)
Group 3	−3.868 (0.786)	−3.798 (0.701)	−4.590 (1.352)	−4.585 (1.303)
Group 4	−3.922 (0.647)	−3.842 (0.577)	−4.259 (0.880)	−4.250 (0.848)
ρ	0.241 (0.055)	0.227 (0.051)	0.214 (0.060)	0.192 (0.055)

Note: Number of litters = 58; mean litter size = 10.5.

and those for γ_{Mk} in (7.2)

$$2\{\partial p_{v,\beta}(q)/\partial \gamma_{Mk}\} = \sum_{i\geq n+1} g_{Mik}(\partial \lambda_i/\partial \xi_{Mi})(1-q_{Mi})\{(d^*_{Mi}-\lambda_i)/\lambda_i^2\} = 0,$$

where n is sample size of y, g_{ik} and g_{Mik} are (i,k)–th elements of model matrices G and G_M, respectively, q_i and q_{Mi} are the i–th and $(n+i)$–th leverages from the augmented GLM Section (6.2.3),

$$T_M(T_M^t \Sigma_M^{-1} T_M)^{-1} T_M^t \Sigma_M^{-1},$$

$d^*_i = d_i/(1-q_i)$ and $d^*_{Mi} = d_{Mi}/(1-q_{Mi})$. They are GLM IWLS estimating equations with response d^*_i (d^*_{Mi}), mean ϕ_i (λ_i), error gamma, link function $h(\phi_i)$ ($h_M(\lambda_i)$), linear predictor $\xi_i = \sum_k g_{ik}\gamma_k$ ($\xi_{Mi} = \sum_k g_{Mik}\gamma_{Mk}$) and prior weight $(1-q_i)/2$ ($(1-q_{Mi})/2$). The prior weight reflects the loss of degrees of freedom in the estimation of the response.

The use of EQLs for component GLMs has an advantage over the use of exact likelihood in that a broader class of models can be fitted and compared in a single framework. The fitting algorithm is summarized in Table 7.3. A quasi–HGLM is composed of four component GLMs. Two component GLMs for the β and u constitute an augmented GLM and are therefore connected by the augmentation. The augmented GLM and the two dispersion GLMs for γ and γ_M are connected and the connections are marked by lines. In consequence all four component GLMs are interconnected. Thus, methods for JGLMs (Chapter 3) and HGLMs (Chapter 6) are combined to produce the algorithm for fitting quasi–HGLMs; the algorithm can be reduced to the fitting of a two-dimensional set of GLMs, one being mean and dispersion, and the other fixed and random effects. Adding a random component v adds two component GLMs, one for u

and the other for λ. Thus, a quasi–HGLM with three random components has eight component GLMs.

In summary, the inferential procedures for GLMs (Chapter 2) can be carried over to this wider class. In each component GLM, we can change the link function, allow various types of terms in the linear predictor and use model selection methods for adding or deleting terms.Furthermore various model assumptions about the components can be checked by applying GLM model checking procedures to the component GLMs. This can be done within the extended-likelihood framework, without requiring prior distributions of parameters or intractable integration.

Table 7.3 *GLM attributes for HGLMs.*

Component	β (fixed)		γ (fixed)
Response	y		d^*
Mean	μ		ϕ
Variance	$\phi V(\mu)$		ϕ^2
Link	$\eta = g(\mu)$		$\xi = h(\phi)$
Linear predictor	$X\beta + Zv$		$G\gamma$
Deviance	d		$\Gamma(d^*, \phi)$
Prior weight	$1/\phi$		$(1-q)/2$

Component	u (random)	λ (fixed)
Response	ψ_M	d_M^*
Mean	u	λ
Variance	$\lambda V_M(u)$	λ^2
Link	$\eta_M = g_M(u)$	$\xi_M = h_M(\lambda)$
Linear predictor	v	$G_M \gamma_M$
Deviance	d_M	$\Gamma(d_M^*, \lambda)$
Prior weight	$1/\lambda$	$(1-q_M)/2$

7.2.3 IWLS equations

An immediate consequence of Table 7.3 is that the extended model can be fitted by solving IWLS estimating equations of three GLMs for the four components as follows:

(a) Given (ϕ, λ), the two components $\omega = (\beta^t, v^t)^t$ can be estimated by IWLS equations (6.7)

$$T_M^t \Sigma_M^{-1} T_M \omega = T_M^t \Sigma_M^{-1} z_{Ma}, \qquad (7.3)$$

which are the equations for the augmented GLM in Section 6.2.3.

(b) Given (ω, λ), we estimate γ for ϕ by the IWLS equations

$$G^t \Sigma_d^{-1} G \gamma = G^t \Sigma_d^{-1} z_d,$$

where $\Sigma_d = \Gamma_d W_d^{-1}$ with $\Gamma_d = \mathrm{diag}\{2/(1 - q_i)\}$,

$$q_i = x_i (X^t W_d X)^{-1} x_i^t,$$

the weight functions $W_d = \mathrm{diag}(W_{di})$ are defined as

$$W_{di} = (\partial \phi_i / \partial \xi_i)^2 / 2\phi_i^2,$$

and the dependent variables are defined as

$$z_{di} = \xi_i + (d_i^* - \phi_i)(\partial \xi_i / \partial \phi_i),$$

with GLM deviance components

$$d_i^* = d_i / (1 - q_i),$$

and

$$d_i = 2 \int_{\widehat{\mu}_i}^{y} (y - s) / V(s) \, ds.$$

This GLM is characterized by a response d^*, gamma error, link function $h()$, linear predictor G, and prior weight $(1 - q)/2$.

(c) Given (ω, ϕ), we estimate γ_M for λ by the IWLS equations

$$G_M^t \Sigma_M^{-1} G_M \gamma_M = G_M^t \Sigma_M^{-1} z_M, \tag{7.4}$$

where $\Sigma_M = \Gamma_M W_M^{-1}$ with $\Gamma_M = \mathrm{diag}\{2/(1 - q_{Mi})\}$; $W_M = \mathrm{diag}(W_{Mi})$ are defined by

$$W_{Mi} = (\partial \lambda_i / \partial \xi_{Mi})^2 / 2\lambda_i^2$$

and z_M by

$$z_{Mi} = \xi_{Mi} + (d_{Mi}^* - \lambda_i)(\partial \xi_{Mi} / \partial \lambda_i)$$

for $d_{Mi}^* = d_{Mi} / (1 - q_{Mi})$, where

$$d_{Mi} = 2 \int_{\widehat{u}_i}^{\psi} (\psi - s) / V_M(s) \, ds$$

and q_M extends the idea of leverage to HGLMs (Lee and Nelder, 2001a). This GLM is characterized by a response d_M^*, gamma error, link function $h_M()$, linear predictor G_M, and prior weight $(1 - q_M)/2$.

7.2.4 Modifications

DEQL gives a uniform algorithm for the estimation of parameters in a broader class of models. However, the use of EQL or PL can cause inconsistency or inefficiency, so that it is better to use exact likelihoods for component GLMs if they exist. This can be done by modifying the GLM leverages (q, q_M) as we saw in Section 3.6.2. The modification of the second-order Laplace approximation can be done similarly (Lee and Nelder, 2001a). For the use of $p_v(h)$ for estimating β in binary data, the IWLS procedures can be used by modifying the augmented responses (Noh and Lee, 2007a).

7.3 Examples

We illustrate how various assumptions of the complete model can be verified by checking components of the corresponding component GLMs.

7.3.1 Integrated circuit data

An experiment on integrated circuits was reported by Phadke *et al.* (1983). The width of lines made by a photoresist nanoline tool were measured in five locations on silicon wafers; measurements were taken before and after an etching process being treated separately. We present the results for the pre-etching data. The eight experimental factors (A through H) were arranged in an L_{18} orthogonal array and produced 33 measurements at each of 5 locations, giving a total of 165 observations. There were no whole-plot (i.e., between-wafer) factors. Wolfinger and Tobias (1998) developed a structured dispersion analysis for a normal HGLM, having wafers as random effects. Let q be the index for wafers and r for observations within wafers. Our final model for the mean is

$$y_{ijkop,qr} = \beta_0 + a_i + b_j + c_k + g_o + h_p + v_q + e_{qr}, \qquad (7.5)$$

where $v_q \sim N(0, \lambda)$, $e_{qr} \sim N(0, \phi)$, and λ and ϕ represent the between-wafer and within-wafer variances, respectively, which can also be affected by the experimental factors (A through H). Our final models for the dispersions are

$$\log \phi_{imno} = \gamma_0^w + a_i^w + e_m^w + f_n^w + g_o^w \qquad (7.6)$$

and

$$\log \lambda_M = \gamma_0^b + e_m^b, \qquad (7.7)$$

where the superscripts w and b refer to within- and between-wafer variances.

From the three component GLMs, it is obvious that the inferences for the means and also the those for the dispersions can be made using ordinary GLM methods. Wolfinger and Tobias (1998) ignored the factor G in the

Table 7.4 *Estimation results of integrated circuit data.*

Model	Factor	Estimate	Standard error	t-value
(7.5)	1	2.453	0.049	49.730
	A(2)	0.378	0.046	8.145
	B(2)	−0.568	0.041	−13.808
	C(2)	0.388	0.043	8.914
	C(3)	0.521	0.052	9.972
	G(2)	−0.176	0.051	−3.462
	G(3)	−0.393	0.045	−8.662
	H(2)	−0.003	0.047	−0.071
	H(3)	0.307	0.051	5.980
(7.6)	1	−4.711	0.329	−14.305
	A(2)	−0.862	0.244	−3.540
	E(2)	−0.016	0.305	−0.052
	E(3)	0.677	0.292	2.317
	F(2)	0.697	0.300	2.324
	F(3)	1.043	0.299	3.488
	G(2)	−0.145	0.290	−0.500
	G(3)	−0.651	0.299	−2.177
(7.7)	1	−4.778	0.666	−7.172
	E(2)	−1.299	1.153	−1.127
	E(3)	1.489	0.863	1.726

dispersion model (7.6) since the deviance $(-2p_{v,\beta}(h))$ contribution of this factor based upon the restricted loglihood is 4.2 with two degrees of freedom (not significant). However, the regression analysis in Table 7.4 shows that the third level of factor G is significant; it shows a deviance contribution 3.98 with one degree of freedom if we collapse the first and second levels of G so the factor should remain in the dispersion model (7.6). This is an advantage of using a regression analysis for dispersion models so that the significance of individual levels can also be tested. Residual plots for component GLMs in Figures 7.1 through 7.3 did not show systematic departures, confirming our final model. Even though Factor E is not significant in the dispersion model (7.7), we include it in the final model to illustrate the model checking plots.

We should set factors (A, E, F, G) to the levels (2, 2, 1, 3) to reduce the variance and use significant factors (B,C,H) in the mean model (7.5) but exclude significant factors (A,G) in the dispersion model (7.6) to

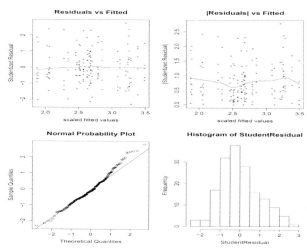

Figure 7.1 *Residual plots for mean model (7.5).*

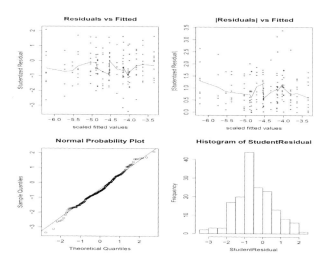

Figure 7.2 *Residual plots for dispersion model (7.6).*

adjust the mean. Phadke *et al.* (1983) originally concluded that A and
F can be used to reduce process variance and B and C to adjust the
mean. By using an efficient likelihood method for the HGLM, we achieve
more significant factors for the means and dispersions. Our conclusion
is slightly different from that of Wolfinger and Tobias (1998) because we
include G in the dispersion model.

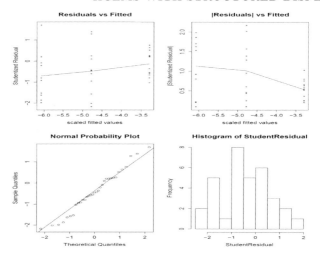

Figure 7.3 *Residual plots for dispersion model (7.7).*

7.3.2 Semiconductor data

This example is taken from Myers *et al.* (2002). It involves an experiment in a semiconductor plant. Six factors are employed, and it is of interest to study the curvatures or cambers of the substrate devices produced in the plant. There is a lamination process, and the camber measurement is made four times on each device produced. The goal is to model the camber taken in 10^{-4} in./in. as a function of the design variables. Each design variable is taken at two levels and the design is a 2^{6-2} fractional factorial. The camber measurement is known to be non-normally distributed. Because the measurements were taken on the same device, they were correlated. Myers *et al.* considered a gamma response model with a log link. They used a GEE approach assuming a working correlation to be AR(1). Because there are only four measurements on each device, the compound symmetric and AR(1) correlation structures may not be easily distinguishable.

First, consider a gamma GLM with log link:

$$\log \mu = \beta_0 + x_1\beta_1 + x_3\beta_3 + x_5\beta_5 + x_6\beta_6.$$

This model has deviances $-2\ell = -555.57$ and $-2p_\beta(\ell) = -534.00$. Next we consider a gamma JGLM with a dispersion model

$$\log \phi = \gamma_0 + x_2\gamma_2 + x_3\gamma_3,$$

which has deviances $-2\ell = -570.42$ and $-2p_\beta(\ell) = -546.30$. Only two

Table 7.5 *Factor levels in semiconductor experiment.*

Run	Lamination Temperature x_1	Lamination Time x_2	Lamination Pressure x_3	Firing Temperature x_4	Firing Cycle Time x_5	Firing Dew Point x_6
1	−1	−1	−1	−1	−1	−1
2	+1	−1	−1	−1	+1	+1
3	−1	+1	−1	−1	+1	−1
4	+1	+1	−1	−1	−1	+1
5	−1	−1	+1	−1	+1	+1
6	+1	−1	+1	−1	−1	−1
7	−1	+1	+1	−1	−1	+1
8	+1	+1	+1	−1	+1	−1
9	−1	−1	−1	+1	−1	+1
10	+1	−1	−1	+1	+1	−1
11	−1	+1	−1	+1	+1	+1
12	+1	+1	−1	+1	−1	−1
13	−1	−1	+1	+1	+1	−1
14	+1	−1	+1	+1	−1	+1
15	−1	+1	+1	+1	−1	−1
16	+1	+1	+1	+1	+1	+1

extra parameters are required and both deviances support the structured dispersion.

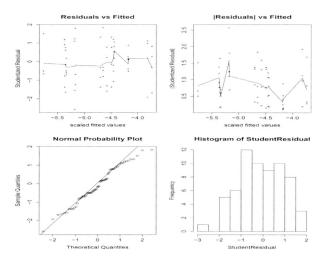

Figure 7.4 *Residual plots of mean model for the semiconductor data.*

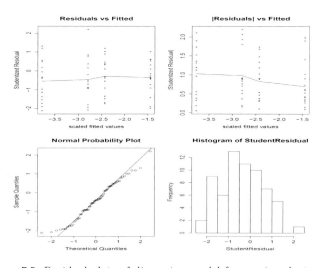

Figure 7.5 *Residual plots of dispersion model for semiconductor data.*

We also consider a gamma HGLM by adding a random effect for the device in the mean model. The variance λ of random effects represents the between-variance, while ϕ represents the within-variance.

Finally, we found no significant effect for the between-variance. This model has deviances $-2p_v(h)(\approx -2\ell) = -573.86$ and $-2p_{v,\beta}(h)$

Table 7.6 *Estimation results of semiconductor data.*

	Covariate	GLM Estimate	SE	t-value	JGLM Estimate	SE	t-value	HGLM Estimate	SE	t-value
$\log(\mu)$	Constant	−4.682	0.045	−103.02	−4.684	0.039	−118.676	−4.711	0.067	−70.568
	x_1	0.180	0.045	3.970	0.254	0.036	7.079	0.209	0.067	3.154
	x_3	0.302	0.045	6.634	0.368	0.039	9.317	0.328	0.067	4.915
	x_5	−0.198	0.045	−4.349	−0.145	0.036	−4.037	−0.174	0.066	−2.626
	x_6	−0.376	0.045	−8.278	−0.324	0.034	−9.410	−0.357	0.066	−5.399
$\log(\phi)$	Constant	−1.994	0.182	−10.962	−2.231	0.182	−12.252	−2.610	0.195	−13.358
	x_2				−0.669	0.182	−3.674	−0.673	0.195	−3.446
	x_3				−0.535	0.182	−2.945	−0.492	0.195	−2.518
$\log(\lambda)$	Constant							−3.014	0.515	−5.854

SE = Standard error.

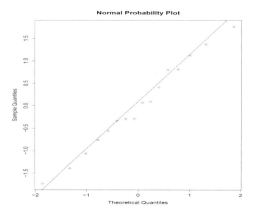

Figure 7.6 *Normal-probability plot of random effect in HGLM for semiconductor data.*

$(\approx -2p_\beta(\ell)) = -555.91$. For testing $\lambda = \text{var}(v) = 0$, which is a boundary value for the parameter space; the critical value for size $\alpha = 0.05$ is $\chi^2_{2\alpha} = 2.71$, so that the deviance difference $(21.91 = 555.91 - 534.00)$ in $-2p_\beta(\ell)$ between the JGLM and HGLM shows that the HGLM should be used (Section 6.5). Results from the three models are in Table 7.6. Residual plots for the mean and dispersion and the normal probability plot for random effects v are in Figures 7.4 through 7.6.

There is no indication of any lack of fit. GLM and JGLM assume independence among repeated measures. Table 7.6 shows that standard errors are larger when the HGLM is used and are most likely better as they account for the correlation. The structured dispersion model shows that to minimize ϕ we need to set the lamination time and pressure at high levels. Then we can adjust the mean by setting appropriate levels of lamination temperature, firing cycle time and firing dew point.

7.3.3 Respiratory data

Tables 7.7 and 7.8 display data from a clinical trial comparing two treatments for a respiratory illness (Stokes *et al.* 1995). In each of two centres, eligible patients were randomly assigned to active treatment (2) or placebo (1). During treatment, respiratory status (poor = 0, good = 1) was determined at four visits. Potential explanatory variables were centre, sex (male = 1, female = 2), and baseline respiratory status (all dichotomous), as well as age (in years) at the time of study entry. There were 111 patients (54 active, 57 placebo) with no missing data for responses or covariates.

Table 7.7 *Respiratory disorder data for 56 subjects from Centre 1.*

Patient	Treatment	Sex	Age	Respiratory Status (0=poor,1=good)				
				Baseline	Visit 1	Visit 2	Visit 3	Visit 4
1	1	1	46	0	0	0	0	0
2	1	1	28	0	0	0	0	0
3	2	1	23	1	1	1	1	1
4	1	1	44	1	1	1	1	0
5	1	2	13	1	1	1	1	1
6	2	1	34	0	0	0	0	0
7	1	1	43	0	1	0	1	1
8	2	1	28	0	0	0	0	0
9	2	1	31	1	1	1	1	1
10	1	1	37	1	0	1	1	0
11	2	1	30	1	1	1	1	1
12	2	1	14	0	1	1	1	0
13	1	1	23	1	1	0	0	0
14	1	1	30	0	0	0	0	0
15	1	1	20	1	1	1	1	1
16	2	1	22	0	0	0	0	1
17	1	1	25	0	0	0	0	0
18	2	2	47	0	0	1	1	1
19	1	2	31	0	0	0	0	0
20	2	1	20	1	1	0	1	0
21	2	1	26	0	1	0	1	0
22	2	1	46	1	1	1	1	1
23	2	1	32	1	1	1	1	1
24	2	1	48	0	1	0	0	0
25	1	2	35	0	0	0	0	0
26	2	1	26	0	0	0	0	0
27	1	1	23	1	1	0	1	1
28	1	2	36	0	1	1	0	0
29	1	1	19	0	1	1	0	0
30	2	1	28	0	0	0	0	0
31	1	1	37	0	0	0	0	0
32	2	1	23	0	1	1	1	1
33	2	1	30	1	1	1	1	0
34	1	1	15	0	0	1	1	0
35	2	1	26	0	0	0	1	0
36	1	2	45	0	0	0	0	0
37	2	1	31	0	0	1	0	0
38	2	1	50	0	0	0	0	0
39	1	1	28	0	0	0	0	0
40	1	1	26	0	0	0	0	0
41	1	1	14	0	0	0	0	1
42	2	1	31	0	0	1	0	0
43	1	1	13	1	1	1	1	1
44	1	1	27	0	0	0	0	0
45	1	1	26	0	1	0	1	1
46	1	1	49	0	0	0	0	0
47	1	1	63	0	0	0	0	0
48	2	1	57	1	1	1	1	1
49	1	1	27	1	1	1	1	1
50	2	1	22	0	0	1	1	1
51	2	1	15	0	0	1	1	1
52	1	1	43	0	0	0	1	0
53	2	2	32	0	0	0	1	0
54	2	1	11	1	1	1	1	0
55	1	1	24	1	1	1	1	1
56	2	1	25	0	1	1	0	1

Table 7.8 *Respiratory disorder data for 55 subjects from Centre 2.*

| Patient | Treatment | Sex | Age | Respiratory Status (0=poor,1=good) | | | | |
				Baseline	Visit 1	Visit 2	Visit 3	Visit 4
1	1	2	39	0	0	0	0	0
2	2	1	25	0	0	1	1	1
3	2	1	58	1	1	1	1	1
4	1	2	51	1	1	0	1	1
5	1	2	32	1	0	0	1	1
6	1	1	45	1	1	0	0	0
7	1	2	44	1	1	1	1	1
8	1	2	48	0	0	0	0	0
9	2	1	26	0	1	1	1	1
10	2	1	14	0	1	1	1	1
11	1	2	48	0	0	0	0	0
12	2	1	13	1	1	1	1	1
13	1	1	20	0	1	1	1	1
14	2	1	37	1	1	0	0	1
15	2	1	25	1	1	1	1	1
16	2	1	20	0	0	0	0	0
17	1	2	58	0	1	0	0	0
18	1	1	38	1	1	0	0	0
19	2	1	55	1	1	1	1	1
20	2	1	24	1	1	1	1	1
21	1	2	36	1	1	0	0	1
22	1	1	36	0	1	1	1	1
23	2	2	60	1	1	1	1	1
24	1	1	15	1	0	0	1	1
25	2	1	25	1	1	1	1	0
26	2	1	35	1	1	1	1	1
27	2	1	19	1	1	0	1	1
28	1	2	31	1	1	1	1	1
29	2	1	21	1	1	1	1	1
30	2	2	37	0	1	1	1	1
31	1	1	52	0	1	1	1	1
32	2	1	55	0	0	1	1	0
33	1	1	19	1	0	0	1	1
34	1	1	20	1	0	1	1	1
35	1	1	42	1	0	0	0	0
36	2	1	41	1	1	1	1	1
37	2	1	52	0	0	0	0	0
38	1	2	47	0	1	1	0	1
39	1	1	11	1	1	1	1	1
40	1	1	14	0	0	0	1	0
41	1	1	15	1	1	1	1	1
42	1	1	66	1	1	1	1	1
43	2	1	34	0	1	1	0	1
44	1	1	43	0	0	0	0	0
45	1	1	33	1	1	1	0	0
46	1	1	48	1	1	0	0	0
47	2	1	20	0	1	1	1	1
48	1	2	39	1	0	1	0	0
49	2	1	28	0	1	0	0	0
50	1	2	38	0	0	0	0	0
51	2	1	43	1	1	1	1	0
52	2	2	39	0	1	1	1	1
53	2	1	68	0	1	1	1	1
54	2	2	63	1	1	1	1	1
55	2	1	31	1	1	1	1	1

Stokes *et al.* (1995) used the GEE method to analyze the data using an independent and unspecified working correlation matrix. The results are in Table 7.9. We consider the following HGLM: for $i = 1, ..., 111$ and $j = 1, ..., 4$

$$\log\{p_{ij}/(1 - p_{ij})\} = x_{ij}^t\beta + y_{i(j-1)}\alpha + v_i,$$

where $p_{ij} = P(y_{ij} = 1|v_i, y_{i(j-1)})$, x_{ij}^t are covariates for fixed effects β, $v_i \sim N(0, \lambda_i)$ and we take the baseline value for the i–th subject as $y_{i(0)}$. The GLM model with additional covariates $y_{i(j-1)}$ was introduced as a transition model by Diggle *et al.* (1994), as an alternative to a random effect model. In this data set, both are necessary for a better modelling of correlation. Furthermore, the random effects have a structured dispersion

$$\log \lambda_i = \gamma_0 + G_i\gamma,$$

where G_i is the covariate age. Heteroscedasity increases with age. Old people with respiratory illness are more variable. For the mean model the centre and sex effects are not significant. The age effect is not significant in the GEE analyses, while it is significant in the HGLM analysis. Furthermore, the previous response is significant, i.e., if a patient had a good status he or she will have more chance of having a good status at the next visit. Model checking plots for structured dispersion are in Figure 7.7. Note that helpful model checking plots are available even for binary data.

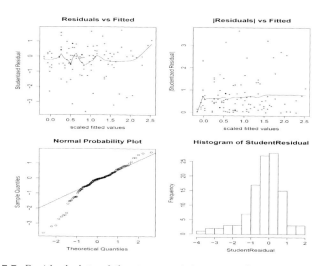

Figure 7.7 *Residual plots of the structured dispersion for the respiratory data.*

Table 7.9 Analyses of respiratory disorder data.

	Covariate	GEE (Independence)			GEE (Unspecified)			HGLM		
		Estimate	SE	t-value	Estimate	SE	t-value	Estimate	SE	t-value
$\log\{p/(1-p)\}$	Constant	−0.856	0.456	−1.88	−0.887	0.457	−1.94	−0.690	0.610	−1.13
	Treatment	1.265	0.347	3.65	1.245	0.346	3.60	1.256	0.415	3.028
	Centre	0.649	0.353	1.84	0.656	0.351	1.87	0.682	0.419	1.626
	Sex	0.137	0.440	0.31	0.114	0.441	0.26	0.261	0.597	0.437
	Age	−0.019	0.013	−1.45	−0.018	0.013	−1.37	−0.035	0.019	−1.889
	Baseline	1.846	0.346	5.33	1.894	0.344	5.51	1.821	0.446	4.079
$\log(\lambda)$	$y_{i(j-1)}$							0.575	0.304	1.891
	Constant							−0.683	0.737	−0.927
	Age							0.047	0.020	2.339

SE = Standard error.

7.3.4 Rat data

Three chemotheraphy drugs were applied to 30 rats that had induced leukemic conditions. White (W) and red blood cell (R) counts were collected as covariates and the response was the number of cancer cell colonies. The data were collected on each rat at four time periods. Data from Myers *et al.* (2002) are in Table 7.10. Among the three covariates, Drug, W and R, the Drug is a between-rat covariate, while W and R are within-rat covariates. The Drug factor has three levels, while W and R are continuous covariates.

We first fitted a Poisson HGLM, which has the scaled deviance

$$D = -2\{\ell(\widehat{\mu} \; ; y|v) - \ell(y; y|v)\} = 28.29$$

with 101.05 degrees of freedom. Thus, Lee and Nelder's (1996) goodness-of-fit test shows no evidence of a lack of fit. This model has the deviance $-2p_v(h) = 621.81$.

We tried the following quasi–Poisson-normal HGLM

$$
\begin{aligned}
\mu_{ij} &= E(y_{ij}|v_i) \\
\text{var}\,(y_{ij}|v_i) &= \phi\mu_{ij} \\
\log\mu_{ij} &= x_{ij}^t\beta + v_i \\
\log\lambda_i &= \gamma_0 + W_i\gamma_1 + W_i^2\gamma_2.
\end{aligned}
$$

The results are in Table 7.11. This model has the deviance $-2p_v(h) = 508.64$, so that this test shows that the quasi–Poisson-normal HGLM is better than the Poisson HGLM. The scaled deviance test for lack of fit is a test for the mean model, so that the deviance test $-2p_v(h)$ is useful for finding a good model for the dispersion. We also tried a quasi–Poisson-gamma HGLM (not shown), but the quasi–Poisson-normal model based upon a likelihood criterion performed slightly better. Overall both results were similar. Note that in the quasi–Poisson-gamma model we have

$$\text{var}(y) = \phi\mu + \lambda\mu^2.$$

In Table 7.11, $\hat{\phi} = \exp(-2.263) \simeq 0.1$ is near zero, implying that a gamma HGLM would be plausible. The results from the gamma HGLM are in Table 7.11. The two HGLMs gave similar results. To select a better model, we computed the deviances $-2p_v(h)$ and cAIC in Section 6.5. For h of the $y|v$ component, we used (3.8). Both deviances show that the Poisson HGLM is not appropriate (Table 7.12) and the quasi–Poisson HGLM or gamma HGLM is better, with the quasi–Poisson model having the advantage. The residual plots for the quasi–Poisson-normal HGLM are given in Figures 7.8 and 7.9.

Table 7.10 *Data from rat leukemia study.*

Subject	Drug	W1	W2	W3	W4	R1	R2	R3	R4	Y1	Y2	Y3	Y4
1	1	15	18	19	24	2	3	2	5	14	14	12	11
2	1	8	11	14	14	2	4	4	5	17	18	18	16
3	1	4	5	6	4	7	5	4	4	23	20	19	19
4	1	16	14	14	12	3	4	4	2	13	12	12	11
5	1	6	4	4	4	7	6	5	2	24	20	20	19
6	1	22	20	21	18	4	3	3	2	12	12	10	9
7	1	18	17	17	16	5	3	5	2	16	16	14	12
8	1	4	7	4	4	8	7	4	4	28	26	26	26
9	1	14	12	12	10	3	4	4	5	14	13	12	10
10	1	10	10	10	10	3	4	5	2	16	15	15	14
11	2	14	14	16	17	6	6	7	6	16	15	15	14
12	2	7	7	6	5	4	4	4	2	36	32	30	29
13	2	9	8	9	11	8	8	7	4	18	16	17	15
14	2	21	20	20	20	3	3	4	3	14	13	13	12
15	2	18	17	17	17	4	4	2	2	19	19	18	17
16	2	3	6	6	2	10	10	8	7	38	38	37	37
17	2	8	9	9	8	3	3	2	2	18	18	17	16
18	2	29	30	29	29	6	6	5	4	8	8	7	6
19	2	8	8	8	7	9	9	8	8	19	19	18	17
20	2	5	4	4	3	8	7	7	7	36	35	30	29
21	3	16	17	17	18	2	3	4	2	15	16	17	15
22	3	13	11	12	12	6	4	5	4	17	16	16	18
23	3	7	8	6	5	3	2	2	3	28	25	27	31
24	3	9	8	9	9	4	5	3	3	29	30	32	30
25	3	18	19	21	20	3	2	5	4	11	12	12	13
26	3	23	25	24	24	5	5	4	4	8	10	9	8
27	3	27	28	27	30	7	6	6	4	7	8	8	7
28	3	30	32	33	35	6	7	8	7	4	5	5	4
29	3	17	19	20	21	4	3	3	2	14	13	13	12
30	3	12	12	13	11	3	5	4	5	17	15	16	16

Table 7.11 *Estimation results of rat data.*

		Quasi–Poisson-normal HGLM			Gamma-normal HGLM		
	Covariate	Estimate	SE	t-value	Estimate	SE	t-value
$\log(\mu)$	Constant	2.709	0.098	27.816	2.718	0.100	27.207
	W	−0.014	0.005	−2.790	−0.015	0.005	−2.883
	R	0.028	0.007	4.213	0.030	0.008	3.915
	DRUG 2	0.166	0.099	1.667	0.165	0.100	1.647
	DRUG 3	0.109	0.095	1.148	0.111	0.094	1.179
$\log(\phi)$	Constant	−2.263	0.144	−15.69	−4.950	0.150	−33.11
$\log(\lambda)$	Constant	1.076	1.254	0.859	0.973	1.194	0.815
	W	−0.597	0.174	−3.43	−0.589	0.169	−3.478
	W^2	0.018	0.005	3.417	0.018	0.005	3.642

SE = Standard error.

Table 7.12 *Deviances of models for rat data.*

	Poisson HGLM	Quasi–Poisson HGLM	Gamma HGLM
$-2p_v(h)$	621.81	508.64	513.64
cAIC	618.06	425.41	437.22

Myers *et al.* (2002) used the GEE approach on a correlated Poisson model. Our analysis shows that using the GEE approach for the gamma model would be better. We prefer our likelihood approach because we can model the correlation and also the structured dispersions and achieve suitable likelihood-based criteria for model selection.

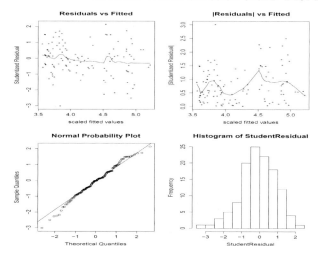

Figure 7.8 *Residual plots of y|v component in quasi–Poisson-normal HGLM for rat data.*

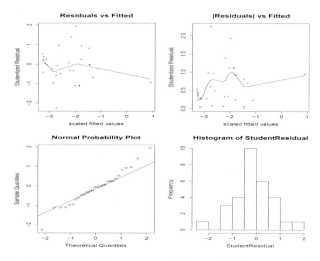

Figure 7.9 *Residual plots of v component in quasi–Poisson-normal HGLM for rat data.*

CHAPTER 8

Correlated random effects for HGLMs

There have been many models and methods proposed for the description and analysis of correlated non-normal data. Our general approach is via the use of HGLMs. Following Lee and Nelder (2001b), we further extend HGLMs in this chapter to cover a broad class of models for correlated data, and show that many previously developed models appear as instances of HGLMs. Rich classes of correlated patterns in non-Gaussian models can be produced without requiring explicit multivariate generalizations of non-Gaussian distributions. We extend HGLMs by adding an additional feature to allow correlations among the random effects.

Most research has focused on introducing new classes of models and methods for fitting them. With HGLMs as summarized in Table 6.1, we can check underlying model assumptions to discriminate between different models. We have illustrated that deviances are useful for comparing a nested sequence of models, and this extends to correlation patterns. Model selection using some deviance criteria can give only relative comparisons, not necessarily providing evidence of absolute goodness of fit, so that suitable model checking plots are helpful. However, these plots can sometimes be misleading, as we shall show. For comparison of non-nested models, AIC-type model selection can be useful. We show how to extend HGLMs to allow correlated random effects. Then by using examples we demonstrate how to use them to analyse various types of data. Likelihood inferences can provide various model selection tools.

8.1 HGLMs with correlated random effects

Let $y_i = (y_{i1}, ..., y_{iq_i})$ be the vector of q_i measurements on the i–th unit and $t_i = (t_{i1}, ..., t_{iq_i})$ be associated information; in longitudinal studies t_i is the corresponding set of times and in spatial statistics the set of locations at which these measurements are made. Consider HGLMs of the following form:

Conditional on random effects v_i, y_i follows a GLM with

$$\mu_i = E(y_i|v_i) \quad \text{and} \quad \text{var}(y_i|v_i) = \phi V(\mu_i)$$

and

$$\eta_i = g(\mu_i) = X_i\beta + Z_i v_i,$$

where $v_i = L_i(\rho)r_i$, $r_i \sim MVN(0, \Lambda_i)$ with $\Lambda_i = \text{diag}(\lambda_{ij})$, and $L_i(\rho)$ is a $p_i \times q_i$ matrix $(p_i \geq q_i)$ with $q_i = rank(L_i(\rho))$. Thus, while the random effects $v_i \sim N_{p_i}(0, L_i(\rho)\Lambda_i L_i(\rho)^t)$ may have a multivariate normal distribution with a singular covariance matrix, the random effects r_i do not.

When $\Lambda_i = \lambda I$ and $L_i(\rho) = I$, we have an HGLM with homogeneous random effects (Chapter 6), while an arbitrary diagonal Λ gives an HGLM with structured dispersion components (Chapter 7). We shall show that various forms of $L(\rho)$ can give rise to a broad class of models.

For simplicity of notation we suppress the subscripts. An arbitrary covariance matrix for v, $\text{var}(v) = L(\rho)\Lambda L(\rho)^t$, can be defined by choosing $L(\rho)$ to be an arbitrary upper- or lower-triangular matrix and a diagonal matrix Λ; see Kenward and Smith (1995) and Pourahmadi (2000). The most general form requires $(q + 1)q/2$ parameters and the number of these increases rapidly with the number of repeated measurements. Use of such a general matrix may cause a serious loss of information when data are limited. An obvious solution is to use models with patterned correlations. Various models for the analysis of correlated data fall into three categories:

- $\Lambda = \lambda I$ and $L(\rho) = L$, a matrix with fixed elements not depending upon ρ.

- Models for the covariance matrix: $\text{var}(v) = \lambda C(\rho)$, where $C(\rho) = L(\rho)L(\rho)^t$

- Models for the precision matrix: $[\text{var}(v)]^{-1} = P(\rho)/\lambda$, where $P(\rho) = (L(\rho)^t)^{-1}L(\rho)^{-1}$

Most previously developed models are multivariate normal. Our generalization, however, is to the wider class of GLMMs that constitute a subset of HGLMs. We could use other conjugate families of distribution for r_i and all the results in this chapter hold. However, $v_i = L_i(\rho)r_i$ may no longer belong to the same family. For example, r_i as a gamma does not imply that v_i is a gamma. We now show in more detail how these models can be written as instances of extended HGLMs.

8.2 Random effects described by fixed L matrices

There are many models using fixed L. One advantage of using such a model is that it is very fast to fit because it is not necessary to estimate ρ. We can use the algorithm developed in the last chapter by fitting

$$\eta_i = g(\mu_i) = X_i\beta + Z_i^* r_i,$$

where $Z_i^* = Z_i L_i$ and $r_i \sim MVN(0, \Lambda_i)$.

These are often defined as random effect models with singular precision matrices. They are more thoroughly studied in the next section.

8.2.1 Temporally correlated errors

The class $\Lambda = \lambda I$ and a fixed $L(\rho) = L$ include many of the state-space models of Harvey (1989) and also those of Besag et al. (1995) and Besag and Higdon (1999). For example,

- $r_i = \Delta v_i \equiv v_i - v_{i-1}$, a random walk model

- $r_i = \Delta^2 v_i \equiv v_i - 2v_{i-1} + v_{i-2}$, a second-order difference model

These models can be described by $r(= Av) \sim N(0, \lambda I)$, where A is a $q \times p$ matrix with rank $q(\leq p)$.

Consider the local trend model used for the seasonal decomposition of time series. In state-space form (e.g., Harvey, 1989) we can write this as

$$y_t = \mu_t + e_t,$$

where

$$\mu_t = \mu_{t-1} + \beta_t + r_t,$$

$$\beta_t = \beta_{t-1} + p_t,$$

with $r_t \sim N(0, \lambda_r)$ and $p_t \sim N(0, \lambda_p)$ being independent. Let $\beta_0 = 0$ and μ_0 be an unknown fixed constant; then this model can be represented as a normal HGLM

$$y_t = \mu_0 + f_t + s_t + e_t,$$

where $f_t = \sum_{j=1}^{t} r_j$ represents a long-term trend, and $s_t = \sum_{j=1}^{t}(t - j + 1)p_j$ is a local trend or seasonal effect and e_t the irregular term. This is another example of a model represented by $r = Av$.

8.2.2 Spatially correlated errors

The random walk and second-order difference have been extended to spatial models by Besag and Higdon (1999). They propose singular multivariate normal distributions, one of which is the intrinsic autoregressive (IAR) model, with kernel

$$\sum_{i \sim j} (v_i - v_j)^2 / \lambda_r,$$

where $i \sim j$ denotes that i and j are neighbours, and another that Besag and Higdon (1999) call locally quadratic representation, with the kernel

$$\sum_{i=1}^{p_1-1} \sum_{j=1}^{p_2-1} r_{i,j}^2 / \lambda_r$$

where $r_{i,j} = v_{i,j} - v_{i+1,j} - v_{i,j+1} + v_{i+1,j+1}$. Here $r = Av$, where r is a $q \times 1$ vector, v is a $p \times 1$ vector and A is a $q \times p$ matrix with $q = (p_1 - 1)(p_2 - 1)$ and $p = p_1 p_2$.

8.2.3 Smoothing splines

We cover smoothing in detail in Chapter 9, but for now we describe only how the idea is covered by correlated random effects. Non–parametric analogues of the parametric modelling approach have been developed. For example, Zeger and Diggle (1994) proposed a semiparametric model for longitudinal data, where the covariate entered parametrically as $x_i \beta$ and the time effect entered non–parametrically as $v_i(t_i)$. Consider a semiparametric model for longitudinal data with covariates enter parametrically as $x_i \beta$ and the time effect non–parametrically as follows,

$$\eta_i = x_i \beta + f_m(t_i),$$

where the functional form of $f_m()$ is unknown. Analogous to the method of Green and Silverman (1994, p. 12), natural cubic splines can be used to fit $f_m(t_i)$, by maximizing the h-likelihood, which is the so-called penalized likelihood in the smoothing literature:

$$\log f(y|v) - v^t P v / (2\lambda),$$

where the variance component λ takes the role of a smoothing parameter and P/λ is the precision matrix of v. Here $v^t P v / \lambda$, proportional to $-2 \log f_\theta(v)$, is the penalty term, symmetric around zero. This means that the penalty term in smoothing can be extended, for example, to a non-symmetric one by using the h-likelihood.

The matrix P is determined by the parameterization of the model, but we give here a general case where v is a vector of $f_m()$ at potentially unequal-spaced t_i. Then

$$P = QR^{-1}Q^t,$$

where Q is the $n \times (n-2)$ matrix with entries $q_{i,j}$ and, for $i = 1, ..., n$ and $j = 1, ..., n-2$,

$$q_{j,j} = 1/h_j, \quad q_{j+1,j} = -1/h_j - 1/h_{j+1}, \quad q_{j+2,j} = 1/h_{j+1},$$
$$h_j = t_{j+1} - t_j,$$

with the remaining elements being zero. R is the $(n-2) \times (n-2)$ symmetric matrix with elements $r_{i,j}$ given by

$$r_{j,j} = (h_j + h_{j+1}), \quad r_{j+1,j} = r_{j,j+1} = h_{j+1}/6,$$

and $r_{i,j} = 0$ for $|i - j| \geq 2$. The model is an HGLM with

$$\eta_i = x_i \beta + v_i(t_i),$$

where $v_i(t_i)$ is a random component with a singular precision matrix P/λ, depending upon t_i. For fitting of this model, see Chapter 9.

8.3 Random effects described by a covariance matrix

Laird and Ware (1982) and Diggle *et al.* (1994) considered random effect models having $\text{var}(v) = \lambda C(\rho)$. Consider the first-order autoregressive AR(1) model, assuming $\text{var}(v) = \lambda C(\rho)$, where $C(\rho)$ is the correlation matrix whose (i, j)–th element is given by

$$\text{corr}(v_i, v_j) = \rho^{|i-j|}.$$

For unequally spaced time intervals t_j, Diggle *et al.* (1994) extended the AR(1) model to the form

$$\text{corr}(v_j, v_k) = \rho^{|t_j - t_k|^u} = \exp(-|t_j - t_k|^u \kappa)$$

with $0 < u < 2$ and $\rho = \exp(-\kappa)$. Diggle *et al.* (1998) studied these autocorrelation models using the variogram. Other useful correlation structures are

- CS (compound symmetric): $\text{corr}(v_j, v_k) = \rho$, for $j \neq k$
- Toeplitz: $\text{corr}(v_j, v_k) = \rho_{|j-k|}$ and $\rho_0 = 1$

In these models we choose $L(\rho)$ to satisfy $C(\rho) = L(\rho)L(\rho)^t$.

8.4 Random effects described by a precision matrix

It is often found that the precision matrix $[\text{var}(v)]^{-1}$ has a simpler form than the covariance matrix $\text{var}(v)$ and models may be generated accordingly. We present two models for which $r = A(\rho)v$ where $A(\rho) = L(\rho)^{-1}$.

8.4.1 Serially correlated errors

AR models can be viewed as a modelling of a precision matrix. Consider the AR(1) model with equal time intervals. Here we have $r_t = v_t - \rho v_{t-1}$, i.e., $r = A(\rho)v$ with $A(\rho) = L(\rho)^{-1} = I - K$ where the non-zero elements of K are

$$\kappa_{i+1,i} = \rho.$$

For unequally spaced time intervals, we may consider a model

$$\kappa_{i+1,i} = \rho/|t_{i+1} - t_i|^u.$$

Antedependence structures form another extension of AR models using the precision matrix rather than the covariance matrix. A process exhibits antedependence of order p (Gabriel, 1962) if $\kappa_{i,j} = 0$ if $|i - j| > p$, which is a generalization of the AR(1) model. For implementation, see Kenward and Smith (1995).

8.4.2 Spatially correlated errors

For spatial correlation with locations t_{ij} Diggle *et al.* (1998) considered a form of autocorrelation

$$\text{corr}(v_j, v_k) = \rho^{|t_j - t_k|^u}.$$

With spatially correlated errors, a more natural model would be that equivalent to the Markov random field (MRF) model of the form

$$[\text{var}(v)]^{-1} = \Lambda^{-1}(I - M(\rho)).$$

Cressie (1993, p 557) considered an MRF model with $[\text{var}(v)]^{-1} = (I - M(\rho))/\lambda$, where $\Lambda = \lambda I$ and the non-zero elements in $M(\rho)$ are given by

$$M_{i+j,i} = \rho/|t_{i+j} - t_i|^u \quad \text{if } j \in N_i,$$

where N_i is the set of neighbours of the i–th location. Cressie (1993) considered multivariate normal models. An MRF model can be immediately extended to non-normal data via Lee and Nelder's (2001b) generalization.

8.5 Fitting and model checking

Multivariate distributions can be obtained from random effect models by integrating out the unobserved latent random effects from the joint density. An important innovation in our approach is the definition of the h-likelihood and its use for inferences from such models, rather than the generation of families of multivariate distributions. Lee and Nelder (2000a) showed that for an HGLM with arbitrary diagonal Λ and $L(\rho) = I$ the h-likelihood provides a fitting algorithm that can be decomposed into the fitting of a two-dimensional set of GLMs, one dimension being mean and dispersion, and the other fixed and random effects, so that GLM codes can be modified for fitting HGLMs. Lee and Nelder demonstrated that the method leads to reliable and useful estimators; these share properties with those derived from marginal likelihood, while having the considerable advantage of not requiring the integrating out of random effects. Their algorithm for fitting joint GLMs can be generalized to extended HGLMs easily as follows:

(a) Given correlation parameters ρ and hence $L_i(\rho)$, apply Lee and Nelder's (2000a) joint GLM algorithm in Chapter 6 to estimate $(\beta, r, \phi, \lambda)$ for the model

$$\eta_i = g(\mu_i) = X_i\beta + Z_i^* r_i$$

where $Z_i^* = Z_i L_i(\rho)$ and $r_i \sim N_{q_i}(0, \Lambda_i)$ with $\Lambda_i = \text{diag}(\lambda_{ij})$.

(b) Given $(\beta, r, \phi, \lambda)$, find an estimate of ρ by the Newton-Raphson method, which maximizes the adjusted profile likelihood, $p_{v,\beta}(h)$.

(c) Iterate (a) and (b) until convergence is achieved.

Inferences can again be made by applying standard procedures for GLMs. Thus the h-likelihood allows likelihood-type inference to be extended to this wide class of models in a unified way.

8.6 Examples

In this section we show the fitting of various models with patterned random effects.

8.6.1 Gas consumption in the UK

Durbin and Koopman (2000) analysed the lagged quarterly demand for gas in the UK from 1960 to 1986. They considered a structural time-series model. As shown in Section 8.2 the so-called local linear trend model with quarterly seasonals can be represented as a normal HGLM

$$y_t = \alpha + f_t + s_t + q_t + e_t$$

where $f_t = \sum_{j=1}^{t} r_j$ and $s_t = \sum_{j=1}^{t}(t - j + 1)p_j$ are random effects for the local linear trend, the quarterly seasonals q_t with $\sum_{j=0}^{3} q_{t-j} = w_t$, and $r_t \sim N(0, \lambda_r)$, $p_t \sim N(0, \lambda_p)$, $w_t \sim N(0, \lambda_w)$, $e_t \sim N(0, \phi_t)$. This model has scaled deviance $D = 31.8$ with $n - p = 108 - p = 31.8$ degrees of freedom, which are the same because the dispersion parameters ϕ_t are estimated. Thus, an AIC based upon the deviance is not meaningful; see Section 6.5. Here the AIC should be based upon the conditional loglihood (Section 6.5)

$$cAIC = -2\log f(y|v) + 2p = -297.2.$$

This model involves four independent random components (r_t, p_t, w_t, e_t) of full size.

We may view f_t, s_t and q_t as various smoothing splines for $f_m(t)$ as illustrated in Section 8.2.3. We consider a linear mixed model by adding a linear trend $t\beta$ to give

$$y_t = \alpha + t\beta + f_t + s_t + q_t + e_t.$$

With this model we found that the random walk f_t was not necessary because $\hat{\lambda}_r$ tends to zero. The model

$$y_t = \alpha + t\beta + s_t + q_t + e_t \tag{8.1}$$

has the scaled deviance $D = 31.0$ with $n - p = 108 - p = 31.0$ degrees of freedom and the conditional loglihood gives

$$cAIC = -2\log f(y|v) + 2p = -298.6.$$

Residual plots for this model shown in Figure 8.1 display apparent outliers. There were disruptions in the gas supply in the third and fourth quarters of 1970. Durbin and Koopman pointed out that these might lead to a distortion in the seasonal pattern when a normality assumption is made for the error component e_t, and they proposed heavy-tailed models, such as those with t-distributions for the error component e_t. This model still involves three independent random components of full size.

Lee (2000) proposed to delete the random quarterly seasonals and add further fixed effects to model the 1970 disruption and seasonal effects

$$\begin{aligned} y_t &= \alpha + t\beta + \alpha_i + t\beta_i + \delta_1(t = 43) + \delta_2(t = 44) \\ &+ \gamma_1 \sin(2\pi t/104) + \gamma_2 \cos(2\pi t/104) + s_t + e_t, \end{aligned} \tag{8.2}$$

where $i = 1, ..., 4$ represents quarters, and δ_1 and δ_2 are for the third and

Table 8.1 Estimates from analyses of gas consumption data.

Coefficient	Mean model (8.2)			Coefficient	Dispersion model (8.3)		
	Estimate	SE	t-value		Estimate	SE	t-value
α	5.079	0.120	41.418	φ	-5.865	0.300	-19.539
α_2	-0.095	0.034	-2.807	ψ_2	0.540	0.419	1.289
α_3	-0.489	0.040	-12.378	ψ_3	0.944	0.421	2.242
α_4	-0.360	0.051	-6.995	ψ_4	1.579	0.419	3.770
β	0.0156	0.009	1.778	$\log(\lambda_p)$	-12.095	0.698	-17.331
β_2	-0.006	0.001	-11.170				
β_3	-0.009	0.001	-15.037				
β_4	0.0005	0.0008	0.570				
δ_1	0.472	0.089	5.290				
δ_2	-0.391	0.121	-3.222				
γ_1	-0.143	0.092	-1.563				
γ_2	-0.061	0.104	-0.590				

SE = Standard error.

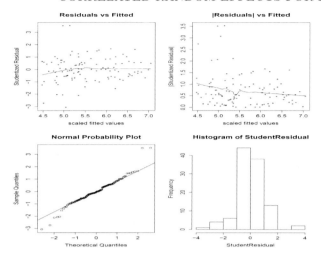

Figure 8.1 *Residual plots of error component e_t for mean model (8.1).*

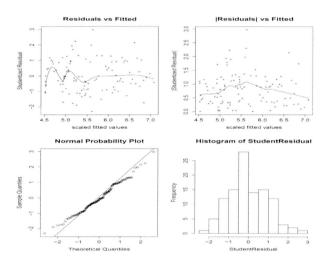

Figure 8.2 *Residual plots of error component e_t for mean model (8.2).*

fourth quarters of 1970. Lee (2000) further found extra dispersion in the third and fourth quarters, which led to a structured dispersion model

$$\log \phi_t = \varphi + \psi_i \qquad (8.3)$$

where ψ_i are the quarterly main effects. Parameter estimates are in Table 8.1. The dispersion model clearly shows that the heterogeneity increases with quarters. Model checking plots are in Figure 8.2, and show that most of the outliers have vanished. The heterogeneity can be explained by adding covariates for the dispersion. The final model has the scaled deviance $D = 92.0$ with $n - p = 108 - p = 92.0$ degrees of freedom, giving

$$cAIC = -2 \log f(y|v) + 2p = -228.1.$$

Thus, both model checking plots and cAICs clearly indicate that the final model is the best among those considered for this data set.

8.6.2 Scottish data on lip cancer rates

Clayton and Kaldor (1987) analysed observed (y_i) and expected numbers (n_i) of lip cancer cases in the 56 administrative areas of Scotland with a view to producing a map that would display regional variations in cancer incidence and yet avoid the presentation of unstable rates for the smaller areas (see Table 8.2). The expected numbers allowed for the different age distributions in the areas by using a fixed effect multiplicative model; these were regarded for the purpose of analysis as constants based on an external set of standard rates. Presumably the spatial aggregation is due in large part to the effects of environmental risk factors. Data were available on the percentage of the work force in each area employed in agriculture, fishing, or forestry (x_i). This covariate exhibits spatial aggregation paralleling that for lip cancer itself. Because all three occupations involve outdoor work, exposure to sunlight, the principal known risk factor for lip cancer, might be the explanation. For analysis, Breslow and Clayton (1993) considered the following Poisson HGLM with the log link

$$\eta_i = \log \mu_i = \log n_i + \beta_0 + \beta_1 x_i/10 + v_i$$

where v_i represented unobserved area-specific log-relative risks. They tried three models:

M1: $v_i \sim N(0, \lambda)$

M2: $v_i \sim$ intrinsic autoregressive model in Section 8.2.2

M3: $v_i \sim$ MRF in which $[\text{var}(v)]^{-1} = (I - \rho M)/\lambda$ where M is the incidence matrix for neighbours of areas

We present plots of residuals (Lee and Nelder, 2000a) against fitted expected values in Figure 8.3. M1 shows a downward linear trend of residuals against fitted values, which is mostly removed in M2. M1 has the scaled deviance $D = 22.3$ with $n - p = 56 - p = 16.1$ degrees of freedom. The conditional loglihood gives an $cAIC = -2 \log f(y|v) + 2p = 310.1$ and deviance $-2p_{v,\beta}(h) = 348.5$. M2 has scaled deviance $D = 30.5$ with $n - p = 27.1$ degrees of freedom, giving an $cAIC = -2 \log f(y|v) + 2p = 296.1$ and the deviance $-2p_{v,\beta}(h) = 321.4$.

Table 8.2 *Lip cancer data from 56 Scottish counties.*

County	y_i	n_i	x	Adjacent Counties
1	9	1.4	16	5, 9, 11, 19
2	39	8.7	16	7, 10
3	11	3	10	6, 12
4	9	2.5	24	18, 20, 28
5	15	4.3	10	1, 11, 12, 13, 19
6	8	2.4	24	3, 8
7	26	8.1	10	2, 10, 13, 16, 17
8	7	2.3	7	6
9	6	2	7	1, 11, 17, 19, 23, 29
10	20	6.6	16	2, 7, 16, 22
11	13	4.4	7	1, 5, 9, 12
12	5	1.8	16	3, 5, 11
13	3	1.1	10	5, 7, 17, 19
14	8	3.3	24	31, 32, 35
15	17	7.8	7	25, 29, 50
16	9	4.6	16	7, 10, 17, 21, 22, 29
17	2	1.1	10	7, 9, 13, 16, 19, 29
18	7	4.2	7	4, 20, 28, 33, 55, 56
19	9	5.5	7	1, 5, 9, 13, 17
20	7	4.4	10	4, 18, 55
21	16	10.5	7	16, 29, 50
22	31	22.7	16	10, 16
23	11	8.8	10	9, 29, 34, 36, 37, 39
24	7	5.6	7	27, 30, 31, 44, 47, 48, 55, 56
25	19	15.5	1	15, 26, 29
26	15	12.5	1	25, 29, 42, 43
27	7	6	7	24, 31, 32, 55
28	10	9	7	4, 18, 33, 45
29	16	14.4	10	9, 15, 16, 17, 21, 23, 25, 26, 34, 43, 50
30	11	10.2	10	24, 38, 42, 44, 45, 56
31	5	4.8	7	14, 24, 27, 32, 35, 46, 47
32	3	2.9	24	14, 27, 31, 35
33	7	7	10	18, 28, 45, 56
34	8	8.5	7	23, 29, 39, 40, 42, 43, 51, 52, 54
35	11	12.3	7	14, 31, 32, 37, 46
36	9	10.1	0	23, 37, 39, 41
37	11	12.7	10	23, 35, 36, 41, 46
38	8	9.4	1	30, 42, 44, 49, 51, 54
39	6	7.2	16	23, 34, 36, 40, 41
40	4	5.3	0	34, 39, 41, 49, 52
41	10	18.8	1	36, 37, 39, 40, 46, 49, 53
42	8	15.8	16	26, 30, 34, 38, 43, 51
43	2	4.3	16	26, 29, 34, 42
44	6	14.6	0	24, 30, 38, 48, 49
45	19	50.7	1	28, 30, 33, 56
46	3	8.2	7	31, 35, 37, 41, 47, 53
47	2	5.6	1	24, 31, 46, 48, 49, 53
48	3	9.3	1	24, 44, 47, 49
49	28	88.7	0	38, 40, 41, 44, 47, 48, 52, 53, 54
50	6	19.6	1	15, 21, 29
51	1	3.4	1	34, 38, 42, 54
52	1	3.6	0	34, 40, 49, 54
53	1	5.7	1	41, 46, 47, 49
54	1	7	1	34, 38, 49, 51, 52
55	0	4.2	16	18, 20, 24, 27, 56
56	0	1.8	10	18, 24, 30, 33, 45, 55

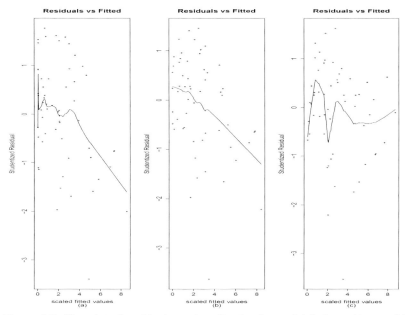

Figure 8.3 *The plot of residuals against fitted values of (a) the ordinary, (b) IAR, and (c) MRF models.*

From the residual plot in Figure 8.3, Lee and Nelder (2001b) chose the model M3 as best because the smoothing line is the flattest. The MRF model with $\rho = 0$ is the M1 model. Here MRF with $\hat{\rho} = 0.173$ provides a suitable model. However, the MRF model has the scaled deviance $D = 31.8$ with $n - p = 24.6$ degrees of freedom, giving an $cAIC = -2\log f(y|v) + 2p = 302.5$ and the deviance $-2p_{v,\beta}(h) = 327.3$. We found that the main difference between M1 and M3 is the prediction for county 49, which has the highest predicted value because it has the largest n_i. This gives a large leverage value of 0.92, for example, under M3. For an observed value of 28, M2 predicts 27.4, while M3 gives 29.8. The leverage exaggerates the differences in predictions. M2 has total prediction error of $P = \sum(y_i - \hat{\mu}_i)^2/\hat{\mu}_i = 24.9$, while M3 has $P = 25.5$, so that M2 is slightly better in prediction. While model checking plots are useful, they can be misleading and objective criteria based upon the likelihood are also required in the model selection.

8.6.3 Factor analysis of law school admission data

Factor analysis and structural equation model (SEM) are widely used in many disciplines, especially in social science and psychology. Exploratory factor analysis reduces the dimension of the data by extracting a few important directions in the data space, while confirmatory factor analysis is a procedure for estimating and testing specific models with latent factors underlying the observed data. Factor analysis and SEM involve random effects called factors.

Consider Bock and Lieberman 's (1970) law school admission data, which consists of six items y_{ij} taking 1 (Correct) or 0 (Not correct) values for a law school admission test with $n = 350$ subjects. Let $\pi_{ij} = P(y_{ij} = 1|\boldsymbol{v}_i)$. They considered a binary factor model

$$\text{logit}(\boldsymbol{\pi}_i) = \boldsymbol{\beta}_0 + \boldsymbol{\Lambda}\boldsymbol{v}_i,$$

where $\boldsymbol{\pi}_i = (\pi_{i1}, \cdots, \pi_{i6})^T$ and $\boldsymbol{\beta}_0 = (\beta_{01}, \cdots, \beta_{06})^T$ are the vectors of π_{ij}'s and β_{0j}'s, respectively,

$$\boldsymbol{\Lambda}^T = \begin{pmatrix} 1 & \lambda_2 & \lambda_3 & 0 & 0 & 0 \\ 0 & 0 & 0 & 1 & \lambda_5 & \lambda_6 \end{pmatrix},$$

and $\boldsymbol{v}_i = (v_{1i}, v_{2i})^T$ follow the bivariate normal distribution with the covariance matrix

$$\text{cov}(\boldsymbol{v}_i) = \begin{pmatrix} \gamma_{11} & \gamma_{12} \\ \gamma_{12} & \gamma_{22} \end{pmatrix}.$$

We reported fitting results in Table 8.3. Here REML(1) is based on the first-order Laplace approximation in Section 6.3.3 and REML(2) is based on the second-order Laplace approximation. Via small simulation studies, Noh *et al.* (2016) showed that REML(1) has the fewest mean squares errors while REML(2) shows the least bias. Among existing methods, the goodness-of-fit test based on the deviance and the degrees of freedom in Section 6.5 show no lack of fit, which allows a lack–of–fit test for factor and SEM models.

Factor and SEM models are widely used in practice. However, these models are primarily normal based. HGLMs with correlated random effects can be used to fit these models. The number of parameters in $\boldsymbol{\Lambda}$ of HGLMs in Section 8.3 is usually one, but in factor and SEM literature efficient algorithms have been developed to fit rich classes of covariance modelling with large numbers of parameters. Thus, it is useful to combine developments in two areas (Noh *et al.*, 2016).

8.6.4 Pittsburgh particulate matter data

Particulate matter (PM) is one of the six constituent air pollutants regulated by United States Environmental Protection Agency. Negative effects of PM on human health have been consistently indicated by much

EXAMPLES
237

Table 8.3 *Parameter estimates (SE) for two–factor model for law school admission data.*

	REML(1)	REML(2)
λ_2	0.48(0.12)	0.50(0.13)
λ_3	0.44(0.10)	0.46(0.11)
λ_5	0.51(0.11)	0.53(0.11)
λ_6	0.42(0.10)	0.42(0.10)
γ_{11}	0.38(0.11)	0.40(0.13)
γ_{12}	0.52(0.08)	0.52(0.08)
γ_{22}	0.56(0.08)	0.58(0.10)
β_{01}	$-1.39(0.08)$	$-1.42(0.09)$
β_{02}	$-0.52(0.05)$	$-0.50(0.05)$
β_{03}	$-0.12(0.07)$	$-0.11(0.07)$
β_{04}	$-0.65(0.05)$	$-0.67(0.05)$
β_{05}	$-1.07(0.06)$	$-1.06(0.06)$
β_{06}	$-0.91(0.05)$	$-0.93(0.06)$

epidemiological research. The current regulatory standards for PM specify two categories: PM10 and PM2.5, which refer to all particles with median aerodynamic diameters smaller than or equal to 10 microns and 2.5 microns, respectively. PM data often exhibit diverse spatio-temporal variation, and other variations induced by the design of the sampling scheme used in collecting the data. Bayesian hierarchical models have often been used in analyzing such PM data.

The data collected from 25 PM monitoring sites around the Pittsburgh metropolitan area in 1996 comprise the 24-hour averages of PM10 observations. Not all monitoring sites reported every day and some sites recorded two measurements on the same day. The number of observations reported ranged from 11 to 33 per day coming from 10 to 25 sites, a total of 6448 observations in 1996. Information on weather at the site was not available and local climatological data (LCD) were used as representative weather information. The LCD comprises 23 daily weather variables, recorded at the Pittsburgh International Airport weather station (PIT).

Cressie *et al.* (1999) and Sun *et al.* (2000) used a log transformation designed to stabilize the mean-variance relationship of the PM values. For the Pittsburgh data, log(PM10) appeared to have seasonal effects measuring higher in the summer season, and higher during the week thanover weekends. Wind blowing from the south significantly increased the amount of log(PM10). The data showed that precipitation had a substantial effect on log(PM10), and the amount of precipitation during the previous day explained slightly more of the variability in log(PM10) than the amount during the current day. Thus, for the PM analysis we could use a gamma HGLM with a log link. However, here we used log-normal linear mixed models to allow the comparison with the Bayesian analysis below.

Daniels *et al.* (2001) considered Bayesian hierarchical models for the PM data. They used non-informative prior distributions for all hyper-parameters. In the mean model, they considered six weather covariates and three seasonal covariates. The six weather covariates comprised three temperature variables: the (daily) average temperature, the dew-point temperature and the maximum difference of the temperature; two wind variables: the wind speed and the wind direction; and one precipitation variable: the amount of precipitation. The three seasonal covariates comprised a linear and a quadratic spline term over the year and a binary weekend indicator variable (1 for a weekend day, 0 for a non-weekend day). The linear spline term for the day increases linearly from 0 to 1 as the day varies from 1 to 366, and the quadratic spline variable for the day has its maximum value 1 at days 183 and 184.

These variables were selected from all available those by testing the models repeatedly. Daniels *et al.*'s (2001) analysis showed both spatial and temporal heterogeneity. When the weather effects were not incorporated in the model, they also detected an isotropic spatial dependence decaying with distance.

Lee *et al.* (2003) showed that an analysis can be made by using HGLMs without assuming priors. The variable log(PM10) is denoted by y_{ijk} for the i–th locations of monitoring sites, j–th day of year 1996, and k–th replication. Because the number of observations varies from 0 to 2 for each day, for a specific site, the PM observations at a site i and a day j have the vector form, $\mathbf{y}_{ij} \equiv (y_{ij1}, \ldots, y_{ijk_{ij}})^t$, depending on the number of observations $k_{ij} = 0, 1, 2$ at the site $i = 1, \ldots, 25$ and the day $j = 1, \ldots, 366$. Consider a linear mixed model:

(a) The conditional distribution of $\mathbf{y}_{ij}$ on the given site-specific random effects, a_i, b_i and c_i follows the normal distribution;

$$\mathbf{y}_{ij}|a_i, b_i, c_i \sim N(\mu_{ij}, \ \sigma_{ij}^2 I_{ij}), \tag{8.4}$$

where $\mu_{ij} = 1_{ij}\alpha + W_{ij}\beta + X_{ij}\gamma + a_i 1_{ij} + W_{ij}b_i + X_{ij}c_i$, 1_{ij} is the $k_{ij} \times 1$ vector of ones, I_{ij} is the $k_{ij} \times k_{ij}$ identity matrix, α, β and γ are corresponding fixed effects, and W_{ij} and X_{ij} are vectors respectively for the six weather and the three seasonal covariates at site i and date j.

$W_{ij} = W_j$ and $X_{ij} = X_j$ for all i and j, because we use the common weather data observed at PIT for all PM monitoring sites, and we assume that the seasonality is the same for all sites.

(b) The random effects b_i and c_i for $i = 1, \ldots, 25$ are assumed to follow normal distributions

$$b_i \sim N_6(0, \lambda_b), \quad c_i \sim N_3(0, \lambda_c),$$

where $\lambda_b = \text{diag}(\lambda_{b1}, \ldots, \lambda_{b6})$, $\lambda_c = \text{diag}(\lambda_{c1}, \ldots, \lambda_{c3})$, and the random effects $a = (a_1, \ldots, a_{25})^t$ to follow normal distributions

$$a \sim N_{25}(0, \lambda_a (I - A(\rho))^{-1}), \tag{8.5}$$

for $\lambda_a \in R^1$. The parameter ρ models the spatial dependence among sites. When the PM monitoring sites h and l are at distance $d_{h,l}$ apart, the (h, l)-th off-diagonal element of the matrix $A = A(\rho)$ is assumed to be $\rho/d_{h,l}$, and all the diagonal terms of A are 0. This distance decaying spatial dependence is popular in literature for Markov random field spatial models (e.g., Cliff and Ord, 1981; Cressie, 1993; Stern and Cressie, 2000).

The error component σ_{ij}^2 in model (8.4) captures the variability of the small-scale processes, which comprise the measurement error and micro-scale atmospheric processes. For modelling the variance, Daniels *et al.* (2001) considered three Bayesian heterogeneity models analogous to the following dispersion models: the homogeneous model,

$$\log(\sigma_{ij}^2) = \tau_0,$$

the temporal heterogeneity model,

$$\log(\sigma_{ij}^2) = \tau_{m(j)}^t,$$

where the index function $m = m(j) \in \{1, \ldots, 12\}$ defines the month in which day j falls, and the spatial heterogeneity model

$$\log(\sigma_{ij}^2) = \tau_i^s.$$

Lee *et al.* (2003) considered a combined heterogeneity model

$$\log(\sigma_{ij}^2) = \tau_{im}^a = \tau_0 + \tau_i^s + \tau_m^t.$$

The parameter estimates and t-values of the fixed effects for the HGLM using weather covariate W_j and for the Bayesian model are listed in Table 8.4. The estimates from the two types of models are very similar for all the heterogeneity models. Table 8.4 presents the results from the spatial heterogeneity model only. The *t*-values of the Bayesian model are calculated from the reported 95% credible intervals based upon large sample theory. All the reported credible intervals are symmetric, except

Table 8.4 *Summary statistics of fixed effect estimates for PM data.*

Variable	HGLM with W_j^* estimate	t-value	HGLM with W_j estimate	t-value	Bayesian Model estimate	t-value
Intercept	3.445	74.44	2.696	68.49	2.700	70.56
Linear spline	−3.117	−9.94	−2.904	−9.35	−2.960	−8.66
Quadratic spline	2.753	11.08	2.942	11.76	2.970	12.52
Weekend	−0.127	−5.77	−0.127	−5.70	−0.140	−6.81
Average temperature	0.025	1.89	0.027	2.07	0.030	14.7
Dew point temperature	−0.009	−0.68	−0.011	−0.84	−0.013	−6.37
Difference of temperature	0.002	0.17	0.004	0.33	0.004	2.61
Wind direction	−0.165	−6.78	−0.163	−6.68	−0.160	−5.75
Wind speed	−0.547	−26.92	−0.061	−4.63	−0.064	−17.92
Precipitation	−0.479	−11.24	−0.270	−10.52	−0.282	−6.91
PMSE	0.136		0.141		0.141	

in the case of wind speed. For wind speed, the t-value of the Bayesian model was calculated by taking the average of the right and left tail values. The estimates of the Bayesian model and the HGLM model with W_j are very similar. However, the t-values of the two models differ noticeably for the temperature variables. In the HGLM analysis, the dew point temperature and the daily difference of the temperature are not significant, while in the Bayesian analysis they are. Consistently over the Bayesian analyses, all types of the heterogeneity models produced the same predicted mean square errors (PMSEs) up to the third decimal of their values in the HGLM. The PMSEs of the Bayesian models were obtained at each iteration of the Gibbs sampler in Daniels *et al.* (2001), while the PMSEs in the HGLM models were straightforwardly obtained from the predicted values

$$\hat{\alpha}1_{ij} + W_{ij}\hat{\beta} + X_{ij}\hat{\gamma} + \hat{a}_i 1_{ij} + W_{ij}\hat{b}_i + X_{ij}\hat{c}_i.$$

Daniels *et al.* (2001) considered three mean models: the full, intermediate and reduced types. The full model had both weather and seasonal covariates, the intermediate model had only seasonal covariates, and the reduced model used intercept only. The full model gave the lowest PMSE regardless of the types of heterogeneity model in both the Bayesian approach and the HGLM approach. In the following we used mainly the full mean models.

The plots of residuals against the fitted values are in Figure 8.4. The left side shows the residual plot from the HGLM (8.4). The group of residuals marked with circles (∘) are distinguishable from the rest of residuals marked with dots (•). This pattern may indicate that the lowest predicted values are too low. By tracing the data corresponding to the separated group of points, we found these residuals came from the observations on the 63rd day (March 3) and 177th day (June 25) in 1996. From the LCD of the area, March 3 and June 25 of 1996 were special in their wind speed and amount of precipitation respectively. On March 3, the wind speed was very high, and on June 25, the precipitation on the day before was the highest of all the days in 1996.

Figure 8.5 explains the reason for such large residuals from the current HGLM. When the precipitation amount and the daily average of the log(PM10) for all 366 days were plotted, the 177th day is the point with exceptional precipitation. The 63rd day has high wind speed, and the daily average of log(PM10) for the day is the highest in the group with higher wind speed. Figure 8.5 shows that the linear fit with the precipitation and wind speed variables resulted in very low predicted values for the high precipitation days and the high wind speed days. In Figure 8.5, fitting curved lines, log(1 + precipitation) and log(1 + wind speed)

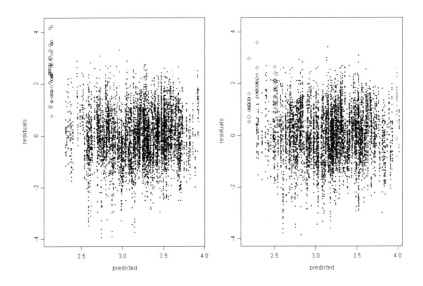

Figure 8.4 *Plots of residuals versus predicted values when the non-transformed weather variables W_j are used (left) and when the transformed weather variables W_j^* are used (right).*

reduces such large residuals. Thus Lee *et al.* (2003) fitted the HGLM using transformed weather covariates W_j^* containing the same variables, but with log-transformed variables $\log(1+\text{wind speed})$ and $\log(1+\text{precipitation})$. The value 1 is added in the log transformations to prevent negative infinite values when the wind speed and precipitation have zero values.

Table 8.3 shows the HGLM model with transformed weather covariates W_j^* indicating a smaller PMSE than the other models. The Bayesian model has almost identical PMSE values to the HGLM model with W_j for all the heterogeneity models. Because of parameter orthogonality between the mean and dispersion parameters, the type of heterogeneity model does not affect the PMSEs from the mean models.

In the Bayesian model, τ_m^t and τ_i^s are assumed to follow $N(\tau, \tau_0^2)$ distribution, and the heterogeneities ($\tau_0^2 > 0$) are tested by Bayes factors using the Savage-Dickey density ratio, $p(\tau_0^2 = 0|Z)/p(\tau_0^2 = 0)$, on the appropriately defined uniform shrinkage prior $\pi(\tau_0^2)$. In HGLMs these parameters in heterogeneity models can be verified by a likelihood test based upon the restricted loglihood $p_{v,\beta}(h)$. From Table 8.5 the loglihood

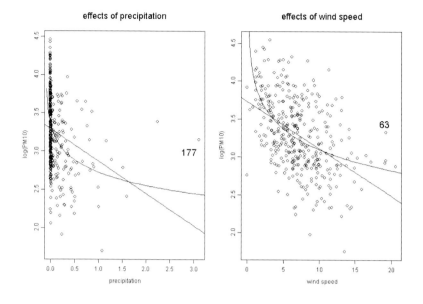

Figure 8.5 *Plots showing the effects of precipitation (left) and wind speed (right) on log(PM10).*

Table 8.5 *Restricted likelihood test for heterogeneity.*

Model	HGLM with W_j^*		HGLM with W_j	
	$-2p_{v,\beta}(h)^*$	df^*	$-2p_{v,\beta}(h)^*$	df^*
Homogeneous	273.76	35	288.45	35
Spatially heterogeneous	121.34	11	126.12	11
Temporally heterogeneous	147.53	24	148.27	24
Additive model	0	0	0	0

* $-2p_{v,\beta}(h)$ and df are computed relative to minimum and maximum values, respectively

test shows that the full heterogeneity model is appropriate, i.e., that both spatial and temporal heterogeneities exist. Daniels *et al.* (2001) did not consider the Bayesian model analogous to the full heterogeneity model. Likelihood inference is available without resorting to simulation methods or assuming priors.

8.7 Twin and family data

8.7.1 Genetic background

There are at least two large areas of statistical application where random effect models have been extremely successful, namely animal breeding and genetic epidemiology. Both involve the use of family data where the latent similarity between family members is modelled as a set of random effects. In these applications, analytical and computational complexity arise rapidly from (a) the correlations between the family members and (b) the large data sets typically obtained. Because our examples are taken from genetic epidemiology, we adopt its terminology here.

Animal breeding motivated many early developments of mixed models, particularly by Henderson (Henderson *et al.*, 1959). Strong interest in finding genetic risk factors of diseases made genetic epidemiology one of the fastest growing areas in genomic medicine, and we present family data from this perspective. The first hint of genetic basis of a disease comes from evidence of familial clustering. Table 8.6 (from Pawitan *et al.*, 2004) shows the distribution of the number of occurrences of pre-eclampsia (PE), a hypertensive condition induced by pregnancy, among women who had two or three pregnancies. A total of 570 women had two pre-eclamptic pregnancies, while we would expect only 68 such women if PEs occurred purely as random Bernoulli events. Among the women who had three pregnancies, 21 were pre-eclamptic during all three, whereas we should expect none if PEs occurred randomly.

Table 8.6 *Familial clustering of pre-eclmpsia, summarized for women who had two or three pregnancies in Sweden between 1987 and 1997.*

Number of pregnancies	Pre-eclampsia incidence	Number of women	Random
2	0	100,590	100,088
2	1	4,219	5,223
2	2	570	68
3	0	20,580	20,438
3	1	943	1,206
3	2	124	24
3	3	21	0

Random = corresponding expected number if pre-eclampsia occurs randomly computed using estimated binomial probabilities.

Since most mothers have children with a common father, it is clear that we cannot separate the maternal and paternal genetic contributions based only on the disease clustering from nuclear families. Furthermore, familial clustering may be due to common environmental effects. Separating these effects requires investigation of larger family structures with appropriate models.

Before considering the modelling in detail and discussing it in the general context of genetic epidemiology, the standard steps in genetic analysis of a disease are worth outlining. For a very brief terminology: a *phenotype* of a subject is an observed trait or simply an outcome variable, and a *genotype* is the genetic make-up of a subject. Genotyping means finding and reading the content of the DNA or chromosomal material from the subject — typically taken from blood or tissue. The whole content of the DNA of a person is called the *genome. Markers* are specific DNA sequences with known locations (loci) on the genome.

- In a *segregation analysis*, we first establish whether a genetic effect is present, by analysing the co-occurrence of the phenotype among family members. A segregation analysis tells us whether a condition is genetic, but it does not tell us which genes are involved or where they are in the genome.

- For this, a *linkage analysis* is needed, where some markers are genotyped and correlated with disease occurrence. Linkage studies are performed on families.

- An *association study* also correlates phenotype and genotype, with the intent of finding the genes involved in a disease as in linkage analysis, but it is often performed on unrelated individuals in a population. Both linkage and association studies are also called gene-mapping studies .

Segregation analysis is a necessary first step in establishing the genetic basis of a disease. The methods we describe here cover segregation analysis only. Linkage analysis requires much more detailed probabilistic modelling of gene transmissions from parents to offspring, and it is beyond our scope. However, the mixed model method is also useful for this purpose, for example in quantitative trait linkage (QTL) analysis; see Amos (1994) and Blangero *et al.* (2001).

All the phenotypes we study using mixed models belong to the so-called complex or *non-Mendelian* phenotypes. A *Mendelian* phenotype is determined by one or two genes that have strong effects, such that the genotype of a person can be inferred simply by looking at the pattern of co-occurrences of the phenotype within a family. In contrast, non–Mendelian phenotypes are determined potentially by many genes, each with typically small effects and possibly caused by the environment.

8.7.2 Twin data

Because of their simplicity, we first describe the twin data. Let $y_i = (y_{i1}, y_{i2})$ be the phenotypes of interest, measured from the twin pair i. The simplest model assumes that y_{ij} is Bernoulli with probability p_{ij} and

$$g(p_{ij}) = \beta + g_{ij} + e_{ij},$$

where $g()$ is the link function, β is a fixed parameter associated with prevalence, the additive genetic effect g_{ij} is $N(0, \sigma_g^2)$ and the common childhood environment effect e_{ij} is $N(0, \sigma_e^2)$. Let $g_i = (g_{i1}, g_{i2})$ and $e_i = (e_{i1}, e_{i2})$, the effects being assumed independent of each other. Between-pair genetic effects are independent, but within-pair values are not. For monozygotic (MZ) twins, it is commonly assumed that

$$\begin{aligned} \mathrm{Cor}(g_{i1}, g_{i2}) &= 1 \\ \mathrm{Cor}(e_{i1}, e_{i2}) &= 1 \end{aligned}$$

and for dizygotic (DZ) twins that

$$\begin{aligned} \mathrm{Cor}(g_{i1}, g_{i2}) &= 0.5 \\ \mathrm{Cor}(e_{i1}, e_{i2}) &= 1. \end{aligned}$$

While it is possible to assume some unknown parameter for the correlation, our ability to estimate it from the available data is usually very limited. *The discrepancy in genetic correlation between MZ and DZ twins allows us to separate the genetic from the common environmental factor.* For the purpose of interpretation, it is convenient to define the quantity

$$h^2 = \frac{\sigma_g^2}{\sigma_g^2 + \sigma_e^2 + 1},$$

known as narrow *heritability*. Since we assume the probit link, the heritability measures the proportion of the variance (of *liability* or predisposition to the phenotype under study) due to additive genetic effects. We follow this definition of heritability, which we can show agrees with the standard definition of heritability in biometrical genetics (Sham, 1998, p. 212). It is common to assume the probit link function in biometrical genetics as this is equivalent to assuming that vulnerability to disease is normally distributed. From the probit model, with the standard normal variate denoted by Z, the model-based estimate of the prevalence is given by

$$
\begin{aligned}
P(Y_{ij} = 1) &= P(Z < \beta + g_{ij} + e_{ij}) \\
&= P(Z - g_{ij} - e_{ij} < \beta) \\
&= \Phi\left(\frac{\beta}{(\sigma_g^2 + \sigma_e^2 + 1)^{1/2}} \right). \qquad (8.6)
\end{aligned}
$$

The logit link could be used, but for most diseases seen in practice, the two link functions will produce very similar results.

For binary outcomes, when there is no covariate, the information from each twin pair is the number of concordances $k = 0, 1, 2$ for the disease. The full data set from n pairs of twins can be summarized as (n_0, n_1, n_2), with $\sum n_k = n$. If there is no genetic or environmental effect, the outcomes are independent Bernoulli with estimated probability

$$
\widehat{p} = \frac{n_1 + 2n_2}{2n}.
$$

A clustering effect can be tested by comparing the observed data (n_0, n_1, n_2) with the expected frequencies

$$
\widehat{n}_0 = n(1 - \widehat{p})^2, \ \ \widehat{n}_1 = n\widehat{p}(1 - \widehat{p}), \ \ \widehat{n}_2 = n\widehat{p}^2.
$$

To express the amount of clustering, a concordance rate c can be computed as the proportion of persons with the disease whose co-twins also have the disease:

$$
c = \frac{2n_2}{n_1 + 2n_2}. \qquad (8.7)
$$

Using the probit model, the marginal likelihood can be computed explicitly using the normal probability as follows. The formula is easily extendable to more general family structures (see Pawitan $et\ al.$, 2004). Since $y_i | g_i, e_i$ is assumed Bernoulli, we first have

$$
\begin{aligned}
P(Y_i = y_i) &= E\{P(Y_{ij} = y_{ij}, \text{ for all } j | g_i, e_i)\} \\
&= E\{\prod_j P(Y_{ij} = y_{ij} | g_i, e_i)\} \\
&= E\{\prod_j p_{ij}^{y_{ij}} (1 - p_{ij})^{1 - y_{ij}}\}. \qquad (8.8)
\end{aligned}
$$

From the model

$$
\begin{aligned}
p_{ij} &= P(Z_j < x_{ij}^t \beta + g_{ij} + e_{ij}) \qquad (8.9) \\
&= P(Z_j - g_{ij} - e_{ij} < x_{ij}^t \beta), \qquad (8.10)
\end{aligned}
$$

where the Z_js are independent standard normal variates. Hence the probability (8.8) can be written as an orthant probability

$$P(Y_i = y_i) = P(l_{ij} < V_{ij} < u_{ij}, \text{ for all } j), \qquad (8.11)$$

where $V_{ij} \equiv Z_j - g_{ij} - e_{ij}$, so the vector $V_i \equiv (V_{i1}, V_{i2})$ is $N(0, \Sigma)$ with

$$\Sigma = \Sigma_g + \Sigma_e + I_2,$$

where Σ_g is the genetic variance matrix, Σ_e the environmental variance matrix and I_2 the identity matrix of size 2. The upper bounds $u_{ij} = x_{ij}^t \beta$ if $y_{ij} = 1$, and $u_{ij} = \infty$ if $y_{ij} = 0$. Similarly, the lower bounds $l_{ij} = -\infty$ if $y_{ij} = 1$, and $l_{ij} = x_{ij}^t \beta$ if $y_{ij} = 0$.

Thus, the marginal loglihood given the data (n_0, n_1, n_2) is

$$\ell(\beta, \sigma_g^2, \sigma_e^2) = \sum_{k=0}^{2} n_k \log p_k,$$

where $p_k = P(Y_i = y_i, \ y_{i1} + y_{i2} = k)$. Since the data have only two degrees of freedom, the parameters are not identifiable. However, if we combine the MZ and DZ twins and assume there are parameters in common, these can be estimated.

Table 8.7 *Concordances of breast and prostate cancers in monozygotic and dizygotic twins*

Cancer	Type	n	n_0	n_1	n_2	Concordance
Breast (women)	MZ	8,437	7,890	505	42	0.14
	DZ	15,351	14,276	1023	52	0.09
Prostate (men)	MZ	7,231	8,098	299	40	0.21
	DZ	13,769	14,747	584	20	0.06

MZ = monozygotic.
DZ = dizygotic.
Concordance rate computed using (8.7).
Source: Lichtenstein *et al.*, 2000.

Table 8.7 shows the number of concordances of breast and prostate cancer summarized from 44,788 Nordic twins (Lichtenstein *et al.*, 2000). The higher concordance rate among the MZ twins is evidence for genetic effects for these cancers. To estimate the parameters, we assume that the cancer prevalences in the MZ and DZ groups are the same. Figure 8.6 shows the likelihood functions of the variance components. The estimated parameters for breast cancer are $\widehat{\sigma}_g^2 = 0.46$ (standard error = 0.20) and $\widehat{\sigma}_e^2 = 0.12$ (standard error = 0.12), so the estimated heritability is $\widehat{h}^2 = 0.27$. For prostate cancer, the estimates are $\widehat{\sigma}_g^2 =$

1.06 (standard error $= 0.17$) and $\widehat{\sigma}_e^2 = 0.0$, and $\widehat{h}^2 = 0.51$. The number of parameters in the model is three (a common intercept, and two variance components), and the number of independent categories is $6 - 2 = 4$, so that one degree of freedom is left for checking the model fit.

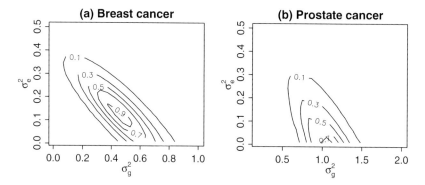

Figure 8.6 *The joint likelihood of genetic and environmental variance components from the twin data in Table 8.7. The confidence level associated with each contour line is 1 minus the contour level, so the outermost contour line defines the 90% confidence region.*

8.7.3 Nuclear family data

For the next order of complexity after the twin model, suppose y_i is an outcome vector measured on the offspring of common parents and $Ey_i = p_i$. We can model

$$g(p_i) = X_i\beta + b_i, \qquad (8.12)$$

where X_i is the matrix obtained by stacking the row vectors x_{ij}, b_i is $N(0, \sigma^2 R)$ and R has the compound symmetric structure

$$\begin{pmatrix} 1 & \rho & \cdots & \rho \\ \rho & 1 & \cdots & \rho \\ \vdots & & \ddots & \vdots \\ \rho & \rho & \cdots & 1 \end{pmatrix}.$$

This model is regularly used in the context of family data, for example, by Curnow and Smith (1976). The correlation is due to genetic and common family environment effects, but these are not separable using the available data and model (8.12).

Subject	Father	Mother
1	u	u
2	u	u
3	1	2
4	1	u
5	4	3
6	u	u
7	5	6

u = unknown.

8.7.4 Arbitrary family structure

For an arbitrary family structure i (not necessarily a nuclear family), let $y_i = (y_{i1}, \ldots, y_{in_i})$ be the outcome vector measured on some (not necessarily all) family members and, conditional on random effects b_i, assume y_{ij} to be independent Bernoulli with parameter p_{ij}. Assume that

$$g(p_{ij}) = x_{ij}^t \beta + z_{ij}^t v_i, \qquad (8.13)$$

where x_{ij} is a vector of fixed covariates, and β a vector of fixed effect parameters. The random effect parameters v typically consist of several components; for example, $v_i = (g_{ij}, c_{ij}, a_{ij})$ represents shared genetic, childhood and adult environmental effects, and

$$z_{ij}^t v_i \equiv g_{ij} + c_{ij} + a_{ij}.$$

The vector $g_i = (g_{i1}, \ldots, g_{in_i})$ of genetic effects is typically assumed to be normal with mean zero and variance $\sigma_g^2 R$, where R is a genetic kinship or relationship matrix induced by the relationships between family members. The vector $c_i = (c_{i1}, \ldots, c_{in_i})$ is assumed normal with zero mean and variance $\sigma_c^2 C$ and $a_i = (a_{i1}, \ldots, a_{in_i})$ is normal with zero mean and variance $\sigma_a^2 A$. It can be readily seen that a realistic genetic model requires multiple variance components.

Relationships between family members can be shown in a pedigree diagram, but this is not feasible for a large family. The most convenient yet powerful representation utilises the so-called pedigree data that identify the parents of each subject. For example, suppose we have seven individuals labelled 1, 2, ..., 7 with the following pedigree table.

Subjects 1, 2 and 6 have unknown parents; 3 is the offspring of 1 and 2; 4 has a known father 1, but no known mother; 5 has known parents 4 and 3; 7 has known parents 5 and 6. In principle, everything we want to know about a family can be computed from the pedigree data, including all variance matrices of the random effects such as R and S above.

The computation of the genetic relationship matrix R is particularly instructive. The environmental matrices C and A can be similarly constructed by assuming that those exposed to the same environment have a correlation of one. Let r_{ij} be the elements of R; the value of r_{ij} as a function of the pedigree data is defined by the following concepts:

1. *For diagonal elements:*

$$r_{ii} = 1 + \frac{1}{2} r_{father,mother}$$

where father and mother are from the i-th subject. The second term on the right-hand side takes account of inbreeding, if there is no inbreeding then $r_{ii} = 1$, and R is a correlation matrix.

2. *For off-diagonal elements r_{ij}, where j is of an older generation than i:*

If both parents of i are known:

$$r_{ij} = \frac{1}{2}(r_{j,father} + r_{j,mother}).$$

If only one parent of i is known:

$$r_{ij} = \frac{1}{2} r_{j,father}$$

or

$$r_{ij} = \frac{1}{2} r_{j,mother}.$$

If both parents of i are unknown:

$$r_{ij} = 0.$$

By definition $r_{ij} = r_{ji}$, so R is a symmetric matrix.

For the sample pedigree data we have

$$
\begin{aligned}
r_{11} &= 1 \\
r_{12} &= 0 \\
r_{13} = r_{31} &= \frac{1}{2}(r_{11} + r_{12}) = \frac{1}{2} \\
r_{14} = r_{41} &= \frac{1}{2} r_{11} = \frac{1}{2} \\
r_{15} = r_{51} &= \frac{1}{2}(r_{14} + r_{13}) = \frac{1}{2} \\
r_{16} &= 0 \\
r_{17} = r_{71} &= \frac{1}{2}(r_{15} + r_{16}) = \frac{1}{4}
\end{aligned}
$$

etc. We can verify that the upper triangle of R is given by

$$R = \begin{pmatrix} 1 & 0 & 0.5 & 0.5 & 0.5 & 0 & 0.25 \\ & 1 & 0.5 & 0 & 0.25 & 0 & 0.125 \\ & & 1 & 0.25 & 0.625 & 0 & 0.3125 \\ & & & 1 & 0.625 & 0 & 0.3125 \\ & & & & 1.125 & 0 & 0.5 \\ & & & & & 1 & 0.5 \\ & & & & & & 1 \end{pmatrix}.$$

8.7.5 Melanoma example

Let y_{ij} be the binary outcome indicating whether member j in family i is diagnosed with melanoma. The family is assumed to be nuclear, consisting of a father, a mother and several children. Conditional on the random effects, y_{ij} is assumed to be independent Bernoulli with probability p_{ij}, and, using probit model,

$$\Phi^{-1}(p_{ij}) = x_{ij}^t \beta + g_{ij} + c_{ij} + a_{ij},$$

where $\Phi(\cdot)$ is the normal distribution function, g_{ij} is the additive genetic effect, c_{ij} is the shared childhood environmental effect and a_{ij} is the shared adult environmental effect. Let g_i, c_i and a_i be vectors of g_{ij}, c_{ij} and a_{ij} respectively. We assume that g_i is $N(0, \sigma_g^2 R_g)$, c_i is $N(0, \sigma_c^2 R_c)$ and a_{ij} is $N(0, \sigma_a^2 R_a)$. We limit the fixed effect covariates to prevalence levels in the different generations; these enter the model as constant terms.

Table 8.8 *Individuals from about 2.6 million nuclear families in Sweden, melanoma cases and concordances between family members.*

Individuals	n	Cases
Parents	5,285,184	21,932
Children	4,491,919	8,488

Relation	Both affected
Parent-child	232
Spouse-spouse	73
Child-child	36

Source: Lindström, *et al.*, 2006.

Table 8.8 shows the number of individuals from around 2.6 million nuclear families in Sweden and the number of melanoma cases from Lindström *et al.* (2006). This table is typical in the sense that we need a large number of families to obtain sufficient number of concordances. From the tabled values, it is difficult to compute exactly the expected numbers of concordances between family members when there is no clustering effect, but we can approximate them. First, the prevalence among parents is

$$p_0 = \frac{21932}{5285184} = 0.00415$$

and the prevalence among children is

$$p_1 = \frac{8488}{4491919} = 0.00189.$$

(The lower prevalence among children is due to the shorter length of followup. A proper analysis of the data requires the techniques of survival analysis, but this is not feasible with such large data sets. Some compromises to deal with the length bias are described in Lindström *et al.* (2006).) Thus, when there is no clustering effect, the expected number of spouse-spouse concordances is $2,642,592 \times 0.00415^2 = 45.5$, while we observed 73. Hence, there is evidence of a shared adult environmental effect. Similarly, assuming that each family has two children (which was not true), under independence we expect $4,491,919 \times 0.00189^2 = 16$, while we observed 36, indicating potentially genetic and childhood environmental effects.

Using a similar method to compute the marginal likelihood as in Section 8.7.2, we obtain

$$\begin{aligned}
\widehat{\sigma}_g^2 &= 0.51 \pm 0.022, \\
\widehat{\sigma}_c^2 &= 0.13 \pm 0.045, \\
\widehat{\sigma}_a^2 &= 0.10 \pm 0.025.
\end{aligned}$$

8.7.6 Pre-eclampsia example

To be able to separate the maternal and paternal effects in Table 8.6 Pawitan *et al.* (2004) considered the pregnancy outcomes from a pair of siblings: brother-brother, brother-sister and sister-sister. From pair i, let y_i be the vector of 0 to 1 outcomes with mean p_i and assume

$$g(p_i) = x_i^t \beta + m_i + f_i + e_i + s_i, \tag{8.14}$$

where $g(\cdot)$ is a known link function, m_i is the vector of maternal effects, f_i the foetal effects, e_i the common family environment effect and s_i the common sibling environment. The common family environment is the unique environment created by the father and the mother, and the sibling environment is the common childhood and adolescent environment experienced by the sisters. We assume that $m_i \sim N(0, \sigma_m^2 R_m)$, $f_i \sim N(0, \sigma_f^2 R_f)$, $e_i \sim N(0, \sigma_e^2 R_e)$ and $s_i \sim N(0, \sigma_s^2 R_s)$. Various correlations exist within each effect, as expressed by the known R matrices, but the effects are assumed independent from each other and between the different pairs of families.

In general, confounding between fixed effect predictors and random effects is possible and is not ruled out by the model. In fact, as expected, we see that the genetic effects are partly mediated by the fixed effect predictors. To compute the entries in the correlation matrices, we make the following usual assumptions in biometrical genetics (see Sham, 1998, Chapter 5): the genetic correlation between full siblings is 0.5 and that between cousins is 0.125. Hence the full set of variance parameters is $\theta = (\sigma_m^2, \sigma_f^2, \sigma_e^2, \sigma_s^2)$.

To give a simple example of the correlation matrices, suppose each mother from a sister-sister pair had two pregnancies, so that the outcome y_i is a binary vector of length 4. Then

$$
R_m = \begin{pmatrix} 1 & 1 & 0.5 & 0.5 \\ 1 & 1 & 0.5 & 0.5 \\ 0.5 & 0.5 & 1 & 1 \\ 0.5 & 0.5 & 1 & 1 \end{pmatrix}
$$

$$
R_f = \begin{pmatrix} 1 & 0.5 & 0.125 & 0.125 \\ 0.5 & 1 & 0.125 & 0.125 \\ 0.125 & 0.125 & 1 & 0.5 \\ 0.125 & 0.125 & 0.5 & 1 \end{pmatrix}
$$

$$
R_e = \begin{pmatrix} 1 & 1 & 0 & 0 \\ 1 & 1 & 0 & 0 \\ 0 & 0 & 1 & 1 \\ 0 & 0 & 1 & 1 \end{pmatrix}, \quad R_s = \begin{pmatrix} 1 & 1 & 1 & 1 \\ 1 & 1 & 1 & 1 \\ 1 & 1 & 1 & 1 \\ 1 & 1 & 1 & 1 \end{pmatrix}.
$$

Using the data in Pawitan *et al.* (2004), we consider all pregnancies in Sweden between 1987 and 1997, and limit the information to the first three known pregnancies for each couple. The distribution of families for different sibling-pairs is given in Table 8.9. We include a total of 239,193 pairs of families, with 774,858 recorded pregnancies, of which 20,358 were pre-eclamptic.

For fixed effect predictors, it is well known that the risk of PE is higher during first pregnancies among diabetics, Nordic, older individuals and non-smokers. Thus we consider whether the pregnancy is first (yes, no) or subsequent (yes, no), the diabetes status of the mother ($1 =$ no diabetes,

2 = pre-gestational diabetes, 3 = gestational diabetes diagnosed near the delivery time), whether the mother is Nordic (born in Sweden, Finland, Iceland, Norway), age of the mother (1 = < 30, 2 = 30–34, 3 = > 35), and smoking status of the mother (0 = non-daily smoker, 1 = daily smoker).

Table 8.9 *Distribution of sibling-pairs from families with pregnancies between 1987 and 1997.*

Sibling-pair	No. of families	No. of pregnancies	No. of PEs
Mother-mother	2×60,875	197,925 (1.63/fam)	5,185
Father-father	2×61,903	200,437 (1.62/fam)	5,206
Mother-father	2×116,415	376,496 (1.62/fam)	9,967
Total	2×239,193	774,858 (1.62/fam)	20,358

As shown in Table 8.10, Nordic, diabetic or older status is significantly associated with higher risk of PE. First pregnancies are also significantly more at risk of PE than the subsequent pregnancies. Note that the fixed effect estimates from the mixed model are not directly comparable to the corresponding values from the standard GLM. The HGLM provides a conditional effect given the random effects, but the GLM gives a marginal or population-averaged effect. Such a comparison, however, is useful to assess confounding between the fixed and random effects.

The marginal effects of the fixed predictors from the HGLM can be found approximately by a simple scaling: see (8.6) and (8.10) for the derivation using probit link. For the logit link, the scale factor is

$$\sqrt{[(\pi^2/3)/\{2.17 + 1.24 + .82 + .00 + (\pi^2/3)\}]} = 0.66,$$

so that, for example, for the first parameter, the marginal effect is $-5.41 \times 0.66 = -3.58$. Compared to the estimates from the GLM, some HGLM estimates are reduced, such as those for diabetes and smoking, indicating that their effects are partly confounded with the genetic effects. However, we might also say that the genetic factors are only fractionally explained by the known risk factors such as diabetes, so that there may be other possibly unknown mechanisms involved. By comparing the variance components, the analysis also shows that familial clustering is explained mostly by maternal and foetal genetic factors.

Table 8.10 *Parameter estimates for GLM and HGLM for pre-eclampsia data.*

Effect	HGLM Estimate	HGLM Standard Error	Scaled Estimate	GLM Estimate	GLM Standard Error
Fixed effect parameters					
First	-5.41	0.095	-3.58	-3.60	0.021
Subsequent	-6.82	0.097	-4.51	-4.62	0.023
Diabetes 2	2.17	0.089	1.44	1.72	0.062
Diabetes 3	1.05	0.101	0.69	0.90	0.067
Nordic	0.75	0.090	0.50	0.51	0.059
Age 2	0.10	0.038	0.07	0.08	0.020
Age 3	0.43	0.047	0.28	0.38	0.025
Smoking	-0.51	0.040	-0.34	-0.49	0.022
Dispersion parameters					
Maternal	2.17	0.054	–	–	–
Fetal	1.24	0.082	–	–	–
Family-environment	0.82	0.095	–	–	–
Sibling	0.00	0.008	–	–	–

8.8 Ascertainment problem

The previous melanoma example shows a very common problem in genetic studies: a large number of individuals is required to observe some disease clustering in families. If disease prevalence is low, say 1%, it seems inefficient to randomly sample 10,000 individuals just to obtain 100 affected cases. Instead, it is more convenient and logistically easier to collect data from families with at least one affected member. Intuitively these genetically loaded families contain most of the information about the genetic properties of the disease.

There is a large literature on non–random ascertainment in genetical studies, starting with Fisher (1934), and later Elston and Sobel (1979) and de Andrade and Amos (2000). Recently, Burton *et al.* (2001) considered the effects of ascertainment in the presence of latent trait heterogeneity and, using examples, claimed that ascertainment-adjusted estimation led to biased parameter estimates. Epstein *et al.* (2002) showed that consistent estimation is possible if a researcher knows and models the ascertainment adjustment properly. However, a simulation study of the logistic variance component model in Glidden and Liang (2002) indicated that estimation using ascertained samples is highly sensitive to model misspecification of the latent variable. This is a disturbing result, since in practice it is unlikely that we will know the distribution of the latent variable exactly.

Noh *et al.* (2005) developed an h-likelihood methodology to deal with ascertainment adjustment and accommodate latent variables with heavy-tailed models. They showed that the latent variable model with heavy tails leaded to robust estimation of the parameters under ascertainment adjustment. Some details of the method are given in Section 10.5.

CHAPTER 9

Smoothing

Smoothing or non–parametric function estimation was one of the largest areas of statistical research in the 1980s, and is now a recognized tool for exploratory data analysis. In regression problems, instead of fitting a simple linear model

$$E(y|x) = \beta_0 + \beta_1 x,$$

we fit a non–parametric smooth or simply a smooth approach to data:

$$E(y|x) = f(x)$$

where $f(x)$ is an arbitrary smooth function. Smoothness of the function is a key requirement, as otherwise the estimate may have excessive variation that masks interesting underlying patterns. The model is non–parametric in that is has no easily interpretable parameters as in a linear model, but as we shall see, the estimation of $f(x)$ implicitly involves some estimation of parameters. One crucial issue in all smoothing problems is how much to smooth; it is a problem that has given rise to a lot of theoretical developments.

In statistical modelling, there are two uses for smoothing terms: (a) removing the effects of an uninteresting nuisance covariate so that other terms in the linear predictor are better estimated, and (b) as a preliminary step in finding the shape of a response to a covariate — the smooth step replaced by a suitable parametric function.

The smoothing literature is enormous and there are a number of monographs, including Silverman (1986), Eubank (1988), Wahba (1990), Green and Silverman (1994), and Ruppert *et al.* (2003). We find the exposition by Eilers and Marx (1996) closest to what we need, because it has both the simplicity and extendability to cover diverse smoothing problems. Our purpose here is to present smoothing from the perspective of classical linear mixed models and HGLMs, showing that all the previous methodology applies naturally and immediately. Furthermore, the well known problem of choosing the smoothing parameter is seen as equivalent to estimating a dispersion parameter.

9.1 Spline models

To state the statistical estimation problem, we have observed data $(x_1, y_1), \ldots, (x_n, y_n)$, and we want to estimate the conditional mean $\mu = E(y|x)$ as

$$\mu = f(x),$$

for arbitrary smooth $f(x)$. The observed x values can be arbitrary, not necessarily equally spaced, but we assume that they are ordered; multiple ys with the same x are allowed. One standard method of controlling the smoothness of $f(x)$ is by minimizing a penalized least squares formula,

$$\sum_i (y_i - f(x_i))^2 + \rho \int |f^{(d)}(x)|^2 dx, \qquad (9.1)$$

where the second term represents a roughness penalty, with $f^{(d)}(x)$ being the d–th derivative of $f(x)$. In practice it is common to use $d = 2$. The parameter ρ is called the smoothing parameter. A large ρ implies more smoothing; the smoothest solution obtained as ρ goes to infinity is a polynomial of degree $(d - 1)$. If $\rho = 0$, we get the roughest solution $\widehat{f}(x_i) = y_i$. The 'smooth' or 'rough' terms are relative. Intuitively we may say that a local pattern is rough if it contains large variation relative to the local noise level. When the level is high, the signal-to-noise ratio is small, so observed patterns in the signal are likely to be spurious. Rough patterns indicate overfitting of the data, and with sets of single covariate data, they are easy to spot.

The use of roughness penalties dates back to Whittaker (1923), who dealt with discrete index series rather than continuous functions and used third differences for the penalty. The penalty was justified as comprising a log-prior density in a Bayesian framework, although in the present context it will be considered as the likelihood of the random effect parameter. Reinsch (1967) showed that the minimizer of (9.1) must (a) be cubic in the interval (x_i, x_{i+1}) and (b) have at least two continuous derivatives at each x_i. These properties are consequences of the roughness penalty, and functions satisfying them are called *cubic splines*.

The simplest alternative to a smooth in dealing with an arbitrary nonlinear function is to use the power polynomial

$$f(x) = \sum_{j=0}^{p} \beta_j x^j,$$

which can be estimated easily. However, this is rarely a good option; a high-degree polynomial is often needed to estimate a nonlinear pattern, but it usually comes with unwanted local patterns that are not easy to control. Figure 9.1 shows the measurements of air ozone concentration versus air temperature in New York in during the summer 1973 (Chambers *et al.*, 1983). There is a clear nonlinearity in the data, which can be reduced but not removed if we log-transform the ozone concentration. Except for a few outliers, the variability of ozone measurements does not change appreciably over the range of temperature, so we shall continue to analyze the data in the original scale. The figure shows second- and eighth-degree polynomial fits. The eighth-degree model exhibits local variation that is well below the noise level. Nevertheless, because it is easily understood, the polynomial model is useful for illustrating the general concepts.

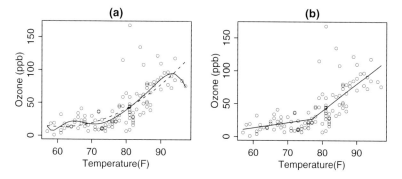

Figure 9.1 *The scatter plot of air ozone concentration (in parts per billion) versus air temperature (in degrees Fahrenheit). (a) Quadratic (dashed) and eighth-degree polynomial (solid) fits to the data. (b) Piecewise linear with a knot at $x = 77$.*

The collection of predictors $\{1, x, \ldots, x^p\}$ forms the *basis functions* for $f(x)$. The basis functions determine the model matrix X based on the data; for example, using a quadratic model $\{1, x, x^2\}$ we have

$$X = \begin{bmatrix} 1 & x_1 & x_1^2 \\ \vdots & \vdots & \vdots \\ 1 & x_n & x_n^2 \end{bmatrix},$$

so each basis function corresponds to one column of X. In general, the problem of estimating $f(x)$ from the data reduces to the usual problem of estimating β in a linear model

$$E(y) = X\beta,$$

with $y = (y_1, \cdots, y_n)$ and $\beta = (\beta_0, \ldots, \beta_p)$.

We shall focus on the so-called *B-spline* basis (de Boor, 1978), which is widely used because of its local properties. The resulting B-spline $f(x)$ is determined only by values at neighbouring points; in contrast, the polynomial schemes are global. The design points $d_1, d_2, \ldots$ for a B-spline are called the *knots*; there are many ways of determining these knots, but here we shall simply set them at equal space intervals within the range of x. The j–th B-spline basis function of degree m is a piecewise polynomial of degree m in the interval (d_j, d_{j+m+1}), and zero otherwise. For example, the B-spline basis of 0 degree is constant between (d_i, d_{i+1}) and zero otherwise. The resulting B-spline of degree 1 is the polygon that connects $(d_i, f(d_i))$; higher-order splines are determined by assuming a smoothness or continuity condition on the derivatives. In practice it is common to use the cubic B-spline to approximate smooth functions (de Boor, 1978), but as we shall see even lower-order splines are sufficient if combined with other smoothness restrictions.

The B-spline models can be motivated simply by starting with a piecewise linear model; see Ruppert *et al.*, (2003). For example,

$$f(x) = \beta_0 + \beta_1 x + v_1 (x - d_1)_+$$

where $a_+ = a$ if $a > 0$ and is equal to zero otherwise. It is clear that the linear function bends at the knot location d_1. At $x = d_1$ the slope of the function changes by the amount v_1. The basis functions for this model are $\{1, x, (x - d_1)_+\}$ and the corresponding model matrix X can be constructed for the purpose of estimating the parameters. This piecewise linear model is a B-spline of order 1, although the B-spline basis is not exactly $\{1, x, (x - d_1)_+\}$, but another equivalent set as described below. In Figure 9.1 a piecewise linear model seems to fit the data well, but it requires estimation of the knot location and whether the change point at $x = 77$ is physically meaningful is not clear.

The exact formulas for the basis functions are tedious to write, but much more instructive to draw. Figure 9.2 shows the B-spline bases of degrees 1, 2 and 3. Each basis function is determined by the degree and the knots. In practice we simply set the knots at equal intervals within the range of the data. Recall that, given the data, each set of basis functions determines a model matrix X, so the problem always reduces to computing regression parameter estimates. If we have three knots $\{0, 1, 2\}$, the bases for B-spline of degree 1 contain three functions,

$$\{(1-x)I(0 < x < 1), xI(0 < x < 1) + (2-x)I(1 < x < 2), (x-1)I(1 < x < 2)\},$$

which can be understood more easily with a plot. As we stated earlier, this set is equivalent to the piecewise linear model basis $\{1, x, (x-1)_+\}$.

An arbitrarily complex piecewise linear regression can be constructed by increasing the number of knots. Such a function $f(x)$ is a B-spline of degree 1, generated by basis functions shown at the top of Figure 9.2. Thus,

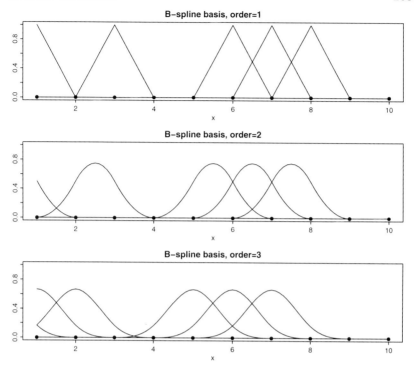

Figure 9.2 *B-spline basis functions of degrees 1, 2 and 3, where the knots are set at* $1, 2, \ldots, 10$. *The number of basis functions is the number of knots plus the degree minus 1. Although there are 10 basis functions of degree 1, but to make the plots clearer, only some are shown. To generate the basis functions at the edges, we need to extend the knots beyond the range of the data.*

in principle, we can approximate any smooth $f(x)$ by a B-spline of order 1 as long as we use a sufficient number of knots. However, when such a procedure is applied to real data, the local estimation will be dominated by noise, so the estimated function becomes unacceptably rough. This is shown in Figure 9.3, where the number of knots is increased from 3 to 5, 11 and 21. The analysis proceeds as follows:

(a) Define the knots at equal intervals covering the range of the data.

(b) Compute the design matrix X.

(c) Estimate β in the model $y = X\beta + e$.

With 21 knots, the estimated function is too rough, since it exhibits more local variation than are warranted by the noise level. A similar problem arises when using polynomial of too high a degree, but now we have one crucial difference: because of the local properties of the B-splines, *it is quite simple to impose smoothness by controlling the coefficients.*

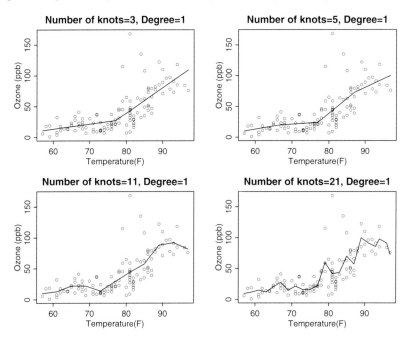

Figure 9.3 *B-splines of degree 1 with various numbers of knots.*

If we use a small number of knots, a B-spline of degree 1 might not be appealing because it is not smooth at the knots. This low-order spline is also not appropriate if we are interested in the derivatives of $f(x)$. This problem is avoided by the higher-order B-splines, where the function is guaranteed smooth by one or two derivatives continuous at the knots. Figure 9.4 shows the B-spline fits of degrees 2 and 3, using 4 and 21 knots. From this example we can see that a large number of knots may lead to serious overfitting of the data and little difference between degrees.

9.2 Mixed model framework

A general spline model with q basis functions can be written as

$$f(x) = \sum_{j=1}^{q} v_j B_j(x),$$

where the basis functions $B_j(x)$s are computed based on the knots $d_1, \ldots, d_p$. In practice, there is never any need to go beyond third degree

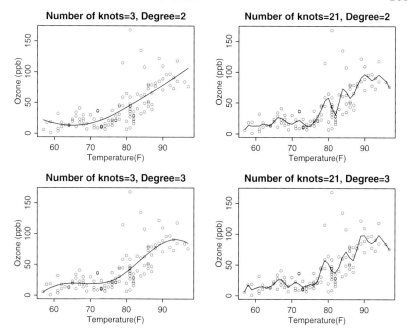

Figure 9.4 *The effects of increasing the numbers of knots for B-splines of degrees 2 and 3.*

and generally quadratic splines seem enough. The explicit formulas for the B-splines are not illuminating; de Boor (1978, Chapter 10) uses recursive formulae to compute $B_j(x)$. Statistical computing environments such as Splus and R have built-in functions to compute B-splines and their derivatives

We now put the estimation problem into a familiar mixed model framework. Given observed data $(x_1, y_1), \ldots, (x_n, y_n)$, where $E(y_i|v) = f(x_i)$, we can write the familiar regression model

$$y_i = \sum_{j=1}^{q} v_j B_j(x_i) + e_i$$

and in matrix form

$$y = Zv + e, \tag{9.2}$$

where the elements of the model matrix Z are given by

$$z_{ij} \equiv B_j(x_i).$$

We use the symbol Z instead of X to conform to our previous notation for mixed models.

The smoothness penalty term in (9.1) can be simplified:

$$\rho \int |f^{(d)}(x)|^2 dx \equiv \rho v^t P v,$$

where the (i, j) element of matrix P is

$$\int B_i^{(d)}(x) B_j^{(d)}(x) dx.$$

This formula simplifies dramatically if we use quadratic B-splines with equally spaced knots, where it can be shown (Eilers and Marx, 1996) that

$$h^2 \sum_j v_j B_j^{(2)}(x, m) = \sum_j \Delta^2 v_j B_j(x, m - 2), \qquad (9.3)$$

where h is the space between knots, $B_j(x, m)$ is a B-spline basis with an explicit degree m and Δ is the usual difference operator, such that

$$\Delta^2 v_j \equiv v_j - 2v_{j-1} + v_{j-2},$$

For $m = 2$ and $d = 2$, we arrive at the 0-degree B-spline bases, which have non-overlapping support, so

$$
\begin{aligned}
\int |f^{(2)}(x)|^2 dx &= \int |\sum_j v_j B_j^{(2)}(x, 2)|^2 dx \\
&= h^{-2} \int |\sum_j (\Delta^2 v_j) B_j(x, 0)|^2 dx \\
&= h^{-2} \sum_j |\Delta^2 v_j|^2 \int |B_j(x, 0)|^2 dx \\
&\equiv c \sum_j |\Delta^2 v_j|^2,
\end{aligned}
$$

where the constant c is determined by how $B_j(x, 0)$ is scaled — usually to integrate to one. Thus the integral penalty form reduces to simple summations. This is the penalty form used by Eilers and Marx (1996) for their penalized spline procedure, although in its general form the authors allow B-spline with arbitrary degrees of differencing. We shall adopt the same penalty here. Even without the exact correspondence with derivatives when $m \neq 2$, the difference penalty is attractive since it is easy to implement and numerically very stable. In effect, we are forcing the coefficients $v_1, \ldots, v_q$ to vary smoothly over their index. This is sensible: from the above derivation of the penalty term, if $v_1, \ldots, v_q$ vary smoothly, the resulting $f(x)$ also varies smoothly. With a second-order penalty ($d = 2$), the smoothest function obtained as $\rho \to \infty$ is a straight line; this follows directly from (9.3) and the fact that $\Delta^2 v_j \to 0$

as $\rho \to \infty$. It is rare that we need more than $d = 2$, but higher-order splines are needed if we are interested in the derivatives of the function. In log-density smoothing we might want to consider $d = 3$, since the smoothest density corresponds to the normal in that case; see Section 9.5.

Thus, the penalized least squares criterion (9.1) takes the much simpler form

$$||y - Zv||^2 + \rho \sum_j |\Delta^2 v_j|^2 \equiv ||y - Zv||^2 + \rho v^t P v, \qquad (9.4)$$

where we have rescaled ρ to absorb the constant c. This can be seen immediately in the form of pieces of the h-loglihood of a mixed model. Thus the alternatives in spline smoothing are:

- Using a few basis functions with careful placement of the knots.
- Using a relatively large number of basis functions where the knots can be set at equal intervals to put a roughness penalty on the coefficients.

Several schemes have been proposed for optimizing the numbers and the positions of the knots (e.g., Kooperberg and Stone 1992), typically applying some form of model selection method on the basis functions. Computationally this is a more demanding method than the second approach, and it does not allow a mixed model specification. In the second approach, which we are adopting here, the complexity of the computation is determined by the number of coefficients q. It is rare to need q larger than 20, so the problem is comparable to a medium-sized regression estimation.

If the data appear at equal intervals or in grid form, as is commonly observed in time series or image analysis problems, we can simplify the problem further by assuming B-splines of degree 0 with the observed xs as knots. We then have no need to set up any design matrix, and the model is simply a discretized model

$$y_i = f(x_i) + e_i \equiv f_i + e_i,$$

and the sequence of function values takes the role of coefficients. The penalized least squares formula becomes

$$\sum_i (y_i - f_i)^2 + \rho \sum_j |\Delta^2 f_j|^2 \equiv \sum_i (y_i - f_i)^2 + \rho f^t P f.$$

The advantage is that we do not have to specify any model matrix Z. In this approach f might be of large size, so the computation will need to exploit the special structure of the matrix P; see below. When the data are not in grid form, it is often advantageous to prebin the data in grid form, which is again equivalent to using a B-spline of degree 0. This method is especially useful for large data sets (Eilers 2004).

In effect, the penalty term in (9.4) specifies that the second differences are iid normal. From $v = (v_1, \ldots, v_q)$ we can obtain $(q - 2)$ second differences, so that by defining

$$
\Delta^2 v \equiv \begin{pmatrix} v_3 - 2v_2 + v_1 \\ v_4 - 2v_3 + v_2 \\ \vdots \\ v_q - 2v_{q-1} + v_{q-2} \end{pmatrix}
$$

to be normal with mean zero and variance $\sigma_v^2 I_{n-2}$, we get

$$
\sum_j |\Delta^2 v_j|^2 = v^t \{ (\Delta^2)^t \Delta^2 \} v
$$

$$
\equiv v^t P v
$$

with

$$
P \equiv (\Delta^2)^t \Delta^2 = \begin{pmatrix} 1 & -2 & 1 & & & & & 0 \\ -2 & 5 & -4 & 1 & & & & \\ 1 & -4 & 6 & -4 & 1 & & & \\ & \ddots & \ddots & \ddots & \ddots & \ddots & & \\ & & & 1 & -4 & 6 & -4 & 1 \\ & & & & 1 & -4 & 5 & -2 \\ 0 & & & & & 1 & -2 & 1 \end{pmatrix}.
$$

Note here that the precision matrix P is singular, because $P1 = 0$ with 1 as the vector of ones.

Taken together, the mixed model specifies that conditional on v, the outcome y is $N(Zv, \sigma^2 I_n)$ and v is normal with mean zero and precision matrix $\sigma_v^{-2} P$. Thus v is a singular normal, in which the joint distribution is determined only from the set of differences. The h-loglihood for the penalized least squares (PLS) method is

$$
h^P = -\frac{n}{2} \log(2\pi\sigma^2) - \frac{1}{2\sigma^2} \|y - Zv\|^2 - \frac{q-2}{2} \log(2\pi\sigma_v^2) - \frac{1}{2\sigma_v^2} v^t P v,
$$

with variance component parameter $\theta = (\sigma^2, \sigma_v^2)$. The smoothing parameter ρ is given by σ^2/σ_v^2, so a significant advantage of the mixed model approach is that we have an established procedure to estimate ρ. Given ρ, the previous formulae for mixed models apply immediately to give

$$
\hat{v} = (Z^t Z + \rho P)^{-1} Z^t y. \tag{9.5}
$$

In Section 9.4, we show how to obtain the predictor $\hat{y} = Z\hat{v}$ from the PLS method by using a random effect model. In that method, the penalty term does not belong to the model and the associated parameter is called the smoothing parameter. However, in random effect models the penalty term corresponds to the distribution of random effects which is part of the model and ρ is the ratio of variance components. Figure 9.5 shows the smoothing of the ozone data using B-splines with 21 knots and various degrees and smoothing parameters ρ. The example indicates that the choice of ρ is more important than the B-spline degree, so for conceptual simplicity first-order B-splines are often adequate.

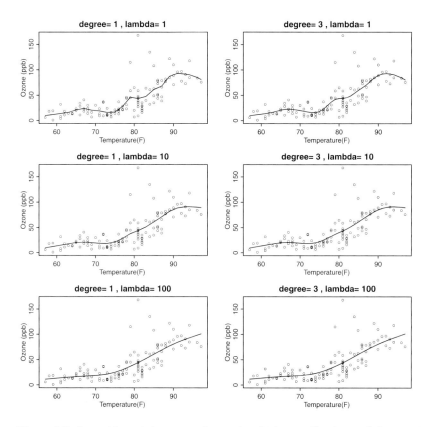

Figure 9.5 *Smoothing of the ozone data using 21 knots, B-splines of degrees 1 and 3, and smoothing parameters ρ equal to 1, 10 and 100.*

A very useful quantity to describe the complexity of a smooth technique is the so-called effective number of parameters or degrees of freedom of the fit. Recall that in the standard linear model $y = X\beta + e$, the fitted value is given by

$$\widehat{y} = X\widehat{\beta} = X(X^tX)^{-1}X^ty \equiv Hy,$$

where H is called the hat matrix. The trace of the hat matrix is given by

$$\text{trace}\{X(X^tX)^{-1}X^t\} = \text{trace}\{X^tX(X^tX)^{-1}\} = p,$$

where p is the number of linearly independent predictors in the model; Thus we have p parameters. The corresponding formula for a non–parametric smooth is

$$\widehat{y} = Z\widehat{v} = Z(Z^tZ + \rho P)^{-1}Z^ty \equiv Hy,$$

where the hat matrix is also known as the smoother matrix. The effective number of parameters of a smooth is defined as

$$\text{df} = \text{trace}(H) = \text{trace}\{(Z^tZ + \rho P)^{-1}Z^tZ\};$$

an identical quantity was also used in Chapter 6 to express the number of parameters in a model. It is a measure of model complexity (e.g. Spiegelhalter *et al.*, 2002) and a useful quantity for comparing the fits from different smoothers. In Figure 9.5, the effective parameters for the fits on the first row are (12.0, 11.4) for $\rho = 1$ and B-splines of degree 1 and 3. The corresponding values for the second and third rows are (7.2, 6.7) and (4.4, 4.3). This confirms our visual impression of little difference between B-splines of various degrees.

9.3 Automatic smoothing

The smoothing parameter ρ can be estimated via the dispersion parameters by optimizing the adjusted profile likelihood

$$p_v(h^P) = h(\theta, \widehat{v}) - \frac{1}{2}\log|I(\widehat{v})/(2\pi)|,$$

where the observed Fisher information is given by

$$I(\widehat{v}) = (\sigma^{-2}Z^tZ + \sigma_v^{-2}P).$$

The resulting procedure is known as automatic smoothing. We can employ the following iterative procedure:

1. Given (σ^2, σ_v^2) and $\rho = \sigma^2/\sigma_v^2$, estimate the B-splines coefficients by

$$\widehat{v} = (Z^tZ + \rho P)^{-1}Z^ty.$$

2. Given $\widehat{v}$, update the estimate of (σ^2, σ_v^2) by optimizing $p_v(h^P)$.

3. Iterate between 1 and 2 until convergence.

The first step is immediate, and at the start only ρ is needed. From our experience, when the signal is not very strong, $\rho = 100$ is a good starting value; since we are using a large number of knots, for the estimated function to be smooth, the signal variance should be about 100 times smaller than the noise variance. To find an explicit updating formula for the second step, first define the error vector $\widehat{e} \equiv y - Z\widehat{v}$, so that

$$\frac{\partial p_v(h^P)}{\partial \sigma^2} = -\frac{n}{2\sigma^2} + \frac{\widehat{e}^t\widehat{e}}{2\sigma^4} + \frac{1}{2\sigma^4}\mathrm{trace}\{(\sigma^{-2}Z^tZ + \sigma_v^{-2}P)^{-1}Z^tZ\}$$

$$\frac{\partial p_v(h^P)}{\partial \sigma_v^2} = -\frac{q-2}{2\sigma_v^2} + \frac{1}{2\sigma_v^4}\widehat{v}^t P\widehat{v} + \frac{1}{2\sigma_v^4}\mathrm{trace}\{(\sigma^{-2}Z^tZ + \sigma_v^{-2}P)^{-1}P\}.$$

Setting these to zero, we obtain rather simple formulas

$$\widehat{\sigma}^2 = \frac{\widehat{e}^t\widehat{e}}{n - \mathrm{df}} \tag{9.6}$$

$$\widehat{\sigma}_v^2 = \frac{\widehat{v}^t P\widehat{v}}{\mathrm{df} - 2}, \tag{9.7}$$

where, as before, the degree of freedom quantity is computed by

$$\mathrm{df} = \mathrm{trace}\{(Z^tZ + \rho P)^{-1}Z^tZ\},$$

using the previous value of ρ.

Figure 9.6 shows the automatic smoothing of the ozone data B-splines of degree 1 and 3. The estimated parameters are $(\widehat{\sigma}^2 = 487.0, \widehat{\sigma}_v^2 = 5.01)$ with $\widehat{\rho} = 97.2$ and df $= 4.4$ for the first-degree B-spline, and $(\widehat{\sigma}^2 = 487.0, \widehat{\sigma}_v^2 = 5.17)$ with $\widehat{\rho} = 94.3$ and df $= 4.3$ for the the third-degree B-spline.

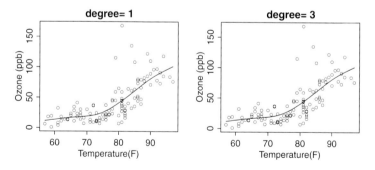

Figure 9.6 *Automatic smoothing of the ozone data using B-splines of degrees 1 and 3; the effective numbers of parameters of the fits are both 4.3.*

AIC and Generalized cross validation

Other methods to estimate the smoothing parameter ρ from the data include the AIC and generalized cross validation approach (Wahba 1979) in which we optimize

$$\text{AIC} = n \log \widehat{\sigma}^2 + 2\text{df}$$

$$\text{GCV} = \frac{\widehat{e}^t \widehat{e}}{(n - \text{df})^2},$$

where $\widehat{\sigma}^2$ and df are computed as above. Figure 9.7 shows that these two criteria are similar; in the ozone example, the optimal smoothing corresponds to about 4.7 (AIC) and 3.7 (GCV) parameters for the fits, fairly comparable to the fit using the mixed model. The advantage of the mixed model approach over the GCV is the extendibility to other response types and an immediate inference using the dispersion parameters. It is well known that the AIC criterion tends to produce less smoothing compared to GCV.

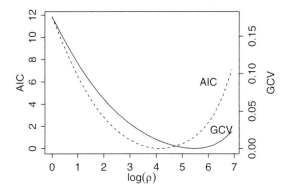

Figure 9.7 *AIC and GCV as functions of the log smoothing parameter for the ozone data using a first-degree B-spline; to make them comparable, the minima are set to zero. The functions are minimized at* $\log \rho$ *around 4.2 and 5.5.*

Confidence band

A pointwise confidence band for the non–parametric fit can be derived from

$$\text{var}(\widehat{v} - v) = I(\widehat{v})^{-1} = (\sigma^{-2} Z^t Z + \sigma_v^{-2} P)^{-1},$$

so the variance of the fitted value $\widehat{\mu} = Z\widehat{v}$ is given by

$$\text{var}(\widehat{\mu}) = Z\text{var}(\widehat{v} - v)Z^t = Z(\sigma^{-2} Z^t Z + \sigma_v^{-2} P)^{-1} Z^t$$

$$= \sigma^2 H,$$

where H is the hat matrix. Hence, 95% confidence interval for $\mu_i = f(x_i)$ is

$$\widehat{\mu}_i \pm 1.96\widehat{\sigma}\sqrt{H_{ii}},$$

where H_{ii} is the i–th diagonal element of the hat matrix. For a large data set H can be a large matrix, but only the diagonal terms are needed, and these can be obtained efficiently by computing the inner products of the corresponding rows of Z and $Z(\sigma^{-2}Z^t Z + \sigma_v^{-2}P)^{-1}$. Two aspects are not taken into account here: (a) the extra uncertainty due to estimation of the smoothing or dispersion parameters and (b) the multiplicity of the confidence intervals over many data points.

Figure 9.8 shows the 95% confidence band plotted around the automatic smoothing of the ozone data. When using a large number of points, it is inefficient to try to get an interval at every observed x. It will be sufficient to compute the confidence intervals at a subset of x values, for example, at equal intervals.

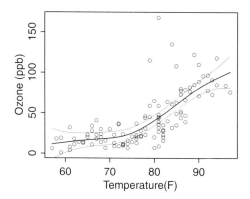

Figure 9.8 *The automatic first-degree B-spline of the ozone data with pointwise 95% confidence intervals.*

9.4 Smoothing via a model with singular precision matrix

In Section 9.3 we presented an iterative procedure for a PLS method. For simplicity of arguments, we assume $Z = I_n$ and allow some fixed effects. Many fitting methods for function estimations such as cubic smoothing splines and spatial models such as intrinsic autoregressive models use a PLS method, minimizing

$$\|y - X_1\beta_1 - v\|^2 + \rho v^t P v,$$

where $v_i = f(t_i)$ for some variable t_i. If the precision matrix P is non-singular, we can use a mixed linear model

$$y = X_1\beta_1 + v + e, \qquad (9.8)$$

with $\mathrm{cov}(v) = \Sigma = \lambda P^{-1}$ with $\rho = \phi/\lambda$ to get an equivalence fit to the PLS method. However, in PLS methods, P is often singular. Suppose P has a dimensionality $n - k$. Then there exists an $(n - k) \times n$ matrix A such that $P = A^T A$ and we may specify the distribution of v by the state transition equation $Av = r \sim N(0, \lambda I_{n-k})$. For an example in a random walk (RW) model, the state-transition equation is

$$v_t - v_{t-1} = r_t, \qquad (9.9)$$

where $r_t \sim N(0, \lambda)$.

These PLS methods have been developed as regularization tools where the penalty does not necessarily originate from the statistical model. The PLS estimators are obtained by minimizing the penalized loglihood

$$-(y - X_1\beta_1 - v)^T(y - X_1\beta_1 - v)/2\phi - v^T P v/2\lambda. \qquad (9.10)$$

To understand the log likelihood for v with a singular P, the procedure in Section 9.3 uses the log density of the completely specified part r, utilizing the relationship $Av = r$

$$
\begin{aligned}
\log f(v) \quad &\equiv \quad -\frac{(n-k)}{2}\log(2\pi\lambda) - \frac{1}{2\lambda}v^T P v \\
&= \quad -\frac{(n-k)}{2}\log(2\pi\lambda) - \frac{1}{2\lambda}r^T r.
\end{aligned}
$$

This leads to the h-loglihood for the PLS

$$h^P = -\frac{n\log(2\pi\phi)}{2} - \frac{(y - X_1\beta_1 - v)^T(y - X_1\beta_1 - v)}{2\phi} + \log f(v). \qquad (9.11)$$

By maximizing the h-likelihood h^P we obtain the estimators $\hat{\beta}_1^P$ and $\hat{v}^P$. The marginal loglihood m^P is

$$m^P = p_v(h^P) = \log \int \exp h^P dv.$$

Because m^P and h^P are proportional under the normality, $\hat{\beta}_1^P$ is the MLE maximizing m^P. The REML estimators can be obtained by maximizing the restricted loglihood $p_{\beta_1,v}(h^P) = p_{\beta_1}(m^P)$.

9.4.1 Choice of the covariance for random effects

Can this PLS fit be obtained from a mixed linear model characterized by $\mathrm{cov}(v) = \Sigma$? When P is singular, any $\Sigma \in S$ can be a possible candidate for cov(v), where

$$S = \{\lambda P^-\} = \{\Sigma | P\Sigma P = \lambda P\} = \{\Sigma | A\Sigma A^T = \lambda I_{n-k}\},$$

where P^- is a generalized inverse of P satisfying $PP^- P = P$.

Consider again a random walk (RW) model for $t = 1, 2, 3$. Suppose we impose a constraint $v_0 = 0$, to give $v_t = r_1 + \cdots + r_t$. This leads to cov$(v) = \lambda G_3$ and

$$G_3 = \begin{pmatrix} 1 & 1 & 1 \\ 1 & 2 & 2 \\ 1 & 2 & 3 \end{pmatrix}. \tag{9.12}$$

Alternatively, we can let $v_4 = 0$ to give $v_t = -r_3 - \cdots - r_t$, leading to cov$(v) = \lambda H_3$ with

$$H_3 = \begin{pmatrix} 3 & 2 & 1 \\ 2 & 2 & 1 \\ 1 & 1 & 1 \end{pmatrix}. \tag{9.13}$$

If we set $v = A^+ r$ where A^+ denotes the Moore-Penrose inverse of A. Then, cov$(v) = \lambda D_3$ with

$$D_3 = 0.111 \times \begin{pmatrix} 5 & -1 & -4 \\ -1 & 2 & -1 \\ -4 & -1 & 5 \end{pmatrix}. \tag{9.14}$$

Note that $\{\lambda D_3, \lambda H_3, \lambda G_3\} \subset S$, so that all of them are possible candidates of covariance matrices corresponding to the singular precision matrix ρP. The diagonal elements of G_3 increase with t, while those of H_3 decrease. Those of D_3 decrease and increase. Our question now is whether these mixed linear models characterized with covariance of random effects will give identical inferences. If not, how do we devise a model which gives the equivalent fit to the PLS method?

Let X_0 be some $n \times k$ matrix, satisfying $PX_0 = 0$ and give (P, X_0) a full column rank of n. For linear equations $Av = r$, a class of solutions for v is

$$v = A^- r + X_0 \beta_0,$$

for any $\beta_0 \in \mathbb{R}^k$. Thus, to get the equivalent fit to the PLS method, we need

$$E(v) = X_0 \beta_0.$$

Because it is common in mixed linear models to assume $E(v) = 0$, we should consider a mixed linear model

$$y = X\beta + v + e = X_0 \beta_0 + X_1 \beta_1 + v + e, \tag{9.15}$$

where $X = (X_0, X_1)$ and $E(v) = 0$ and cov$(v) = \Sigma$ for some $\Sigma \in S$.

Let $\hat{\beta}_0, \hat{\beta}_1$ and $\hat{v}$ be the estimators from this model. Lee and Lee (2013) showed that

$$\hat{\beta}_1 = \hat{\beta}_1^P \text{ and } \hat{v}^P = X_0\hat{\beta}_0 + \hat{v}$$

to give an identical prediction. This means that $\hat{\beta}$ and $\hat{v}$ are uniformly BLUEs and BLUPs for all $\Sigma \in S$. For the dispersion parameter estimation for $\tau = (\phi, \lambda)$, they proposed the REML estimator because it does not depend upon the choice of $\Sigma \in S$. However, the MLE for τ depends upon the choice of Σ, so that care is necessary. Following Lee $et\ al.$ (2016), we propose to use the random effect model (9.15) with $\Sigma = \lambda P^+$. With this choice we have $X_0^t \hat{v} = 0$ to allow a meaningful interpretation of β_0. Similarly, in regression models $1^t \hat{e} = 0$ gives a meaningful interpretation for the intercept.

Example 1: Consider the random walk model for $i = 1, 2$,

$$y_i = \beta_0 + v_i + e_i,$$

where $v_2 - v_1 = r_1 \sim N(0, 1)$, $E(v_i) = 0$, and $e_i \sim N(0, 1)$. The state-transition equation $v_2 - v_1 = r_1$ gives

$$A = (-1, 1), \quad P = \begin{pmatrix} 1 & -1 \\ -1 & 1 \end{pmatrix},$$

so that we have $X_0 = (1, 1)^t$. Suppose we have the responses $y = (2, 5)^t$. Consider three covariances for the random effects v where

$$G_2 = \begin{pmatrix} 1 & 1 \\ 1 & 2 \end{pmatrix}, H_2 = \begin{pmatrix} 2 & 1 \\ 1 & 1 \end{pmatrix}, \text{ and } D_2 = \frac{1}{4}\begin{pmatrix} 1 & -1 \\ -1 & 1 \end{pmatrix}.$$

$\{G_2, H_2, D_2\} \subset S$, and G_2 and H_2 are nonsingular, while D_2 is singular. Let $\hat{\beta}_0^i$ and $\hat{v}^i$ ($i = 1, 2, 3$) be BLUE and BLUP under G_2, H_2 and D_2, respectively. Then $\hat{\beta}_0^1 = 3$, $\hat{v}^1 = (0, 1)^t$, $\hat{\beta}_0^2 = 4$, $\hat{v}^2 = (-1, 0)^t$, and $\hat{\beta}_0^3 = 3.5$, $\hat{v}^3 = (-0.5, 0.5)^T$. We see that the individual estimators are not the same and see the common fit $\hat{y}^1 = \hat{y}^2 = \hat{y}^3 = (3, 4)^t$. Now consider a random walk model without an intercept for $i = 1, 2$,

$$y_i = v_i + e_i.$$

The PLS estimate is $\hat{v}^P = (3, 4)^t$, which gives the common fit $\hat{y}^P = (3, 4)^t$ above. Assume instead that we fit the random effect models under a certain covariance $\Sigma \in S$. Under G_2, $\hat{v}^1 = (1.8, 3.4)^t$ is BLUP, under H_2, $\hat{v}^2 = (2.2, 2.4)^t$ is BLUP, and under D_2, $\hat{v}^3 = (-0.5, 0.5)^t$ is BLUP, illustrating that different choices of covariance no longer lead to the common fit $\hat{y}^1 = \hat{v}^1 \neq \hat{y}^2 = \hat{v}^2 \neq \hat{y}^3 = \hat{v}^3$.

Example 2: Consider the following curve fitting problem: for $i = 1, 2, 3$,

$$y_i = f(t_i) + e_i,$$

where $t = (0.1, 0.2, 0.3)^t$, $y = (2, 4, 3)^t$ and $e_i \sim N(0, 1)$. For simplicity, we assume that $\lambda = 1$. Consider the PLS estimators first. Following Green and Silverman (1994, p. 13), P for the cubic smoothing spline is determined by t and given by

$$P = 1500 \times \begin{pmatrix} 1 & -2 & 1 \\ -2 & 4 & -2 \\ 1 & -2 & 1 \end{pmatrix}.$$

Since $P1 = Pt = 0$, we may choose $X_0 = (1, t)$. The PLS prediction is

$$\hat{v}^P = (I_3 + P)^{-1} y = (\hat{f}(t_1), \hat{f}(t_2), \hat{f}(t_3))^t = (2.5, 3, 3.5)^t.$$

Consider the following two covariances for the random effects v where

$$P^+ = \frac{1}{54000} \times \begin{pmatrix} 1 & -2 & 1 \\ -2 & 4 & -2 \\ 1 & -2 & 1 \end{pmatrix} \text{ and } \Sigma = P^+ + 0.1 X_0 X_0^T.$$

Note that $\{P^+, \Sigma\} \subset S$, and P^+ is singular, while Σ is non–singular. Let $\hat{\beta}_0^L$ and $\hat{v}^L$ be the estimators from (9.15) with $\mathrm{cov}(v) = P^+$. Then, $\hat{\beta}_0^L = (2, 5)^T$ and $\hat{v}^L = -5.55 \times 10^{-5} \times (1, -2, 1)^t$ are obtained, which are the BLUE and BLUP under P^+. For Σ, $\hat{\beta}_0$ and $\hat{v}$ are obtained from (9.15) with $\mathrm{cov}(v) = \Sigma$, and they are BLUE and BLUP under Σ. These give a common fit

$$\hat{y}^L = X_0 \hat{\beta}_0^L + \hat{v}^L = X_0 \hat{\beta}_0 + \hat{v} = (2.5, 3, 3.5)^T = \hat{v}^P.$$

Example 3: Consider the RW model (9.8) with $X_1 = (0, 0, 1)^t$ and $e \sim N(0, I_3)$. For $i = 1, 2$, $v_{i+1} - v_i = r_i \sim N(0, \lambda)$. Thus, we have $X_0 = 1$, $A = \begin{pmatrix} -1 & 1 & 0 \\ 0 & -1 & 1 \end{pmatrix}$ and $P = \begin{pmatrix} 1 & -1 & 0 \\ -1 & 2 & -1 \\ 0 & -1 & 1 \end{pmatrix}$. Suppose we have the responses $y = (2, 7, 10)^t$. We consider the PLS method and two random effect models (9.15) with $\Sigma = \lambda P^+$ and λG_3. The ML and REML estimates for λ from the random effect models are in Table 9.4.1. We see that the PLS method and the random effect model with $\mathrm{cov}(v) = \lambda P^+$ give the same ML estimate, but the model with $\mathrm{cov}(v) = \lambda G_3$ gives a different ML estimate. However, all three models give the common prediction.

We now return to the PLS method in Section 9.4, minimizing

$$||y - Zv||^2 + \rho v^t P v,$$

where P is a $q \times q$ singular precision matrix. We see that it can be fitted by using the mixed linear model,

$$y = X_0 \beta_0 + v + e, \tag{9.16}$$

where $X_0 = Z\tilde{X}_0$, $\tilde{X}_0^t P = 0$, $(\tilde{X}_0, P)$ have a full column rank q and $v \backsim N(0, P^+)$. Let h^P be the h-loglihood of the PLS method in Section 9.2 and h be for the random effect model (9.16). Then $p_v(h^P) = p_{\beta_0, v}(h)$, so that $p_v(h^P)$ gives the REML estimation for $p_{\beta_0, v}(h)$.

We see that the PLS methods implicitly assume fixed effects. Thus, for some methods the loss of information would be large due to implicitly assuming large numbers of fixed effects. However, if we use a model without these fixed effects, care is necessary because inferences depend upon a particular choice of Σ, i.e., unformly BLUP interpretation for all $\Sigma \in S$ is no longer valid.

ML and REML estimates of λ for random walk example

	ML	REML
Model (9.8) with singular precision	10.42	23
Model (9.15) with $\Sigma = \lambda P^+$	10.42	23
Model (9.15) with $\Sigma = \lambda G_3$	5.75	23

9.5 Non–Gaussian smoothing

From the linear normal mixed model framework, extension of smoothing to non–Gaussian responses is straightforward. Some of the most important areas of application include the smoothing of count data, non–parametric density and hazard estimation. On observing the independent paired data $(x_1, y_1), \ldots, (x_n, y_n)$, we assume that y_i comes from the GLM family

$$\log f(y_i) = \frac{y_i \theta_i - b(\theta_i)}{\phi} + c(y_i, \phi),$$

where $\mu_i = E(y_i)$, and assuming a link function $g()$, we have

$$g(\mu_i) = f(x_i)$$

for some unknown smooth function $f()$. As an example of non–Gaussian smoothing, consider an HGLM where conditional on random parameters v, we have

$$g(\mu) = Zv,$$

where Z is the model matrix associated with the quadratic B-spline basis functions in (9.3) and the coefficients $v = (v_1, \ldots, v_q)$ are normal with mean zero and precision matrix $\sigma_v^{-2}P$. The h-loglihood is given by

$$h(\phi, \sigma_v^2, v) = \sum_i \left\{ \frac{y_i\theta_i - b(\theta_i)}{\phi} + c(y_i, \phi) \right\} - \frac{(q-2)}{2} \log(2\pi\sigma_v^2) - \frac{1}{2\sigma_v^2} v^t P v.$$

As before, the smoothing parameter is determined by the dispersion parameters (ϕ, σ_v^2).

If $c(y_i, \phi)$ is not available, the EQL approximation in Section 3.5 will be needed to obtain an explicit formula for the likelihood. Recalling the previous formulae, the h-loglihood is given by

$$h(\phi, \sigma_v^2, v) = \sum \left\{ -\frac{1}{2} \log(\phi V(y_i)) - \frac{1}{2\phi} d(y_i, \mu_i) \right\} - \frac{q-2}{2} \log(2\pi\sigma_v^2) - \frac{1}{2\sigma_v^2} v^t P v,$$

where $d(y_i, \mu_i)$ as in Chapter 4 is given by

$$d_i \equiv d(y_i, \mu_i) = 2 \int_{\mu_i}^{y_i} \frac{y_i - u}{V(u)} du.$$

The computational methods from HGLM apply immediately. In particular, we use the following iterative algorithm:

1. Given a fixed value of (ϕ, σ_v^2), update the estimate of v using the IWLS algorithm.

2. Given $\hat{v}$, update the estimate of (ϕ, σ_v^2) by maximizing the adjusted profile likelihood $p_v(h)$.

3. Iterate between 1 and 2 until convergence.

For the first step, given a fixed value of (ϕ, σ_v^2), we compute the working vector Y with elements

$$Y_i = z_i^t v^0 + \frac{\partial g}{\partial \mu_i}(y_i - \mu_i^0),$$

where z_i is the ith row of Z. Define Σ as the diagonal matrix of the variance of the working vector with diagonal elements

$$\Sigma_{ii} = \left(\frac{\partial g}{\partial \mu_i^0} \right)^2 \phi V_i(\mu_i^0),$$

where $\phi V_i(\mu_i^0)$ is the conditional variance of y_i given v. The updating formula is the solution of

$$(Z^t \Sigma^{-1} Z + \sigma_v^{-2} P) v = Z^t \Sigma^{-1} Y.$$

Also, by analogy with the standard regression model, the quantity

$$df = \text{trace}\{ (Z^t \Sigma^{-1} Z + \sigma_v^{-2} P)^{-1} Z^t \Sigma^{-1} Z \}$$

represents the degrees of freedom or the effective number of parameters associated with the fit.

For the second step, we need to maximize the adjusted profile likelihood

$$p_v(h) = h(\phi, \sigma_v^2, \widehat{v}) - \frac{1}{2} \log |I(\widehat{v})/(2\pi)|,$$

where

$$I(\widehat{v}) = Z^t \Sigma^{-1} Z + \sigma_v^{-2} P.$$

By using the EQL approximation, we can follow the previous derivation, where given $\widehat{v}$ we compute the set of deviances d_is and update the parameters using analogous formulae to (9.6) and (9.7):

$$\widehat{\phi} = \frac{\sum_i d_i}{n - \mathrm{df}} \qquad (9.17)$$

$$\widehat{\sigma}_v^2 = \frac{v^t P v}{\mathrm{df} - 2}. \qquad (9.18)$$

As with GLMs, we might want to set the dispersion parameter ϕ to one, for example, for modelling Poisson or binomial data where we believe there is no overdispersion. If we use $\phi = 1$ when in fact $\phi > 1$, we are likely to undersmooth the data.

A pointwise confidence band for the smooth can be computed first on the linear predictor scale for $g(\mu)$, i.e.,

$$\widehat{g(\mu)} \pm 1.96 \sqrt{H_{ii}},$$

where the hat matrix H is given by

$$H = Z(Z^t \Sigma^{-1} Z + \sigma_v^{-2} P)^{-1} Z^t \Sigma^{-1};$$

it is then transformed to the μ scale.

Smoothing count data

Suppose $y_1, \ldots, y_n$ are independent count data with mean

$$\mu_i = E y_i = P_i f(x_i)$$

or

$$\log \mu_i = \log P_i + f(x_i)$$

for some smooth $f(x_i)$, where P_i is a known offset term and $\mathrm{var}(y_i) = \phi \mu_i$. The previous theory applies with a little modification in the computation of the working vector to deal with the offset:

$$Y_i = \log P_i + z_i^t v^0 + \frac{y_i - \mu_i^0}{\mu_i^0}.$$

The scattered points in Figure 9.9 are the raw rates of breast cancer per 100 person-years in Sweden in 1990. The outcome y_i is the number of breast cancers in a small age-window of length 0.1 years, and the offset term is the number of persons at risk inside the age window. For example, between the ages 40.2 and 40.3, there were $y_i = 6$ observed cancers, and 6,389 women at risk in this age group. Using the general algorithm above, the estimated dispersion parameters are $\widehat{\phi} = 0.79$ and $\widehat{\sigma}_v^2 = 0.00094$, with 4.2 effective parameters for the fit. In this example, if we set $\phi = 1$, we obtain $\widehat{\sigma}_v^2 = 0.00096$ and an indistinguishable smooth. The smoothed rate shows a flattening of risk at about the age of 47, which is about the age of menopause. (This is a well known pattern observed in almost all breast cancer data around the world.)

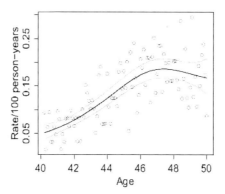

Figure 9.9 *Smoothing of breast cancer rates as a function of age. The estimated dispersion parameters are* $\widehat{\phi} = 0.79$ *and* $\widehat{\sigma}_v^2 = 0.00094$ *and the effective number of parameters of the fit is about 4.2.*

Density estimation

The histogram is an example of a non–parametric density estimate. When there is enough data, histogram is useful to convey the shape of a distribution. The weakness of a histogram is that it has too much local variability (if the bins are too small) or low resolution (if the bins are too large). We consider the smoothing of high-resolution histograms.

The kernel density estimate is often suggested when a histogram is considered too crude. Given data $x_1, \ldots, x_n$, and kernel $K(\cdot)$, the estimate of the density $f(\cdot)$ at a particular point x is

$$\widehat{f}(x) = \frac{1}{n\sigma} \sum_i K\left(\frac{x_i - x}{\sigma}\right).$$

SMOOTHING

$K(\cdot)$ is a standard density such as the normal density function; the scale parameter σ, proportional to the bandwidth of the kernel, controls the amount of smoothing. There is a large literature on kernel smoothing, particularly on the optimal choice of the bandwidth, which we shall not consider further here. There are several weaknesses in the kernel density estimate: (a) it is computationally very inefficient for large data sets, (b) finding the optimal bandwidth (or σ in the above formula) requires special techniques, and (c) there is a large bias at the boundaries. These are overcome by the mixed model approach.

First we prebin the data, so we have equispaced midpoints $x_1, \ldots, x_n$ with corresponding counts $y_1, \ldots, y_n$; there is a total of $N = \sum_i y_i$ data points. This step makes the procedure highly efficient for large data sets, but it has little effect on small data sets since the bin size can be made small. The interval Δ between points is assumed small enough that the probability of an outcome in the i–th interval is $f_i \Delta$; for convenience we set $\Delta \equiv 1$. The loglihood of $f = (f_1, \ldots, f_n)$ is based on the multinomial probability

$$\ell(f) = \sum_i y_i \log f_i,$$

where f satisfies $f_i \geq 0$ and $\sum_i f_i = 1$. Using the Lagrange multiplier technique we want an estimate f that maximizes

$$Q = \sum_i y_i \log f_i + \psi(\sum_i f_i - 1).$$

Taking the derivatives with respect to f_i we obtain

$$\frac{\partial Q}{\partial f_i} = y_i/f_i + \psi.$$

Setting $\frac{\partial Q}{\partial f_i} = 0$, so $\sum f_i(\partial Q/\partial f_i) = 0$, we find $\psi = -N$; hence f is the maximizer of

$$Q = \sum_i y_i \log f_i - N(\sum_i f_i - 1).$$

Defining $\mu_i \equiv N f_i$, the expected number of points in the i–th interval, the estimate of $\mu = (\mu_1, \ldots, \mu_n)$ is the maximizer of

$$\sum_i y_i \log \mu_i - \sum_i \mu_i,$$

exactly the loglihood from Poisson data and we no longer have to worry about the sum-to-one constraint. Computationally, non–parametric density estimation is equivalent to non–parametric smoothing of Poisson data, and the general method in the previous section applies immediately.

The setting $m = 3$ and $d = 3$ is an interesting option for smoothing (log-)densities: (a) the smoothest density is log-quadratic, so it is Gaussian, and (b) the mean and variance from the smoothed density are the same as the variance of the raw data regardless of the amount of smoothing (Eilers and Marx, 1996).

Figure 9.10 shows the scaled histogram counts scaled to integrate to one and the smoothed density estimate of the eruption time of the Old Faithful geyser, using the data provided in R. There are $N = 272$ points in the data, and we first prebin it into 81 intervals; there is very little difference if we use different numbers of bins as long as they are large enough. If we set $\phi = 1$, we get $\widehat{\sigma}_v^2 = 0.18$ and 9.3 parameters for the fit. If we allow ϕ to be estimated from the data, we obtain $\widehat{\phi} = 0.89$ and $\widehat{\sigma}_v^2 = 0.20$, and 9.6 parameters for the fit. This result is indistinguishable from the previous fit.

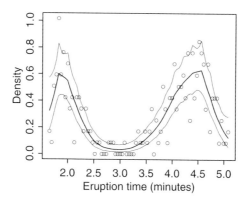

Figure 9.10 *The scaled histogram counts (circles) and smoothed density estimate (solid line) of the eruption times of the Old Faithful geyser. The pointwise 95% confidence band appears as thin lines around the estimate.*

Smoothing the hazard function

For a survival time T with with density $f(t)$ and survival distribution $S(t)$,
$$S(t) = P(T > t),$$
the hazard function is defined as
$$\lambda(t) = \frac{f(t)}{S(t)},$$
and interpreted as the rate of events among those still at risk at time t. For example, if T follows an exponential distribution with mean θ, the

hazard function of T is constant at

$$\lambda(t) = 1/\theta.$$

Survival data are naturally modeled in hazard form, and the likelihood function can be computed based on the following relationships:

$$\lambda(t) = -\frac{d \log P(T > t)}{dt} \tag{9.19}$$

$$\log P(T > t) = -\int_0^t \lambda(u) du \tag{9.20}$$

$$\log f(t) = \log \lambda(t) - \int_0^t \lambda(u) du. \tag{9.21}$$

Consider censored survival data $(y_1, \delta_1), \ldots, (y_n, \delta_n)$, where δ_i is the event indicator, and the underlying t_i has density $f_\theta(t_i)$. This underlying t_i is partially observed in the sense that $t_i = y_i$ if $\delta_i = 1$, but if $\delta_i = 0$, then only $t_i > y_i$ is known.

The loglihood contribution of (y_i, δ_i) is

$$\begin{aligned}
\log L_i &= \delta_i \log f_\theta(y_i) + (1 - \delta_i) \log P_\theta(y_i) \\
&= \delta_i \log \lambda(y_i) - \int_0^{y_i} \lambda(u) du, \tag{9.22}
\end{aligned}$$

where the parameter θ_i is absorbed by the hazard function. Thus only uncensored observations contribute to the first term.

It is instructive to follow a heuristic derivation of the likelihood via a Poisson process, since it shows how we can combine data from different individuals. First, we partition the time axis into tiny intervals of length dt, and let $y(t)$ be the number of events that fall in the interval $(t, t+dt)$. Then the time series $y(t)$ is an independent Bernoulli series with (small) probability $\lambda(t) dt$, which is approximately Poisson with mean $\lambda(t) dt$. Observing (y_i, δ_i) is equivalent to observing a series $y(t)$, which is all zero except for the last value, which is equal to δ_i. Hence, given (y_i, δ_i), we obtain the likelihood

$$\begin{aligned}
L(\theta) &= \prod_t P(Y_t = y_t) \\
&= \prod_t \exp\{-\lambda(t) dt\} \lambda(t)^{y(t)} \\
&\approx \exp\{-\sum_t \lambda(t) dt\} \lambda(y_i)^{\delta_i} \\
&\approx \exp\{-\int_0^{y_i} \lambda(t) dt\} \lambda(y_i)^{\delta_i},
\end{aligned}$$

giving the loglihood contribution

$$\ell_i = \delta_i \log \lambda(y_i) - \int_0^{y_i} \lambda(u) du,$$

as we have just seen.

Survival data from independent subjects can be combined directly to produce hazard estimates. For an interval $(t, t+dt)$ we can simply compute the number of individuals $N(t)$ still at risk in this interval, so the number of events $d(t)$ in this interval is Poisson with mean

$$\mu(t) = N(t)\lambda(t)dt.$$

This means that a non–parametric smoothing of the hazard function $\lambda(t)$ follows immediately from Poisson smoothing discussed above, simply by using $N(t)dt$ as an offset term. If the interval dt is in years, the hazard has a natural unit of the number of events per person-year. The only quantities that require a new type of computation here are the numbers at risk $N(t)$ and the numbers of events $d(t)$, but these are readily provided by many survival analysis programs. Thus, assuming that $0 < y_1 < \cdots < y_n$, the required steps are;

1. From censored survival data $(y_1, \delta_1), \ldots, (y_n, \delta_n)$, compute the series $(y_i, \Delta y_i, N(y_i), d_i)$s, where $\Delta y_i = y_i - y_{i-1}$, $y_0 \equiv 0$ and d_i is the number of events in the interval (y_{i-1}, y_i).
2. Apply Poisson smoothing using the data (y_i, d_i) with offset term $N(y_i)\Delta y_i$.

It is instructive to recognize the quantity $d_i/(N(y_i)\Delta_i)$ as the raw hazard rate. Some care is needed when the data set is large: the observed intervals Δy_i can be so small that the offset term creates unstable computation because it generates wild values of raw hazard. In this case it is sensible to combine several adjoining intervals simply by summing the corresponding outcomes and offset terms.

Figure 9.11(a) shows the Kaplan–Meier estimate of the survival distribution of 235 lymphoma patients following their diagnosis (Rosenwald et al., 2002). The average follow-up is 4.5 years, during which 133 died. The plot shows a dramatic death rate in the first two or three years of follow-up. The smoothed hazard curve in Figure 9.11(b) shows this high early mortality rate more clearly than does the survival function. For this fit, the estimated random effect variance is $\hat{\sigma}_v^2 = 0.016$, which corresponds to 3.7 parameters. (b) shows the scattered points of raw hazard rates that are too noisy to be interpretable.

Figure 9.12 shows the smoothed hazard of breast cancer in twins following the breast cancer of the index twins. At least one of the twins

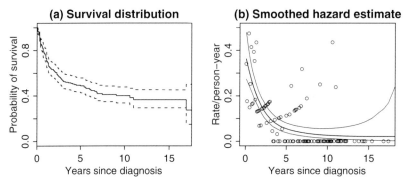

Figure 9.11 *(a) Survival distribution (solid) of 235 lymphoma patients from Rosenwald et al. (2002) with its pointwise 95% confidence band. (b) The smoothed hazard and pointwise 95% confidence band. The scattered points are the raw hazard rates.*

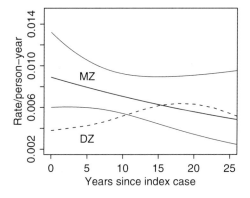

Figure 9.12 *The smoothed hazards of breast cancer of monozygotic (MZ, solid line) and dizygotic (DZ, dashed line) twins, where the follow-up time is computed from the cancer diagnosis of the index twin. The pointwise 95% confidence band for the MZ twins is given around the estimate.*

had breast cancer; the first one in calendar time is called the index case. What is of interest is the hazard function, which is an incidence rate, of breast cancer in the second twin from the time of the first breast cancer.

The data came from 659 monozygotic (MZ) twins, of whom 58 developed breast cancer during the followup, and 1253 dizygotic (DZ) twins, of whom 132 had breast cancer. For the MZ data, the estimated variance $\widehat{\sigma}_v^2 \approx 0$ with two parameters for the fit — this is the smoothest result for $d = 2$ —, and for the DZ data we obtain $\widehat{\sigma}_v^2 = 0.01$ with about 3.3

parameters for the fit. It appears that compared to the DZ twins, the MZ twins had higher rates of breast cancers in the first 10 years following the cancers of their co-twins which is evidence of genetic effects in breast cancer.

CHAPTER 10

Double HGLMs

HGLMs can be further extended by allowing additional random effects in their various components. Lee and Nelder (2006a) introduced a class of double HGLMs (DHGLMs) in which random effects can be specified in both the mean and the residual variances. Heteroscedasticity between clusters can be modelled by introducing random effects in the dispersion model as heterogeneity between clusters in the mean model. HGLMs (Chapter 6) were originally developed from an initial synthesis of GLMs, random effect models, and structured dispersion models (Chapter 7) and extended to include models for temporal and spatial correlations (Chapter 8). Now it is possible to have robust inference against outliers by allowing heavy-tailed distributions. Abrupt changes among repeated measures arising from the same subject can also be modelled by introducing random effects in the dispersion. We shall show how assumptions about skewness and kurtosis can be altered by using such random effects. Many models can be unified and extended further by the use of DHGLMs. These include models in the finance area such as autoregressive conditional heteroscedasticity (ARCH) models (Engel, 1995), generalized ARCH (GARCH), and stochastic volatility (SV) models (Harvey *et al.*, 1994). Models can be further extended by introducing random effects in the variance terms.

In the synthesis of the inferential tools needed for fitting this broad class of models, the h-likelihood (Chapter 4) plays a key role and gives a statistically and numerically efficient algorithm. The algorithm for fitting HGLMs can be easily extended for this larger class of models and requires neither prior distributions of parameters nor quadrature for integration. DHGLMs can be decomposed into a set of interlinked GLMs.

10.1 Model description

Suppose that conditional on the pair of random effects (a, u), the response y satisfies

$$E\left(y|a, u\right) = \mu \quad \text{and} \quad \text{var}\left(y|a, u\right) = \phi V\left(\mu\right),$$

where ϕ is the dispersion parameter and $V()$ is the variance function. The key extension is to introduce random effects into the component ϕ.

(a) Given u, the linear predictor for μ takes the HGLM form

$$\eta = g(\mu) = X\beta + Zv, \qquad (10.1)$$

where $g()$ is the link function, X and Z are model matrices, $v = g_M(u)$ for some monotone function $g_M()$ are the random effects and β are the fixed effects. Dispersion parameters λ for u have the HGLM form

$$\xi_M = h_M(\lambda) = G_M\gamma_M, \qquad (10.2)$$

where $h_M()$ is the link function, G_M is the model matrix and γ_M are fixed effects.

(b) Given a, the linear predictor for ϕ takes the HGLM form

$$\xi = h(\phi) = G\gamma + Fb, \qquad (10.3)$$

where $h()$ is the link function, G and F are model matrices, $b = g_D(a)$ for some monotone function $g_D()$ are the random effects and γ are the fixed effects. Dispersion parameters α for a have the GLM form with

$$\xi_D = h_D(\alpha) = G_D\gamma_D, \qquad (10.4)$$

where $h_D()$ is the link function, G_D is model matrix and γ_D are fixed effects. Here the labels M and D stand for mean and dispersion respectively.

The model matrices allow both categorical and continuous covariates. The number of component GLMs in (10.2) and (10.4) equals the number of random components in (10.1) and (10.3) respectively. DHGLMs become HGLMs with structured dispersion (Chapter 7) if $b = 0$. We can also consider models that allow random effects in (10.2) and (10.4). Consequences are similar and we shall discuss these later.

10.1.1 Models with heavy-tailed distributions

Outliers are observed in many physical and sociological phenomena, and it is well known that the normal-based models are sensitive to outliers. To have robust estimation against outliers, heavy-tailed models have often been suggested. Consider a simple linear model

$$y_{ij} = X_{ij}\beta + e_{ij} \qquad (10.5)$$

where

$$e_{ij} = \sigma_{ij}z_{ij},$$

$z_{ij} \sim N(0,1)$ and $\phi_{ij} = \sigma_{ij}^2$. Here the kurtosis of e_{ij} (or y_{ij}) is

$$E(e_{ij}^4)/\text{var}(e_{ij})^2 = 3E(\phi_{ij}^2)/E(\phi_{ij})^2 \geq 3,$$

where equality holds if and only if ϕ_{ij} are fixed constants. Thus, by introducing a random component in ϕ_{ij}, e.g.,

$$\log \phi_{ij} = \gamma + b_i, \tag{10.6}$$

we can make the distribution of e_{ij} heavier tailed than the normal.

Example 10.1: If $a_i = \exp(b_i) \sim k/\chi_k^2$, the error term $e_i = (e_{i1}, ...e_{im})^t$ follows a multivariate t-distribution (Lange *et al.*, 1989) with

$$E(\phi_{ij}) = \text{var}(y_{ij}) = k \exp(\gamma)/(k-2),$$

for $k > 2$. When $k = 1$ this becomes a Cauchy distribution. This model allows an explicit form for the marginal distribution of e_{ij}, but is restricted to a single random effect model. Such a restriction is not necessary to produce heavy tails.

In our models, we assume $\exp(b_i)$ follows an inverse gamma with $E(\exp b_i) = 1$ and $\text{var}(\exp b_i) = 1/k$, so that we have

$$E(\phi_{ij}) = \text{var}(y_{ij}) = \exp(\gamma).$$

If b_i is Gaussian, correlations can be introduced easily. The use of a multivariate t-model gives robust estimation, reducing to the normal model with $\text{var}(y_{ij}) = \exp(\gamma)$ when $k \to \infty$ or $\text{var}(b_i) = 0$. We prefer this multivariate t-model to that of (Lange *et al.*, 1989), because $\text{var}(y_{ij})$ does not depend upon the degrees of freedom, i.e., the parameterization of covariance matrix remains the same. This is true for other distributions for b_i, which reduce to a normal model when $\text{var}(b_i) = 0$. Generally, tails become heavier with $1/k$. □

By introducing random effects v_i in the mean, we can describe heterogeneity of means between clusters. By introducing random effects b_i in the dispersion we can describe effects of dispersions between clusters; these can in turn describe abrupt changes among repeated measures. However, in contrast with v_i in the model for the mean, b_i in the dispersion does not necessarily introduce correlations. Because $\text{cov}(y_{ij}, y_{il}) = 0$ for $j \neq l$ for independent z_{ij} (defined just under (10.5)), there are two alternative ways of expressing correlation in DHGLMs: by introducing a random effect in the mean linear predictor $X_{ij}\beta + v_i$ or by assuming correlations between z_{ij}. The latter is the multivariate t-model, which can be fitted by using a method similar to that in Chapter 8. The current DHGLMs allow both approaches. Furthermore, by introducing a correlation between v_i and b_i we can also allow asymmetric distributions for y_{ij}, which is important for selection of animals in breeding programs (Felleki *et al.*, 2012). However, in this book, we restrict our attention to models with independent v_i and b_i.

It is actually possible to have heavy-tailed distributions with current

HGLMs without introducing a random component in ϕ. Yun *et al.* (2006) have shown that a heavy-tailed Pareto distribution can be generated as exponential inverse gamma HGLMs, in which the random effects have variance var$(u) = 1/(\alpha - 1)$. The random effects have infinite variance when $0 < \alpha \leq 1$. The authors were nevertheless able to fit the model to Internet service times with α in this interval and found that the estimated distribution had indeed a long tail.

An advantage of using a model with heavy tails such as the t-distribution is that it has bounded influence, so that the resulting ML estimators are robust against outliers. Noh and Lee (2007b) showed that this result can be extended to GLM classes via DHGLMs. See Wang *et al.* (2016) for an extension of t-distribution to t-process.

10.1.2 Altering skewness and kurtosis

In GLMs the higher-order cumulants of y are given by the iterative formula

$$\kappa_{r+1} = \kappa_2(d\kappa_r/d\mu), \quad \text{for} \quad r \geq 2, \tag{10.7}$$

where $\kappa_2 = \phi V(\mu)$. With HGLMs, y can have different cumulants from those for the $y|u$ component. For example, the negativebinomial distribution (equivalent to a Poisson-gamma HGLM) can be shown to satisfy

$$\kappa_1 = \mu_0, \ \kappa_2 = \mu_0 + \lambda\mu_0^2 \ \text{ and } \ \kappa_{r+1} = \kappa_2(d\kappa_r/d\mu) \ \text{ for } \ r = 2, 3, \cdots ;$$

thus it still obeys the rules for GLM skewness and kurtosis, although it does not follow the form of cumulants from a one-parameter exponential family. Thus, random effects in the mean may provide different skewness and kurtosis from those for the $y|v$ component, although they may mimic similar patterns to those from a GLM family of given variance.

By introducing random effects in the model for the residual variance, we can produce models with different cumulant patterns from the GLMs. Consider DHGLMs with no random effects for the means. Let κ_i^* be cumulants for the GLM family of $y|a$ and κ_i be those for y. Then,

$$
\begin{aligned}
\kappa_1 &\equiv E(y_{ij}) = \kappa_1^* \equiv E(y_{ij}|b_i) = \mu, \\
\kappa_2^* &\equiv \text{var}(y_{ij}|b_i) = \phi V(\mu), \\
\kappa_2 &\equiv E\{\text{var}(y_{ij}|b_i)\} = E(\kappa_2^*) = E(\phi)V(\mu)
\end{aligned}
$$

so that

$$
\begin{aligned}
\kappa_3 &= E(\kappa_3^*) = E(\phi^2)V(\mu)dV(\mu)/d\mu \\
&\geq E(\phi)^2 V(\mu)dV(\mu)/d\mu \\
&= \kappa_2(d\kappa_2/d\mu), \\
\kappa_4 &= E(\kappa_4^*) + 3\mathrm{var}(\phi)V(\mu)^2 \\
&\geq E(\kappa_4^*) \\
&= \{E(\phi^2)/E(\phi)^2\}\kappa_2(d\kappa_3/d\mu) \\
&\geq \kappa_2(d\kappa_3/d\mu).
\end{aligned}
$$

Thus, higher-order cumulants no longer have the patterns of those from a GLM family.

Consider now the model (10.5) but with $X_{ij}\beta + v_i$ for the mean. Here, even if all the random components are Gaussian, y_{ij} can still have a skewed distribution because

$$
E\{y_{ij} - E(y_{ij})\}^3 = E(e_{ij}^3) + 3E(e_{ij}^2 v_i) + 3E(e_{ij}v_i^2) + E(v_i^3)
$$

is non-zero if (b_i, v_i, z_{ij}) are correlated. When $v_i = 0$, $\kappa_3 = E(e_{ij}^3) \neq 0$ if z_{ij} and ϕ_{ij} (and hence b_i) are correlated. Thus, in DHGLMs we can produce various skewed distributions by taking non-constant variance functions.

10.2 Models for finance data

Brownian motion is the Black-Scholes model for the logarithm of an asset price and much of modern financial economics is built on this model. However, from an empirical view, this assumption is far from perfect. The changes in many assets show fatter tails than the normal distribution and hence more extreme outcomes happen than would be predicted by the basic assumption. This problem has relevant economic consequences in risk management, pricing derivatives and asset allocation.

Time series models of changing variance and covariance known as volatility models are used in finance to improve the fitting of the returns on a portfolio with a normal assumption. Consider a time series y_t, where

$$
y_t = \sqrt{\phi_t} z_t
$$

with z_t a standard normal variable and ϕ_t a function of the history of the process at time t. In financial models, the responses are often mean corrected to assume null means (e.g., Kim et $al.$, 1998). The simplest autoregressive conditional heteroscedasticity of the order 1 (ARCH(1)) model (Engel, 1995) takes the form

$$\phi_t = \gamma_0^* + \gamma y_{t-1}^2.$$

This is a DHGLM with $\mu = 0$, $V(\mu) = 1$ and $b = 0$, which is in fact the JGLM described in Chapter 4. The ARCH(1) model can be extended to the generalized ARCH (GARCH) model by assuming

$$\phi_t = \gamma_0^* + \gamma_2 \phi_{t-1} + \gamma_1 y_{t-1}^2,$$

which can be written as

$$\phi_t = \gamma_0 + b_t,$$

where $\gamma_0 = \gamma_0^*/(1 - \rho)$, $\rho = \gamma_1 + \gamma_2$, $b_t = \phi_t - \gamma_0 = \rho b_{t-1} + r_t$ and $r_t = \gamma_1(y_{t-1}^2 - \phi_{t-1})$. The exponential GARCH is given by

$$\xi_t = \log \phi_t = \gamma_0 + b_t$$

where $b_t = \xi_t - \gamma_0$. Here the logarithm appears as the GLM link function for the dispersion parameter ϕ. If $r_t \sim N(0, \alpha)$, $i.e.$,

$$b_t = \rho b_{t-1} + r_t \sim AR(1)$$

becomes the popular stochastic volatility (SV) model (Harvey et $al.$, 1994).

If we take positive-valued responses y^2, all these models become mean models. For example, SV models become gamma HGLMs with temporal random effects (see Section 5.4), satisfying

$$E(y^2|b) = \phi \quad \text{and} \quad \text{var}(y^2|b) = 2\phi^2,$$

which is equivalent to assuming $y^2|b \sim \phi\chi_1^2$. Thus, the HGLM method can be used directly to fit these SV models.

Castillo and Lee (2006) showed that various Lévy models can be unified by DHGLMs, in particular, the variance gamma model introduced by Madan and Seneta (1990), the hyperbolic model introduced by Eberlein and Keller (1995), the normal inverse Gaussian model by Barndorff-Nielsen (1997), and the generalized hyperbolic model by Barndorff-Nielsen and Shephard (2001). Castillo and Lee found that the h-likelihood method is numerically more stable than the ML method because fewer runs diverge.

10.3 Joint splines

Compared to the non–parametric modelling of the mean structure in Chapter 9, non–parametric covariance modelling has received little attention. With DHGLMs, we can consider a semiparametric dispersion structure. Consider the model

$$y_i = x\beta + f_M(t_i) + e_i,$$

where $f_M()$ is an unknown smooth function and $e_i \sim N(0, \phi_i)$, with $\phi_i = \phi$. Following Section 9.4, the spline model for $f_M(t_i)$ can be obtained by fitting

$$y = x\beta + Lr + e,$$

where $r \sim N(0, \lambda I_{n-2})$ and $e \sim N(0, \phi I_n)$.

Now consider a heterogeneity in ϕ_i. Suppose

$$\log \phi_i = x\gamma + f_D(t_i)$$

with an unknown functional form for $f_D()$. This can be fitted similarly by using a model

$$\log \phi = x\gamma + La,$$

where $a \sim N(0, \alpha I_{n-2})$. In this chapter, we show how to estimate the smoothing parameters (α, λ) jointly by treating them as dispersion parameters.

10.4 H-likelihood procedure for fitting DHGLMs

For inferences from DHGLMs, we propose use of an h-loglihood in the form

$$h = \log f(y|v, b; \beta, \phi) + \log f(v; \lambda) + \log f(b; \alpha),$$

where $f(y|v, b; \beta, \phi)$, $f(v; \lambda)$ and $f(b; \alpha)$ denote the conditional density functions of y given (v, b), and those of v and b respectively. In forming the h-likelihood, we use the scales of (v, b) as those on which the random effects occur linearly on the linear predictor scale, as we always do with HGLMs. The marginal likelihood $L_{v,b}$ can be obtained from h via an integration,

$$L_{v,b} = \log \int \exp(h) dv db = \log \int \exp L_v db = \log \int \exp L_b dv,$$

where $L_v = \log \int \exp(h) dv$ and $L_b = \log \int \exp(h) db$. The marginal likelihood $L_{v,b}$ provides legitimate inferences about the fixed parameters. However, for general inferences it is not enough because it is uninformative about the unobserved random parameters (v, b).

As estimation criteria for DHGLMs Lee and Nelder (2006a) proposed the use of h for (v, β), $p_\beta(L_v)$ for (b, γ, γ_M) and $p_{\beta,\gamma}(L_{b,v})$ for γ_D. Because L_v and $L_{b,v}$ often involve intractable integration, we propose $p_{v,\beta}(h)$ and $p_{v,\beta,b,\gamma}(h)$, instead of $p_\beta(L_v)$ and $p_{\beta,\gamma}(L_{b,v})$. The whole estimation

Table 10.1 *Estimation scheme for DHGLMs.*

Criterion	Argument	Estimated	Eliminated	Approximation
h	$v, \beta, b, \gamma, \gamma_M, \gamma_D$	v, β	None	h
$p_\beta(L_v)$	$b, \gamma, \gamma_M, \gamma_D$	b, γ, γ_M	v, β	$p_{v,\beta}(h)$
$p_{\beta,\gamma}(L_{b,v})$	γ_D	γ_D	v, β, b, γ	$p_{v,\beta,b,\gamma}(h)$

scheme is summarized in Table 10.1. For the binary data, we may use $p_v(h)$ for estimating β; see Chapter 6.

The h-likelihood procedure gives statistically satisfactory and numerically stable estimations. For simulation studies, see Yun and Lee (2006) and Noh *et al.* (2005). Alternatively, we may use numerical integration such as Gauss-Hermite quadrature (GHQ) for the likelihood inferences. However, it is difficult to apply this numerical method to models with general random effect structures, for example, crossed and multivariate random effects and those with spatial and temporal correlations. Yun and Lee (2006) used the SAS NLMIXED procedure to fit a DHGLM; for the adaptive GHQ with 20 (25) quadrature points it took more than 35 (57) hours on a PC with a Pentium 4 processor and 526 Mbytes of RAM, while the h-likelihood procedure took less than 8 minutes. In 2016, it took just a few seconds.

10.4.1 *Component GLMs for DHGLMs*

Use of h-likelihood leads to the fitting of interconnected GLMs, where some are augmented. In consequence, GLMs serve as basic building blocks to define and fit DHGLMs. The GLM attributes of a DHGLM are summarized in Table 10.2, showing the overall structures of the extended models. We define components as fixed or random parameters. Each component has its own GLM, so that the development of inferential procedures for the components is straightforward. For example, if we are interested in model checking for the component ϕ (γ) in Table 10.2, we can use the procedures for the GLM having response d^*. Even further extensions are possible; for example, if we allow random effects in the λ component, the corresponding GLM becomes an augmented GLM.

Table 10.2 *GLM attributes for DHGLMs.*

	Aug. GLM	GLM	Aug. GLM	GLM
Component	β (fixed)		γ (fixed)	
Response	y		d^*	
Mean	μ		ϕ	
Variance	$\phi V(\mu)$		ϕ^2	
Link	$\eta = g(\mu)$		$\xi = h(\phi)$	
Linear predictor	$X\beta + Zv$		$G\gamma + Fb$	
Deviance	d		$\Gamma(d^*,\phi)$	
Prior weight	$1/\phi$		$(1-q)/2$	
Components	u (random)	λ (fixed)	a (random)	α (fixed)
Response	ψ_M	d_M^*	ψ_D	d_D^*
Mean	u	λ	a	α
Variance	$\lambda V_M(u)$	λ^2	$\alpha V_D(a)$	α^2
Link	$\eta_M = g_M(u)$	$\xi_M = h_M(\lambda)$	$\eta_D = g_D(a)$	$\xi_D = h_D(\alpha)$
Linear predictor	v	$G_M\gamma_M$	b	$G_D\gamma_D$
Deviance	d_M	$\Gamma(d_M^*,\lambda)$	d_D	$\Gamma(d_D^*,\alpha)$
Prior weight	$1/\lambda$	$(1-q_M)/2$	$1/\alpha$	$(1-q_D)/2$

$d_i = 2\int_{\mu_i}^{y}(y-s)/V(s)\,ds.$

$d_{Mi} = 2\int_{u_i}^{\psi_M}(\psi_M-s)/V_M(s)\,ds.$

$d_{Di} = 2\int_{a_i}^{\psi_D}(\psi_D-s)/V_D(s)\,ds.$

$d^* = d/(1-q_0).$

$d_M^* = d_M/(1-q_M).$

$d_D^* = d_D/(1-q_D).$

$\Gamma(d^*,\phi) = 2\{-\log(d^*/\phi) + (d^*-\phi)/\phi\}.$

(q, q_M, q_D) = leverages described in Section 7.2.2.

10.4.2 IWLS

Following the estimation scheme in Table 10.1, we have the following procedure:

(a) For estimating $\psi = (\gamma^t, b^t)^t$ the IWLS equations (6.7) are extended to those for an augmented GLM

$$T_D^t \Sigma_{Da}^{-1} T_D \psi = T_D^t \Sigma_{Da}^{-1} z_{Da}, \qquad (10.8)$$

where $T_D = \begin{pmatrix} G & F \\ 0 & I_q \end{pmatrix}$, $\Sigma_{Da} = \Gamma_{Da} W_{Da}^{-1}$ with $\Gamma_{Da} = \mathrm{diag}(2/(1-q), \Psi)$, $\Psi = \mathrm{diag}(\alpha_i)$ and the weight functions $W_{Da} = \mathrm{diag}(W_{D0}, W_{D1})$ are defined as

$$W_{D0i} = (\partial\phi_i/\partial\xi_i)^2/2\phi_i^2$$

for the data d_i^*, and

$$W_{D1i} = (\partial a_i/\partial b_i)^2 V_D(a_i)^{-1}$$

for the quasi–response ψ_D, while the dependent variates $z_{Da} = (z_{D0}, z_{D1})$

are defined as

$$z_{D0i} = \xi_i + (d_i^* - \phi_i)(\partial \xi_i / \partial \phi_i)$$

for the data d_i^*, and

$$z_{D1i} = b_i + (\psi_D - b_i)(\partial b_i / \partial a_i)$$

for the quasi–response ψ_D.

For example, in stochastic volatility models, $q = 0$ because there is no $(\beta^t, v^t)^t$, and

$$z_{D0i} = \xi_i + (d_i^* - \phi_i)/\phi_i$$

for log link $\xi_i = \log \phi_{ii}^*$, and

$$z_{D1i} = b_i + (\psi_D - b_i)(\partial b_i / \partial a_i) = 0$$

for $\psi_D = 0$ and $b_i = a_i$.

(b) For estimating γ_D we use the IWLS equations

$$G_D^t \Sigma_{D1}^{-1} G_D \gamma_D = G_D^t (I - Q_D) \Sigma_{D1}^{-1} z_D, \qquad (10.9)$$

where $Q_D = \mathrm{diag}(q_{Di})$, $\Sigma_{D1} = \Gamma_D W_D^{-1}, \Gamma_D = \mathrm{diag}\{2/(1 - q_D)\}$, the weight functions $W_D = \mathrm{diag}(W_{Di})$ are defined by

$$W_{Di} = (\partial \alpha_i / \partial \xi_{Di})^2 / 2\alpha_i^2,$$

the dependent variates by

$$z_{Di} = \xi_{Di} + (d_{Di}^* - \alpha_i)(\partial \xi_{Di} / \partial \alpha_i),$$

and the deviance components by

$$d_{Di} = 2 \int_{a_i}^{\psi_D} (\psi_D - s)/V_D(s)\, ds.$$

The q_D is the leverage described in Section 7.2.2. For estimating ρ, we use Lee and Nelder's (2001b) method in Section 5.4. This algorithm is equivalent to that for gamma HGLMs with responses y_t^2.

To summarize the joint estimation procedure:

(a) For estimating $\omega = (\beta^t, v^t)^t$, use the IWLS equations (7.3) in Chapter 7.

(b) For estimating γ_M, use the IWLS equations (7.4) in Chapter 7.

(c) For estimating $\psi = (\gamma^t, b^t)^t$, use the IWLS equations (10.8).

(d) For estimating γ_D, use the IWLS equations (10.9).

This completes the explanation of the fitting algorithm for DHGLMs in Table 10.2. For sparse binary data we use $p_v(h)$ for β by modifying the dependent variates in (7.3) (Noh and Lee, 2007a).

10.5 Random effects in the λ component

We can also introduce random effects in the linear predictor (10.2) of the λ component.(Lee and Noh, 2012) The use of a multivariate t-distribution for the random effects in μ is a special case, as shown in Section 10.1.1. Wakefield $et\ al.$ (1994) found in a Bayesian setting that the use of the t-distribution gave robust estimates against outliers. Noh $et\ al.$ (2005) noted that allowing random effects for the λ component eliminated the sensitivity of the parameter estimates from the choice of random effect distribution.

There has been concern about the choice of random effect distributions, because of the difficulty in identifying them from limited data, especially binary data. The non–parametric maximum likelihood (NPML) estimator can be fitted by assuming discrete latent distributions (Laird, 1978) and its use was recommended by Heckman and Singer (1984) because the parameter estimates in random effect models can be sensitive to misspecification; see also Schumacher $et\ al.$ (1987). However, its use has been restricted by the difficulties in fitting discrete latent models, for example, in choosing the number of discrete points (McLachlan, 1987) and in computing standard errors for NPML (McLachlan and Krishnan, 1997; Aitkin and Alfo, 1998). By introducing random effects into the linear predictor of the λ component we can avoid such sensitivity.

Consider a binomial HGLM, where for $i = 1, ..., n$, $y_i|v_i \sim \text{binomial}(5, p_i)$ and

$$\log\left\{\frac{p_i}{1 - p_i}\right\} = \beta + v_i. \tag{10.10}$$

In a simulation study, Noh $et\ al.$ (2005) took $n = 100,000$, $\beta = -5.0$, $\lambda = \text{var}(v_i) = 4.5$, with v_i coming from one of the following six distributions (the parameters are chosen so that they all have equal variances of 4.5):

(a) $N(0, 4.5)$.

(b) Logistic with mean 0 and variance 4.5.

(c) $N(0, \lambda_i)$, where $\log(\lambda_i) = \log(2.7) + b_i$ with $\exp(b_i) \sim 5/\chi_5^2$.

(d) $N(0, \lambda_i)$, where $\log(\lambda_i) = \log(2.7) + b_i$ with $b_i \sim N\{0, 2\log(4.5/2.7)\}$.

(e) Mixture-normal distribution (MN): a mixture of two normal distributions that produces a bimodal distribution, $0.5N(-\sqrt{3}, 1.5) + 0.5N(\sqrt{3}, 1.5)$.

(f) Gamma distribution: $\sqrt{2.25}(w_i - 2)$, where w_i follows a gamma distribution with shape parameter 2 and scale parameter 1.

As previously discussed, the distributions in cases (c) and (d) lead to heavy-tailed models for the random effects. In (c) $v_i \sim \sqrt{2.7}t(5)$, where $t(5)$ is the t-distribution with 5 degrees of freedom. Model (d) will be called the NLND (normal with log-normal dispersion) model. The distributions in (e) and (f) are bimodal and skewed distributions respectively.

Table 10.3 shows the performances of the the h-likelihood estimators based upon 200 simulated data sets (Noh $et~al.$, 2005). In the studies, we allowed the possibility that the distribution of random effects might be misspecified. For each of the true distributions of random effects that generated the data, we examined the performances of the estimates under three assumed distributions: (a) normal, (b) t and (c) NLND.

The normal random effect leads to a GLMM, in which Noh $et~al.$ (2005) found that the GHQ method yielded results almost identical to those of the h-likelihood method, so that there was no advantage in using the GHQ method. The biases of GLMM estimators have the most impact if the true distribution is skewed and are worse for the dispersion estimate. The use of heavy-tailed distributions such as t and NLND distributions avoids such sensitivity. The biases are also undesirable in the skewed case. Either t or the NLND distribution may be used to achieve robust estimation, and the two models have very similar performances.

As is well known, a robust procedure tends to perform well when the data are well behaved. This is seen in Table 10.3 when the true distribution is normal. Noh $et~al.$ (2005) studied cases for non-randomly ascertained samples, where the GLMM estimator showed a serious bias if the true distribution was not normal. However, the use of heavy-tailed models yields parameter estimates that are remarkably robust over a wide range of true distributions.

10.6 Examples

We give several examples of data analyses using DHGLMs. A conjugate DHGLM has

$$V_M(u) = V(u) \quad \text{and} \quad V_D(a) = a^2.$$

For example, the conjugate normal DHGLM is the normal-normal-inverse gamma DHGLM, in which the first distribution is for the $y|(u,a)$ component, the second for the u component and the third for the a component. Note that the inverse gamma appears as the conjugate of the gamma distribution.

Table 10.3 *Results from 200 simulations assuming normal random effects and heavy-tailed random effects in logistic variance-component model.*

Parameter	True model	Normal mean (S.D.)	t-distribution mean (S.D.)	NLND mean (S.D.)
$\beta = -5.0$	Normal	$-5.01(0.027)$	$-5.01(0.028)$	$-5.01(0.028)$
	Logistic	$-5.45(0.034)$	$-5.02(0.033)$	$-5.02(0.034)$
	$\sqrt{2.7}t(5)$	$-5.71(0.038)$	$-5.04(0.034)$	$-5.06(0.037)$
	NLND	$-5.67(0.035)$	$-5.03(0.034)$	$-5.03(0.034)$
	MN	$-4.19(0.031)$	$-4.74(0.026)$	$-4.73(0.025)$
	Gamma	$-6.62(0.049)$	$-5.63(0.038)$	$-5.61(0.038)$
$\lambda = 4.5$	Normal	$4.48(0.083)$	$4.48(0.085)$	$4.48(0.085)$
	Logistic	$5.83(0.115)$	$4.53(0.091)$	$4.52(0.090)$
	$\sqrt{2.7}t(5)$	$6.62(0.133)$	$4.57(0.094)$	$4.56(0.095)$
	NLND	$6.47(0.130)$	$4.55(0.093)$	$4.55(0.091)$
	MN	$2.96(0.065)$	$4.98(0.112)$	$4.96(0.110)$
	Gamma	$14.00(0.275)$	$4.83(0.108)$	$4.81(0.107)$

NLND = normal with log-normal dispersion.
MN = mixture normal.

10.6.1 Data on schizophrenic behaviour

Rubin and Wu (1997) analysed schizophrenic behaviour data from an eye-tracking experiment with a visual target moving back and forth along a horizontal line on a screen. The outcome measurement is called the gain ratio, which is eye velocity divided by target velocity, and it was recorded repeatedly at the peak velocity of the target during eye-tracking under three conditions. The first condition was plain sine (PS), which means the target velocity was proportional to the sine of time and the colour of the target was white. The second condition was colour sine (CS), which means the target velocity was proportional to the sine of time as for the PS, but the colours kept changing from white to orange or blue. The third condition was triangular (TR), in which the target moved at a constant speed equal to the peak velocity of PS, back and forth, but the colour remained white.

The study involved 43 non-schizophrenic subjects, 22 females and 21 males, and 43 schizophrenic subjects, 13 females and 30 males. Each subject was exposed to five trials, usually three PS, one CS, and one TR. Each trial had 11 cycles, and a gain ratio was recorded for each cycle. However, for some cycles, the gain ratios were lost because of eye blinks, so that there were on average 34 observations of 55 cycles for each subject.

For the moment, we ignore missingness and describe full treatment in Chapter 12. For observed responses y_{ij}, gain ratios for the j–th measurement of the i–th subject, first consider the following HGLM:

$$y_{ij} = \beta_0 + x_{1ij}\beta_1 + x_{2ij}\beta_2 + t_j\beta_3 + sch_i\beta_4 + sch_i \cdot x_{1ij}\beta_5 + sch_i \cdot x_{2ij}\beta_6 + v_i + e_{ij} \tag{10.11}$$

where $v_i \sim N(0, \lambda_1)$ is the random effect, $e_{ij} \sim N(0, \phi)$ is a white noise, sch_i is equal to 1 if a subject is schizophrenic and 0 otherwise, t_j is the measurement time, x_{1ij} is the effect of PS versus CS and x_{2ij} is the effect of TR versus the average of CS and PS. Rubin and Wu (1997) did not consider the time covariate t_j. However, we found this to be necessary, as will be seen later. We found that the sex effect and sex-schizophrenic interaction were not necessary in the model. Figure 10.1(a) is a normal probability plot for the residuals of the HGLM and shows large negative outliers.

Table 10.4 shows repeated measures of three schizophrenics, having the three largest outliers, which showed abrupt changes after repetition. The observations corresponding to the three largest outliers are marked by superscript a. To explain these abrupt measurements, Rubin and Wu (1997) considered an extra-component mixture model.

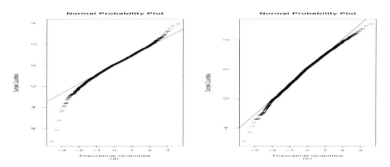

Figure 10.1 *Normal probability plots of the mean models for the schizophrenic behaviour data. (a) HGLM model (10.11). (b) DHGLM model extended to include (10.12).*

Psychological theory suggests a model in which schizophrenics suffer from attention deficits on some trials, as well as general motor reflex retardation; both aspects lead to relatively slower responses for schizophrenics, with motor retardation affecting all trials and attentional deficiency only some. Also, psychologists have known for a long time about large

Table 10.4 *Repeated measures and abrupt changes showed by three schizophrenics.*

ID	trt	1	2	3	4	5	6	7	8	9	10	11
25	PS	.916	.831	.880	.908	.951	.939	.898	.909	.939	.896	.826
	PS	.887	.900	.938	.793	.794	.935	.917	.882	.635	.849	.810
	CS	.836	.944	.889	.909	.863	.838	.844	.784	*	*	*
	PS	.739	.401[a]	.787	.753	.853	.731	.862	.882	.835	.862	.883
129	CS	.893	.702	.902	*	*	.777	*	*	*	*	*
	PS	*	*	*	.849	.774	*	*	*	*	*	.209[a]
207	PS	*	*	.862	.983	*	*	*	.822	.853	*	.827
	CS	.881	.815	.886	.519[a]	*	.657	*	.879	*	*	.881
	CS	.782	*	*	*	.840	*	.837	*	*	.797	*

$*$ =missing. [a] =abrupt change.

variations in within-schizophrenic performance on almost any task (Silverman, 1967). Thus, abrupt changes among repeated responses may be peculiar to schizophrenics and such volatility may differ for each patient. Such heteroscedasticity among schizophrenics can be now modelled by a DHGLM, introducing a random effect in the dispersion. We assume the HGLM (10.11), but conditionally on b_i, $e_{ij} \sim N(0, \phi_i)$ and

$$\log(\phi_i) = \gamma_0 + sch_i\gamma_1 + sch_ib_i, \qquad (10.12)$$

where $b_i \sim N(0, \lambda_2)$ are random effects in dispersion. Given the random effects (v_i, b_i), the repeated measurements are independent, and v_i and b_i are independent. Thus the i–th subject has a dispersion $\exp(\gamma_0 + \gamma_1 + b_i)$ if he or she is schizophrenic and $\exp(\gamma_0)$ otherwise.

Figure 10.1 (b) is the normal probability plot for the DHGLM. Many noticeable outliers in Figure 10.1(a) have disappeared. Table 10.5 shows the analysis from the DHGLM, which is very similar to that from the HGLM. However, the DHGLM gives slightly smaller standard errors, reflecting the efficiency gain from using a better model.

10.6.2 Crack growth data

Hudak *et al.* (1978) presented crack growth data, listed in Lu and Meeker (1993). Each of 21 metallic specimens was subjected to 120,000 loading cycles, with the crack lengths recorded every 10^4 cycles. We take $t =$ no. cycles$/10^6$ here, so $t_j = j/100$ for $j = 1, ..., 12$. The crack increment sequences look irregular. Let l_{ij} be the crack length of the i–th specimen at the j–th observation and let $y_{ij} = l_{ij} - l_{ij-1}$ be the corresponding increment of crack length, which always has a positive value. Models that describe the process of deterioration or degradation of units or systems are of interest and are also key ingredients in processes that model failure events. Lu and Meeker (1993) and Robinson and Crowder (2000) proposed non–linear models with normal errors. Lawless and Crowder (2004) proposed a gamma process with independent increments in discrete time. Their model is similar to the conjugate gamma HGLM, using the covariate t_j. We found that total crack size is a better covariate for crack growth, so that the resulting model has non-independent increments. From the normal probability plot in Figure 10.2(a) for the HGLM without a random component in the dispersion, we can see outliers caused by abrupt changes among repeated measures. Our final model is a conjugate gamma DHGLM with $V_M(u) = u^2$ and

Table 10.5 *Estimation results for schizophrenic behaviour data.*

Covariate		HGLM			DHGLM		
		Estimate	SE	t-value	Estimate	SE	t-value
μ	Constant	0.811	0.014	59.024	0.812	0.014	59.554
	x_1	0.006	0.005	1.420	0.003	0.004	0.800
	x_2	−0.121	0.005	−24.635	−0.117	0.004	−27.522
	$time$	−0.002	0.0004	−5.574	−0.002	0.0004	−6.137
	sch	−0.036	0.020	−1.849	−0.0372	0.0194	−1.914
	$sch \cdot x_1$	−0.0290	0.007	−4.134	−0.0195	0.0059	−3.331
	$sch \cdot x_2$	−0.007	0.008	−0.923	−0.0085	0.0066	−1.297
$\log(\phi)$	Constant	−5.194	0.027	−194.970	−5.461	0.087	−62.553
	sch				0.298	0.125	2.389
$\log(\lambda_1)$	Constant	−4.838	0.157	−30.914	−4.853	0.157	−30.980
$\log(\lambda_2)$	Constant				−1.123	0.083	−13.535

SE=standard error.

$V_D(a) = a^2$, and

$$
\begin{aligned}
\eta_{ij} &= \log \mu_{ij} = \beta_0 + l_{ij-1}\beta_l + v_i, \\
\xi_M &= \log \lambda = \gamma_M, \\
\xi_{ij} &= \log \phi_{ij} = \gamma_0 + t_j\gamma_t + b_i, \\
\xi_{Di} &= \log \alpha_i = \gamma_D.
\end{aligned}
$$

Now we want to test a hypothesis H_0: $\mathrm{var}(b_i) = 0$ (i.e., no random effects in the dispersion). Note that such a hypothesis is on the boundary of the parameter space, so the critical value for a deviance test is $\chi^2_{1,0.1} = 2.71$ for a size 0.05 test (Chernoff, 1954). Here the difference of deviance $(-2p_{v,b,\beta}(h))$ is 50.31 and thus a heavy-tail model is necessary. From Figure 10.2(b) for the DHGLM, we see that most outliers caused by abrupt changes among repeated measures disappear when we introduce a random effect in the dispersion.

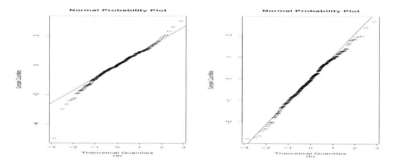

Figure 10.2 *Normal probability plots for (a) HGLM and (b) DHGLM of the crack growth data.*

Table 10.6 shows the results from the DHGLM and submodels. In this data set, the regression estimators β are insensitive to the dispersion modelling. The DHGLM has the smallest standard errors, a gain from use of proper dispersion modelling.

10.6.3 Epilepsy data

Thall and Vail (1990) presented longitudinal data from a clinical trial of 59 epileptics randomized to a new drug or a placebo ($T = 0$ or $T = 1$). Baseline data available at the start of the trial included the logarithm of the average number of epileptic seizures recorded in the 8-week period preceding the trial (B), the logarithm of age (A), and visit (V:

Table 10.6 *Summaries of analyses of crack growth data.*

	HGLM Estimate	s.e.	HGLMSD* Estimate	s.e.	DHGLM Estimate	s.e.
β_0	-5.68	0.09	-5.71	0.09	-5.67	0.08
β_l	2.38	0.07	2.41	0.07	2.36	0.05
γ_0	-3.32	0.09	-2.72	0.20	-3.00	0.20
γ_t			-10.59	2.85	-10.23	2.24
$\log \lambda$	-3.37	0.33	-3.45	0.33	-3.39	0.33
$\log \alpha$					-1.10	0.30
$-2p_{v,b,\beta}(h)$	-1507.48		-1520.76		-1571.07	

* HGLMSD=HGLM with structured dispersion. s.e.=standard error.

a linear trend, coded $(-3, -1, 1, 3)/10$. A multivariate response variable consisted of the seizure counts during 2-week periods before each of four visits to the clinic. Either random effects or extra-Poisson variation ($\phi > 1$) could explain the overdispersion among repeated measures within a subject. Thall and Vail (1990), Breslow and Clayton (1993), Diggle *et al.* (1994) and Lee and Nelder (1996) analysed these data using various Poisson HGLMs. Lee and Nelder (2000) showed that both types of overdispersion are necessary to give an adequate fit to the data. Using residual plots, they showed their final model to be better than other models they considered. However, those plots still showed apparent outliers as in Figure 10.3(a).

Consider a model in the form of a conjugate Poisson DHGLM ($V_M(u) = u$ and $V_D(a) = a^2$) as follows:

$$\eta_{ij} = \beta_0 + T\beta_T + B\beta_B + T*B\beta_{T*B} + A\beta_A + V\beta_V + v_i,$$
$$\xi_{Mi} = \log \lambda_i = \gamma_M,$$
$$\xi_{ij} = \log \phi_{ij} = \gamma_0 + B\gamma_B + b_i, \quad \text{and}$$
$$\xi_{Di} = \log \alpha_i = \gamma_D.$$

The difference in deviance for the absence of a random component $\alpha = \text{var}(b_i) = 0$ between DHGLM and HGLM with structured dispersion is 114.07, so that the component b_i is necessary. From the normal probability plot for the DHGLM (Figure 10.3(b)), we see that an apparent outlier vanishes.

Table 10.7 shows the results from the DHGLM and submodels. The regression estimator β_{T*B} is not significant at the 5% level under HGLM and quasi–HGLM, while significant under HGLM with structured disper-

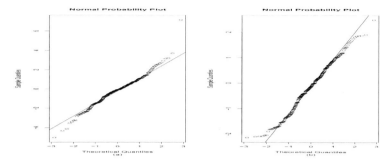

Figure 10.3 *Normal probability plots for (a) HGLM and (b) DHGLM for epilepsy data.*

sion and DHGLM. Again the DHGLM produced the smallest standard errors. Thus, proper dispersion modelling gives a more powerful result.

10.6.4 *Pound-dollar exchange rate data*

We analysed daily observations of the weekday close exchange rates for the Great Britain pound (GBP) / U.S. dollar (USD) from 1 October 1981 through 28 June 1985. We followed Harvey *et al.* (1994) in using as the response, the 936 mean-corrected returns:

$$y_t = 100 * \{\log(r_t/r_{t-1}) - \sum \log(r_i/r_{i-1})/n\},$$

where r_t denotes the exchange rate at time t. Harvey *et al.* (1994), Shephard and Pitt (1997), Kim *et al.* (1998) and Durbin and Koopman (2000) fitted the SV model

$$\log \phi_t = \gamma_0 + b_t, \tag{10.13}$$

where $b_t = \rho b_{t-1} + r_t \sim AR(1)$ with $r_t \sim N(0, \alpha)$. The efficiency of Harvey *et al.*'s (1994) estimator was improved by Shephard and Pitt (1997) using a MCMC method, and improved estimator speed by Kim *et al.* (1998). Durbin and Koopman (2000) developed an importance sampling method for both the ML and Bayesian procedures. The DHGLM estimates are

$$\log \phi_t = -0.894(0.203) + b_t, \quad \text{and} \quad \log \alpha = -3.515(0.416).$$

Table 10.8 shows the results. The parameterization

$$\sigma_t = \sqrt{\phi_t} = \kappa \exp(b_t/2), \quad \text{where} \quad \kappa = \exp(\gamma_0/2)$$

allows a clearer economic interpretation (Kim *et al.*, 1998). SV models

Table 10.7 Summary of analyses of epilepsy data.

	HGLM Estimate	s.e.	HGLMQ* Estimate	s.e.	HGLMSD* Estimate	s.e.	DHGLM Estimate	s.e.
β_0	-1.36	1.24	-2.06	0.80	-1.39	1.01	-1.47	0.89
β_B	0.88	0.14	0.92	0.09	0.88	0.11	0.88	0.10
β_T	-0.94	0.42	-0.99	0.30	-0.91	0.35	-0.91	0.31
β_{T*B}	0.34	0.21	0.40	0.14	0.33	0.17	0.34	0.15
β_A	0.48	0.36	0.69	0.23	0.50	0.30	0.52	0.26
β_V	-0.29	0.10	-0.29	0.15	-0.29	0.10	-0.29	0.10
γ_0			0.78	0.10	-0.05	0.26	-0.35	0.27
γ_B					0.48	0.14	0.40	0.14
$\log \lambda$	-1.26	0.21	-3.13	0.19	-1.77	0.23	-2.12	0.21
$\log \alpha$							-0.33	0.21
$-2p_{v,b,\beta}(h)$	1341.74		1284.31		1250.87		1136.80	

HGLMQ=quasi-HGLM. HGLMSD=HGLM with structured dispersion. s.e.=standard error.

have attracted much attention as a way of allowing clustered volatility in asset returns. Despite their intuitive appeal, SV models have been used less frequently than ARCH-GARCH models in applications. This is partly due to the difficulty associated with estimation in SV models (Shephard, 1996), where the use of marginal likelihood involves intractable integration, the integral being n-dimensional (total sample size). Thus, computationally intensive methods such as Bayesian MCMC and simulated EM algorithms have been developed.

The two previous analyses based on a Bayesian MCMC method report on the $\exp(\gamma_0/2)$ and $\sqrt{\alpha}$ scales, while we report our DHGLM procedure on the γ_0 and $\log \alpha$ scales. For comparison purposes, we use a common scale. The MCMC method assumes priors while the likelihood method does not, so that results may not be directly comparable. Note first that the two Bayesian MCMC analyses by Kim *et al.* (1998) and Shephard and Pitt (1997) are similar, and that the two likelihood analyses by our h-likelihood method and Durbin and Koopman's (2000) importance-sampling method are also similar. The table shows that the 95% confidence bound from our method contains all the other estimates. Thus, all these methods provide compatible estimates.

An advantage of DHGLM is a direct computation of standard error estimates from the Hessian matrix. Furthermore, these models for finance data can now be extended in various ways, allowing mean drift, non-constant variance functions and other operations. With other data having additional covariates such as days, weeks and months, we have found that weekly or monthly random effects are useful for modelling ϕ_t and further studies of these new models for finance data would be interesting.

Table 10.8 *Summaries of analyses of daily exchange rate data.*

| MCMC | Kim[1] | SP[2] | DK[3] | DHGLM | | |
	Estimate	Estimate	Estimate	Estimate	LI	UI
$\exp(\gamma_0/2)$	0.649	0.659	0.634	0.640	0.524	0.780
$\sqrt{\alpha}$	0.158	0.138	0.172	0.172	0.115	0.259

MCMC = Markov chain Monte Carlo.
LI = 95% lower confidence bound.
UI = 95% upper confidence bound.
[1] Kim, Shephard and Chib (1998).
[2] Shephard and Pitt (1997).
[3] Durbin and Koopman (2000).

10.6.5 Joint cubic splines

In Chapter 9 we studied non-parametric function estimation, known as smoothing, for the mean. In previous chapters, we observed that a better dispersion fit gives a better inference for the mean, so that it is natural to consider non-parametric function estimation for the dispersion. DHGLMs provide a straightforward extension.

Suppose the data are generated from a model, for $i = 1, ..., 100$

$$y_i = f_M(x_i) + \sqrt{f_D(x_i)} z_i,$$

where $z_i \sim N(0,1)$. Following Wahba (1990, page 45) we assume

$$f_M(x_i) = \exp[4.26\{\exp(x_i) - 4\exp(-2x_i) + 3\exp(-3x_i)\}]$$

and take

$$f_D(x_i) = 0.07 \exp\{-(x_i - \bar{x})^2\}.$$

In the estimation, the actual functional forms of $f_M(x_i)$ and $f_D(x_i)$ are assumed unknown. For fitting the mean and dispersion we use a DHGLM with joint splines described in Section 10.3, replacing t_i with x_i, giving

$$\mu_i = \beta_0 + x_i \beta_1 + v_i(x_i)$$

and

$$\log \phi_i = \gamma_0 + x_i \gamma_1 + b_i(x_i).$$

Curve fittings for the mean and variance for 100 observations are given in Figure 10.4. With DHGLM extensions to nonGaussian data, e.g., in the form of count or proportion, are immediate.

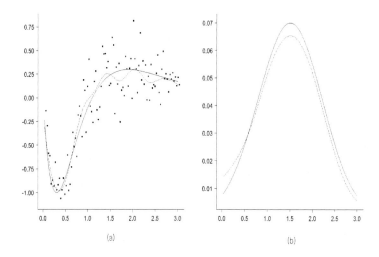

Figure 10.4 *Cubic splines for (a) mean and (b) variance. True values (solid lines) and estimates (dashed lines).*

CHAPTER 11

Variable selection and sparsity models

A key activity in regression modelling known as variable selection is the process of choosing which variables to keep and which to exclude in the final model. The trade-off is clear: with more variables we can potentially explain more of the systematic variation, but we may also bring in more noise. Keeping only the relevant variables in the model is a crucial step with several potential goals: better estimation, better prediction and better interpretation. These statistical goals are in line with a general principle in science — the so-called Occam's razor stating that among competing models of equal fit to the data, we should choose one with the fewest assumptions or parameters. When there are many potential predictors with equal status, i.e., no prior preferences among them, having as few predictors as possible in the model would often help interpretation. When we have a large number of potential predictors, overfitting also becomes a serious problem: it is far too easy to produce models that fit well only on the training or discovery data set, but not on the validation data set.

If there are several variables of special interest — let us call these exposures —, inferences on them can be improved by inclusion of the necessary covariates and degraded by inclusion of the unnecessary ones. The necessary covariates called confounders in epidemiology are correlated with both the exposures and the outcome. Their inclusion reduces both bias and residual variance. On the other hand, the unnecessary covariates are correlated only with the exposures but not with the outcome. In such cases, their inclusion does not reduce the bias of the exposure, but only leads to variance inflation. In practice, the variables of prior interest may include both exposures and well known confounders such as age and sex, but the above considerations remain. In general, when we perform statistical variable selection it should be understood that the selection is performed among variables of equal interest or status.

There are many classical methods of variable selection, often known as

313

subset selection, for example, stepwise forward selection, stepwise backward elimination or best subset selection. The literature is large and can be summarized briefly. The stepwise methods are fast and convenient, but have inferior performance compared to the best subset method. The latter is preferable, but very quickly becomes impractical because, with p predictors, we need to compare 2^p models. In general the subset selection methods are discrete and highly variable processes; they cannot be easily adapted to newer applications where the number of variables p is much greater than the sample size.

11.1 Penalized least squares

Statistically we can see immediately the need for a variable selection method from the fact that can always improve the apparent fit of a model as measured by the standard R^2 by adding pure noise as new predictors into the model. In general the standard model fit criterion alone cannot be used for subset selection. Mallow's Cp, introduced in Chapter 2, is defined as

$$Cp = RSS(p) + 2p\sigma^2,$$

where RSS is the residual sum of squares of the model, and σ^2 is the error variance. The Cp criterion can be interpreted as a penalized least squares (PLS) in Chapter 9, so that adding a new variable does not always reduce the RSS. Because it is a discrete process, model selection using Cp carries the problems stated in the previous paragraph.

Subset selection is of course not the only way to perform variable selection. Prediction accuracy is often improved by shrinking (Chapter 4) or simply setting some coefficients to zero by thresholding. In this chapter we discuss a general approach to simultaneously perform variable selection and estimation of the regression coefficients via random effect models. The problem of too many predictors leads to overfitting, so variable selection can be seen as a regularization problem. Introducing random effects provides a flexible framework and a likelihood-based solution.

Consider the regression model

$$y_i = x_i^t \beta + e_i, \quad i = 1, \ldots, n, \tag{11.1}$$

where β is a $p \times 1$ vector of fixed unknown parameters and the e_i's are iid with mean 0 and variance ϕ. Many variable selection procedures can be described as PLS estimation methods that minimizes,

$$Q_\lambda(\beta) = \frac{1}{2}\sum_{i=1}^{n}(y_i - x_i^t\beta)^2 + \sum_{j=1}^{d}p_\lambda(|\beta_j|), \qquad (11.2)$$

where $p_\lambda(\cdot)$ is a penalty function controlling model complexity.

With the L_0-penalty, namely $p_\lambda(|\beta_j|) = \lambda I(|\beta_j| \neq 0)$, the PLS becomes

$$Q_\lambda(\beta) = \frac{1}{2}\sum_{i=1}^{n}(y_i - x_i^t\beta)^2 + \lambda|M|,$$

where $M = \sum_{j=1}^{d}I(|\beta_j| \neq 0)$ denotes the size of the candidate model. Except for the tuning parameter λ, this is equivalent to the Cp criterion. To overcome the difficulties of the traditional subset selection, several other penalties have been proposed.

With the L_1-penalty, the PLS estimator becomes the least absolute shrinkage and selection operator (LASSO):

$$Q_\lambda(\beta) = \frac{1}{2}\sum_{i=1}^{n}(y_i - x_i^t\beta)^2 + \lambda\sum_{j=1}^{p}|\beta_j|,$$

which automatically sets to zero those predictors with small estimated OLS coefficients, thus performing simultaneous estimation and variable selection. Tibshirani (1996) introduced and gave a comprehensive overview of LASSO as a method of PLS. LASSO has been criticized on the ground that it typically selects too many variables to prevent over-shrinkage of the regression coefficients (Radchenko and James, 2008); otherwise, regression coefficients of selected variables are often over-shrunken. To improve LASSO, various other penalties have been proposed. Fan and Li (2001) proposed the smoothly clipped absolute deviation (SCAD) penalty for oracle estimators, and Zou (2006) proposed the adaptive LASSO. Zou and Hastie (2005) noted that the prediction performances of LASSO can be poor in cases where variable selection is ineffective; they proposed the elastic net penalty, which improves the prediction of LASSO.

With the L_2-penalty, we have

$$Q_\lambda(\beta) = \frac{1}{2}\sum_{i=1}^{n}(y_i - x_i^t\beta)^2 + \lambda\sum_{j=1}^{p}|\beta_j|^2,$$

and the PLS estimator becomes the ridge regression. In this case, all variables are kept in the model, but the resulting estimates are the shrunken versions of the OLS estimates. Ridge regression often achieves good prediction performance, but it cannot produce a parsimonious model. This can be acceptable if the model is used in a black-box fashion for prediction purposes, but it is an issue if a biological interpretation of the model is also desired.

The general version of the PLS is the penalized likelihood criterion

$$Q_\lambda(\beta) = \ell(\beta) - p_\lambda(\beta), \tag{11.3}$$

where $\ell(\beta) = \sum_{i=1}^{n} \log f_\phi(y_i|\beta)$ is the data loglihood and $p_\lambda(\beta)$ is the sparseness penalty. We can in general use variable selection of any GLM-based regression model in this framework.

11.2 Random effect variable selection

Following Lee and Oh (2009, 2014), we now describe a random effect model that generates a family of penalties, including the normal-type (bell-shaped L_2), LASSO-type (cusped L_1) and a new unbounded penalty at the origin. Suppose that conditional on u_j, we have

$$\beta_j | u_j \sim N(0, u_j\theta), \tag{11.4}$$

where θ is a fixed dispersion parameter, and u_j's are iid samples from the gamma distribution with a parameter w such that

$$f_w(u_j) = (1/w)^{1/w} \frac{1}{\Gamma(1/w)} u_j^{1/w-1} e^{-u_j/w},$$

having $\mathrm{E}(u_j) = 1$ and $\mathrm{Var}(u_j) = w$. In this random effect model, sparseness or selection is achieved in a transparent way, since if $u_j \approx 0$ then $\beta_j \approx 0$. Note that $\theta u_j = (a\theta)(u_j/a)$ for any $a > 0$, which means θ and u_j are not separately identifiable. Thus, we constrain the mean $\mathrm{E}(u_j) = 1$ for all w (Section 6.1). This imposes a constraint on random effect estimates such that $\sum_{j=1}^{p} \hat{u}_j/p = 1$.

The marginal model is heavy tailed:

$$
\begin{aligned}
f_{w,\theta}(\beta_j) &= \int f_\theta(\beta_j|u_j) f_w(u_j) du_j \\
&= \frac{w^{-1/w}}{\Gamma(1/w)\sqrt{2\pi\theta}} \int u_j^{1/w-3/2} e^{-\beta_j^2/(2u_j\theta)-u_j/w} du_j,
\end{aligned}
$$

but this is too complicated to deal with directly, and more importantly, this marginal form does not give us individual estimates of u_j's that are needed for variable selection.

Model (11.4) can be re-written as $\beta_j = \sqrt{\tau_j} e_j$, with $e_j \sim N(0,1)$ and

$$\log \tau_j = \log \theta + v_j.$$

Note that $v_j \equiv \log u_j$ is a weak canonical scale (see Chapter 6) which together with (11.1) defines a DHGLM (Chapter 10), with h-loglihood $h = h_1 + h_2$ given by

$$h_1 = \sum_{i=1}^{n} \log f_\phi(y_i|\beta) = -\frac{n}{2}\log(2\pi\phi) - \frac{1}{2\phi}\sum_{i=1}^{n}(y_i - x_i^t\beta)^2,$$

$$h_2 = \sum_{j=1}^{p}\{\log f_\theta(\beta_j|u_j) + \log f_w(v_j)\}$$

$$\log f_\theta(\beta_j|u_j) = -\frac{1}{2}\left\{\log(2\pi\theta) + \log u_j + \beta_j^2/(\theta u_j)\right\}$$

$$\log f_w(v_j) = -\log(w)/w - \log\Gamma(1/w) + v_j/w - \exp(v_j)/w.$$

The outline of the estimation scheme using IWLS as follows:

- Given (β, w, ϕ, θ), estimate u_j's by solving

$$\partial h/\partial u = 0,$$

which gives the random effect estimator

$$\widehat{u}_j \equiv \widehat{u}_j(\beta) = [\{8w\beta_j^2/\theta + (2-w)^2\}^{1/2} + (2-w)]/4. \qquad (11.5)$$

- Then, given $\widehat{u}$, we update β based on the model (11.1) with β satisfying (11.4). This is a purely random effect model $Y = X\beta + e$, where $e \sim N(0, \Sigma \equiv \text{diag}\{\phi\})$, and $\beta \sim N(0, D \equiv \text{diag}\{\widehat{u}_j\theta\})$. From the mixed model equation (5.23) in Chapter 5 we update β by solving

$$(X^t\Sigma^{-1}X + D^{-1})\beta = X^t\Sigma^{-1}y,$$

which in this case simplifies to

$$(X^tX + W_\lambda)\beta = X^ty, \qquad (11.6)$$

where $W_\lambda \equiv \text{diag}\{\lambda/\widehat{u}_j\}$ and $\lambda = \phi/\theta$.

From the model (11.4) it is clear that $\widehat{\beta}_j = 0$ when $\widehat{u}_j = 0$, which is how we achieve sparseness. We can allow thresholding by setting small $\widehat{u}_j$ to zero, but then the corresponding weight $1/\widehat{u}_j$ in W_λ is undefined. We could exclude the corresponding predictors from (11.6), but instead we employ a perturbed random effect estimate $\widehat{u}_{\delta,k} = \lambda(|\beta_k| + \delta)/|p'_\lambda(|\beta_k|)|$ for a small positive $\delta = 10^{-8}$; then, the weight is always defined. See (11.10) below for explanation of this formula. As long as δ is small, the solution is nearly identical to the original IWLS. Note that this algorithm is identical to that of Hunter and Li (2008) for the improvement of local quadratic approximation.

To complete the computations, we need estimates of the dispersion parameters (ϕ, θ), though in practice it is more common to estimate the tuning parameter λ directly using the cross validation procedure. This is discussed in Section 11.5. The parameter w is a crucial determinant of the shape of the sparseness penalty, which we shall discuss next.

11.3 Implied penalty functions

Given fixed parameters (w, ϕ, θ), the estimator of β is obtained by maximizing the profile h-loglihood

$$h_p = (h_1 + h_2)|_{u=\hat{u}},$$

where $\hat{u}$ solves $dh/du = 0$. Since h_1 is the classical loglihood, the procedure corresponds to a penalized loglihood with implied penalty

$$p_\lambda(\beta) = -\phi h_2|_{u=\hat{u}},$$

where $\hat{u}_j$ is computed in the first step of the IWLS above. Specifically, for fixed w, taking only terms that involve β_j and $\hat{u}_j$, the j–th term of the penalty function is

$$p_\lambda(\beta_j) = \frac{\phi}{2\theta} \frac{\beta_j^2}{\hat{u}_j} + \frac{\phi(w-2)}{2w} \log \hat{u}_j + \frac{\phi}{w}\hat{u}_j. \qquad (11.7)$$

Thus the random effect model leads to a family of potentially unbounded penalty functions $p_\lambda(\beta)$ indexed by w and parameterized by ϕ of (11.1) and θ of (11.4). The penalty is a statistical model with an extra parameter w, which includes the ridge and LASSO regression as special cases:

- $w \to 0$: ridge penalty
- $w = 2$: LASSO penalty
- $w > 2$: penalty with infinite value and derivative at 0

To show these, first as $w \to 0$ from (11.5) and (11.7) we get $\hat{u}_j \to 1$ and obtain the L_2-penalty

$$p_\lambda(\beta) \to \phi/(2\theta) \sum_{j=1}^{p} \beta_j^2 + \text{constant}.$$

For convenience, for this $w \to 0$ case we simply write $w = 0$. For $w = 2$, again using (11.5) and (11.7) we have $\hat{u}_j = |\beta_j|/\sqrt{\theta}$ and

$$p_\lambda(\beta) = (\phi/\sqrt{\theta}) \sum_{j=1}^{p} |\beta_j| + \text{constant}$$

becomes the L_1-penalty.

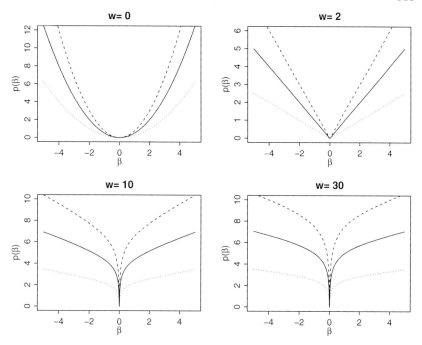

Figure 11.1 *Penalty function* $p_\lambda(\beta)$ *at different values of* w, *for* $\lambda = 1$ *(solid),* $\lambda = 1.5$ *(dashed) and* $\lambda = 0.5$ *(dotted). In general, larger values of* λ *are associated with larger penalties, hence more shrinkage and more sparseness.*

The penalty functions $p_\lambda(\cdot)$ at $w = 0$, 2, 10 and 30 with $\lambda = 1$ (solid), 1.5 (dashed) and 0.5 (dotted) are shown in Figure 11.1. As the concavity near the origin increases, the sparsity of local solutions increases, and as the slope becomes flat, the amount of shrinkage lessens. From Figure 11.1, we see that HL controls the sparsity and shrinkage amount by choosing the values of w and λ simultaneously. The form of the penalty changes from a quadratic shape ($w = 0$) for ridge regression to a cusped form ($w = 2$) for LASSO and then to an unbounded form ($w > 2$) at the origin. Quadratic penalties correspond to ridge or shrinkage estimates, which often lead to better prediction (Chapter 5), while cusped ones lead to simultaneous variable selection and estimation of LASSO and the smoothly clipped absolute deviation (SCAD) penalty (Fan, 1997). Given $w > 2$, the amount of shrinkage becomes larger as λ increases, and becomes unbiased at $\lambda = 0$ (Chapter 6). Given $\lambda > 0$, the sparsity increases as w becomes larger. By controlling sparsity and shrinkage simultaneously, the HL has much better chances of selecting the correct models without losing prediction accuracy than the other methods (Kwon *et al.*, 2016).

Singularities in LASSO and SCAD imply that their derivatives are not defined at the origin. Given λ, however, both penalties satisfy $p_\lambda(0) < \infty$ and $|p'_\lambda(0)| < \infty$, while the unbounded penalty has $p_\lambda(0) = -\infty$ and $|p'_\lambda(0)| = \infty$. The penalty $p_\lambda(|\beta_j|) = \lambda|\beta|^p$ at the origin for $0 < p < 1$ has also been considered which, although not differentiable at the origin, still has a finite range.

While the shape of the implied penalty function changes dramatically as the parameter w moves from 0 to 2, beyond $w = 10$, the shape of the penalty and the performance of the corresponding sparse procedure are largely similar. Following Lee and Oh (2014), for simplicity of arguments in this chapter we shall use $w = 30$.

11.4 Scalar β case

It is really fruitful to see the estimation in the simplest case with a single parameter β, where z is the OLS estimate of β. Even more specifically, think of β as the population mean and z the sample mean. Here we can draw the penalized criterion, and illustrate various variable selection procedures in their effects on thresholding and shrinkage of the OLS estimate to zero. The IWLS step (11.6) gives

$$\hat{\beta} = z/[1 + \lambda/\hat{u}]. \tag{11.8}$$

The corresponding PLS criterion is

$$Q_\lambda(\beta) = \frac{1}{2}(z - \beta)^2 + p_\lambda(\beta). \tag{11.9}$$

Figure 11.2 shows the penalized likelihood surfaces at different values of z. Given λ as z approaches zero (when $z \leq 2$), there is only one maximum at zero, so in this case the estimate is zero and the corresponding predictor is not selected in the model. Otherwise, bimodality occurs. In this case, from the figures, it can be seen that the likelihood surface appears to support the non-null maximum value (selecting the corresponding predictor as necessary) because a perturbation caused by the singularity at the origin is negligible. Thus, we found that this singularity at the origin does not pose a numerical difficulty in finding non–zero local maxima.

Note that the implied penalized likelihood $Q_\lambda(\beta)$ is non–convex, but the model can be expressed hierarchically as (a) $y_i|\beta$ is normal, and (b) $\beta_j|u_j$ is normal with (c) gamma u_j; all three models are convex. Thus,

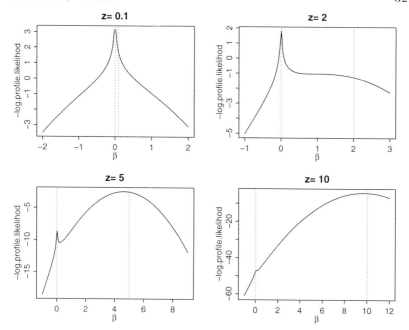

Figure 11.2 *Implied penalized log-likelihood functions equal to* $-Q_\lambda(\beta)$ *in (11.9) at different values of* z *and fixed* $\lambda = 1$.

the proposed IWLS algorithm overcomes the difficulties of a non–convex optimization by solving three interlinked convex optimizations.

Equalizing the score equations for β from (11.8) and from the PLS (11.9) we have

$$\beta(1 + \lambda/\widehat{u}) - z = \partial Q_\lambda/\partial \beta = -(z - \beta) + p'_\lambda(\beta),$$

and get a useful general formula

$$\widehat{u}(\beta) = \lambda\beta/p'_\lambda(\beta). \tag{11.10}$$

This formula allows us to obtain results for LASSO, SCAD or the so-called adaptive LASSO (Zou, 2006) by using different random effect estimates $\widehat{u}$ in the IWLS of (11.8). Examples of the penalty derivatives for some methods are given in Table 11.1. For the LASSO, $p_\lambda(\beta) = \lambda|\beta|$, so $\widehat{u} = |\beta|$. The adaptive LASSO corresponds to $p_\lambda(\beta) = 2\lambda|\beta|/|z|$, so the estimate can be obtained by IWLS using

$$\widehat{u} = |\beta||z|/2. \tag{11.11}$$

The SCAD estimate can be computed using

$$\widehat{u} = |\beta|/\left\{ I(|\beta| \le \lambda) + \frac{(a\lambda - |\beta|)_+}{(a-1)\lambda} I(|\beta| > \lambda) \right\}, \qquad (11.12)$$

for some $a > 2$.

Table 11.1 *Derivative of penalty function for some methods.*

Types	$p'_\lambda(\beta)$
LASSO	$\lambda \, \mathrm{sign}(\beta)$
SCAD	$\lambda \, \mathrm{sign}(\beta) \left\{ I(|\beta| < \lambda) + \frac{(a\lambda-|\beta|)_+}{(a-1)\lambda} I(|\beta| > \lambda) \right\}$
HL	$\lambda\beta/\{w\{(2/w - 1) + \kappa_j\}/4\}$
	where $\kappa_j = \{8\beta^2/(w\theta) + (2/w - 1)^2\}^{1/2}$

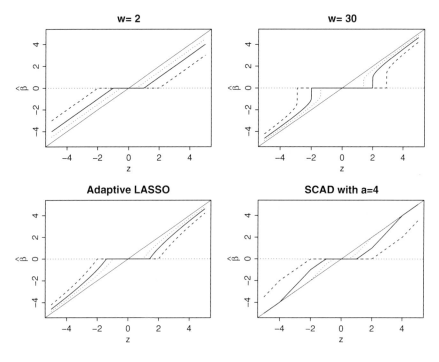

Figure 11.3 *Different IWLS solutions (11.8) as a function of z at fixed $\lambda = 1$ (solid), $\lambda = 2$ (dashed) and $\lambda = 0.5$ (dotted). The formula for $\widehat{u}$ is given by (11.5) for $w = 2$ and 30, and by (11.11) and (11.12) for the adaptive LASSO and SCAD estimates, respectively.*

Figure 11.3 provides the solutions (11.8) at $w = 2$ and 30, and the corresponding adaptive LASSO and SCAD estimates. In all cases we set the tuning parameter $\lambda = 1$ (solid), $\lambda = 2$ (dashed) and $\lambda = 0.5$ (dotted). We can see that larger values of λ are associated with more sparseness and more shrinkage. Among the methods, the random effect solution at $w = 30$ and the adaptive LASSO are closest. SCAD does not shrink when $|z|$ is large.

11.5 Estimating the dispersion and tuning parameters

In random effect models, we can use the ML or REML estimates for (w, ϕ, θ), and compute the tuning parameter λ as the ratio ϕ/θ. However, in the variable selection literature it is common to estimate the tuning parameter λ by using the K-fold cross validation method. This extra method is needed because in the PLS procedures λ is not a model parameter. However, cross validation is a useful general method and worth describing in detail. First, the error variance ϕ can be estimated from the OLS regression,

$$\hat{\phi} = \frac{1}{n-p} \sum_{i=1}^{n} (y_i - \mathbf{x}_i^t \hat{\beta}_{OLS})^2.$$

The data are split into K parts, then for a fixed value of λ,

- Use $(K-1)$ parts as a training set to estimate the sparse model using formulas (11.5) and (11.6) at parameter value λ. The parameter θ needed in (11.5) is set to $\hat{\phi}/\lambda$.
- Use the estimated model to predict the left-out part as test data.
- Repeat this procedure K times, going around each part as the test data once.
- Accumulate the prediction errors over the K parts.

When $K = n$ we get the familiar leave-one-out cross validation, which is computationally demanding, because the procedure has to be done for each value of λ. A common choice in practice is $K = 5$ or 10. To illustrate the procedure with $K = 10$, denote the full data set by $\mathcal{T}$, and the cross-validation training and test data sets by $\mathcal{T} - \mathcal{T}^s$ and $\mathcal{T}^s$ for $s = 1, 2, \ldots, 10$, respectively. For each λ, obtain the estimator $\hat{\beta}_\lambda^s$ using $\mathcal{T} - \mathcal{T}^s$ as the training data, where for clarity we made the dependency on λ explicitly. Therefore, we compute the cross-validated error

$$CV(\lambda) = \sum_{s=1}^{10} \sum_{(y_k, x_k) \in \mathcal{T}^s} (y_k - \mathbf{x}_k^t \hat{\beta}_\lambda^s)^2.$$

The optimal λ is the minimizer of $CV(\lambda)$. A one-dimensional optimization can be applied to $CV(\lambda)$, although a simple grid search could suffice.

11.6 Example: diabetes data

We first analysed a data set on the disease progression of diabetes in Efron *et al.* (2004). There were 10 predictive variables: age, sex, body mass index (bmi), blood pressure (bp), and six serum measurements (triglycerides (tc), low density lipid (ldl), hish density lipid (hdl), total cholesterol (tch), low tension glaucoma (ltg), glucose (glu)) obtained from 442 diabetes patients. The response of interest is a quantitative measure of disease progression one year after baseline. We considered a quadratic model having $p = 64$ predictive variables, including 10 original predictive variables, 45 interactions, and 9 squares, where all predictive variables were standardized. The three methods computed and compared were LASSO, SCAD and HL($w = 30$).

Table 11.2 *Analysis of diabetes data (n = 442, p = 64).*

Method	LASSO	SCAD	HL
Number of variables	15	10	11
Cross-validated error	1379.65	1373.18	1354.16

As shown in Table 11.2, the numbers of variables selected by the three methods are similar, varying from 10 to 15, though the HL method has the smallest cross-validated error. The estimated coefficients are given in Table 11.3. If we look at estimates of main effects, the LASSO estimators are shrunk the most and the SCAD estimators the least. We see that all methods include the age:sex interaction in their final model, consistent with the known result that diabetes progression behaves differently in women after menopause.

As seen in Table 11.3, the HL solution shows interaction terms of age and a quadratic term of glu, but age and glu themselves have not been chosen. This violates the functional marginality rule discussed in Section 2.1. The same problem is encountered with the other methods. With an automatic variable selection method with large p, the marginality rule will be easily violated, i.e., higher-order terms can appear without necessary lower-order interactions. A systematic way of handling such a problem is structured variable selection as we shall study.

Table 11.3 *Estimated coefficients of diabetes data.*

Variables	LASSO	SCAD	HL
sex	−122.19	−233.73	−202.12
bmi	501.16	527.46	505.65
bp	257.48	319.86	308.22
hdl	−194.22	−275.01	−255.54
ltg	468.91	491.04	492.65
glu	21.83		
age			
glu^2	73.24	58.13	111.03
bmi^2	40.92		63.61
age^2	12.64	13.75	
bmi:bp	88.14	96.33	113.16
age:sex	111.67	136.42	162.06
age:bp	30.38	9.81	48.14
age:ltg	10.69		41.06
age:glu	10.48		
sex:bp	3.96		

11.7 Numerical studies

There are several more aspects to consider when assessing a variable selection method than when assessing a standard parameter estimation method, particularly in many-dimensional cases. These aspects include variable selection performance and estimation and prediction performance. We illustrate these by comparing the empirical performances of several methods via simulation studies of the standard regression model (11.1). Several examples are considered, including cases of aggregation and sparse situations with or without grouped variables; some have been previously studied by Tibshirani (1996), Fan and Li (2001), and Zou (2006). We use $\phi^{1/2} = 3$ in all examples except 3 and 6, where we use $\phi^{1/2} = 2$ and $\phi^{1/2} = 10$, respectively. The first three examples have sample size $n = 40$, and the last three, $n = 100$.

- **Example 1.** A few large effects with $\beta = (3, 1.5, 0, 0, 2, 0, 0, 0)^t$: The predictors x_i are iid standard normal vectors. The correlation between x_i and x_j is $\rho^{|i-j|}$ with $\rho = 0.5$. The signal-to-noise ratio is approximately 5.7.

- **Example 2.** Many small effects with $\beta = (0.85, 0.85, \cdots, 0.85)^t$. The rest of the settings are the same as in Example 1. The signal-to-noise ratio is approximately 1.8. The ridge regression is expected to perform well.

- **Example 3.** Single large effect with $\beta = (5,0,0,0,0,0,0,0)^t$. The remaining settings are the same as in Example 1. This represents a typical case in which the significant predictors are very sparse. The signal-to-noise ratio is approximately 7. Example 1 may represent the middle ground between Example 2 and Example 3.

- **Example 4.** Inconsistent LASSO path with $\beta = (5.6, 5.6, 5.6, 0)^t$. The predictor variables x_i are iid $\mathrm{N}(\mathbf{0}, C)$, where

$$
C = \begin{pmatrix}
1 & \rho_1 & \rho_1 & \rho_2 \\
\rho_1 & 1 & \rho_1 & \rho_2 \\
\rho_1 & \rho_1 & 1 & \rho_2 \\
\rho_2 & \rho_2 & \rho_2 & 1
\end{pmatrix}
$$

with $\rho_1 = -0.39$ and $\rho_2 = 0.23$. Zou (2006) studied this case and showed that LASSO does not work.

- **Example 5.** A few large grouped effects. This is a relatively large problem with grouped variables. This setting is interesting because only a few grouped variables are significant such that the variables are sparse in terms of groups, but the variables within a group all show the same effects. The true coefficients are

$$
\beta = (\underbrace{0, \ldots, 0}_{10}, \underbrace{2 \ldots, 2}_{10}, \underbrace{0, \ldots, 0}_{10}, \underbrace{2, \ldots, 2}_{10})^t.
$$

The pairwise correlation between x_i and x_j is 0.5 for all i and j.

- **Example 6.** Consider Example 5 to have a low signal-to-noise ratio. Zou and Hastie (2005) introduced this case because no variable selection method works properly despite many null effects. With this example they showed that LASSO has a poor prediction performance.

For comparisons we consider the ridge ($w = 0$), LASSO ($w = 2$) and SCAD (11.12) with $a = 3.7$. The results are based on 100 simulated data sets. More complete comparisons, including more complex versions of the HL-based procedure are given in Lee and Oh (2014). For each method, the tuning parameter λ is estimated using the 10-fold cross-validation as described in Section 11.5.

Now consider several measures to evaluate the methods covering variable selection, estimation and prediction performances:

- Proportion of cases where the true models are selected (i.e., all null coefficients are correctly identified).

Table 11.4 *Proportion of selecting true model.*

Example	Ridge	LASSO	SCAD	HL
1	0.00	0.03	0.04	0.59
	(0.0)	(0.0)	(0.0)	(0.0)
2	1.00	0.53	0.54	0.00
	(0.0)	(0.0)	(0.5)	(4.0)
3	0.00	0.15	0.15	0.70
	(0.0)	(0.0)	(0.0)	(0.0)
4	0.00	0.23	0.28	0.87
	(0.0)	(0.0)	(0.0)	(0.0)
5	0.00	0.00	0.00	0.29
	(0.0)	(0.0)	(0.0)	(0.0)
6	0.00	0.00	0.00	0.00
	(0.0)	(3.0)	(2.0)	(9.0)

Note: Parentheses indicate median number of incorrect zeros.

- Median number of incorrect zeros are selected (depicting the median number of coefficients wrongly set to zeros).
- Estimation mean-square error (MSE)

$$\text{MSE}(\hat{\beta}) = \frac{1}{p} \sum_{j=1}^{p} (\beta_j - \hat{\beta}_j)^2.$$

- Prediction mean-squared error (PMSE)

$$\text{PMSE} = \frac{1}{n} \sum_{i=1}^{n} (y_i - \hat{y}_i)^2.$$

Note that $\hat{y}_i$ is a prediction based on the estimated coefficients from the training set ($n = 100$), but the outcome y_i is from an independent test set ($n = 200$).

From Tables 11.4 and 11.5, we obtain the following empirical observations:

- With regard to variable selection, the performances of LASSO and SCAD are similar, and HL improves on the other methods for all sparse cases. In Example 6, no variable selection and estimation methods worked.

Table 11.5 *Median MSEs from 100 simulated data sets.*

Example	Ridge	LASSO	SCAD	HL
1	0.26 (0.12)	0.19 (0.15)	0.19 (0.15)	0.14 (0.16)
2	0.13 (0.067)	0.27 (0.12)	0.27 (0.11)	0.57 (0.18)
3	0.14 (0.086)	0.043 (0.041)	0.043 (0.039)	0.013 (0.018)
4	0.23 (0.22)	0.21 (0.25)	0.20 (0.25)	0.13 (0.13)
5	0.15 (0.030)	0.12 (0.029)	0.12 (0.038)	0.11 (0.024)
6	0.71 (0.16)	1.12 (0.28)	1.13 (0.27)	1.85 (0.35)

Note: Parentheses indicate median absolute deviation.

Table 11.6 *Median PMSEs from test set with the.*

	Ridge	Lasso	SCAD	HL
Ex 1	9.49 (1.03)	9.32 (0.92)	9.32 (0.92)	9.29 (1.06)
Ex 2	9.40 (0.92)	9.59 (1.00)	9.59 (1.00)	10.29 (0.86)
Ex 3	4.31 (0.47)	4.14 (0.42)	4.14 (0.42)	4.09 (0.50)
Ex 4	9.57 (0.99)	9.54 (0.99)	9.54 (0.99)	9.50 (1.06)
Ex 5	11.91 (1.40)	11.72 (1.69)	11.41 (1.33)	11.50 (1.51)
Ex 6	117.44 (10.93)	125.30 (17.34)	121.72 (13.16)	138.34 (19.71)

Note: Parentheses indicate median absolute deviation.

- With regard to estimation and prediction performance, HL outperformed the others in all cases except in Examples 2 and 6, where the ridge regression produced the best results. It is possible to improve HL to match the ridge regression by allowing w to vary; see Lee and Oh (2014), Kwon *et al.* (2016) and Ng, Oh and Lee (2016). Elastic-net is a compromise between $L_1(w = 2)$, LASSO, and $L_2(w = 0)$ and Ridge, while the choice of HL penalty can be made among $0 < w < \infty$. In finite samples, Kwon *et al.* (2016) showed that HL tended to select the correct model with larger probability without losing prediction accuracy if we allowed both λ and w to vary simultaneously.

11.8 Asymptotic property of HL method

Fan and Li (2001) proved that the SCAD achieves the oracle property: the resulting estimator is asymptotically equivalent to the ordinary least squares estimator obtained by using the relevant covariates only. The results hold for HL estimators for high-dimensional models (Kwon *et al.*, 2016). However, Fan and Li 's oracle property implies that at least one of the SCAD's local solutions is consistent with the true model. Ng, Oh and Lee (2016) established the selection consistency of all local solutions of HL, which implies the uniqueness of local solutions under certain conditions. Recently, Ng, Lee and Lee (2016) reformulated the change-point problem as a PLS approach. They showed that the HL estimator achieves consistent estimation of the number of change points, and their locations and sizes, while the existing methods such as LASSO and SCAD cannot satisfy such a property.

11.9 Sparse multivariate methods

The random effect method has been applied to produce sparse versions of the classical multivariate techniques, such as the principal component analysis (PCA), canonical covariance analysis (CCA), partial least squares for Gaussian outcomes and partial least squares for survival outcomes (Lee *et al.*, 2010, Lee *et al.*, 2011a, Lee *et al.*, 2011b, Lee *et al.*, 2013). We are going to explain the method clearly for the sparse PCA (SPCA) method, but for the other techniques we refer the readers to the original papers.

Suppose X is an $n \times p$ data matrix centred across the columns, where n and p are the number of observations and the number of variables, respectively. Also, let $S_X = X^t X/n$ be the sample covariance matrix of X. In PCA, the interest is to find the linear combination $z_k = X v_k$, for $k = 1, \ldots, p$, that maximizes

$$z_k^t z_k/n = v_k^t S_X v_k, \qquad (11.13)$$

with the constraints $v_k^t v_k = 1$ and $v_k \perp v_h$ for all $h < k$. PCA can be computed through the singular-value decomposition (SVD) of X. Let the SVD of X be

$$X = UDV^t, \qquad (11.14)$$

where D is $n \times p$ matrix with singular value d_i on the (i, i)–th element and zero otherwise; the columns of $Z = UD = XV$ are the principal component scores, and the columns of the $p \times p$ matrix V are the corresponding loadings. The vector v_k in (11.13) is the k–th column of V.

Each principal component score in (11.14) is a linear combination of p loadings or variables. Simple interpretation and subsequent usage of PCA results often depend on the ability to identify subsets with non–zero loadings, but this effort is hampered by the fact that the standard PCA yields non–zero loadings on all variables. If the low-dimensional projections are relatively simple, many loadings are not statistically significant, so the non–zero values reflect the high variance of the standard method. In applications involving high-dimensional or whole-genome gene expression data, biologically we expect only a small portion of the genes to be expressed in any tissue, and an even smaller fraction to be involved in a particular process. If we want to represent the molecular variability in lower dimensions by the PCA method, sparse PCA (SPCA) becomes a necessity (Lee *et al.*, 2010).

Standard algorithms for SVD (e.g., Golub and Reinsch, 1971) give the PCA loadings, but if p is large and we only want to obtain a few singular vectors, the computation to obtain the whole set of singular vectors may be impractical. Furthermore, with these algorithms how to impose sparsity on the loadings is not obvious . Höskuldsson (1988) described the so-called NIPALS algorithm that works like a power method for obtaining the largest eigenvalue of a matrix and its associated eigenvector. The NIPALS algorithm computes only a singular vector at a time and is efficient if we only want to extract a few singular vectors. More importantly, the steps are recognizable in regression terms, so the algorithm is immediately amenable to random effect modification as needed to obtain the sparse version.

First we review the ordinary NIPALS algorithm. We set the initial value of z_1 as the first column of X, then:

1. Find v_1: $v_1 \leftarrow X^t z_1 / (z_1^t z_1)$.
2. Normalize v_1: $v_1 \leftarrow v_1 / \sqrt{v_1^t v_1}$.
3. Find z_1: $z_1 \leftarrow X^t v_1$.
4. Repeat steps 1 to 3 until convergence.

To obtain the second largest singular value and the associated vectors, we first compute residual $X_2 = X - z_1 v_1^t$, then apply the NIPALS algorithm above by replacing X by X_2.

To impose sparseness on the PCA loadings we first introduce the regression framework into step 1 of the NIPALS algorithm. Denoting X_j as the j–th column of X, we have

$$X_j = z_1 v_{1j} + e_j,$$

where v_{1j} is the j–th element of the $p \times 1$ vector v_1 (the first loading vector), and e_j is an error term that has mean zero and variance ϕ. If z_1 is assumed to be known, the OLS estimate for v_1 is given by

$$\hat{v}_1 = (z_1^t z_1)^{-1} X^t z_1,$$

exactly as given in Step 1 of NIPALS.

Now, to impose sparsity, let v_{1j} be a random variable such that

$$v_{1j}|u_j \sim N(0, u_j\theta), \qquad (11.15)$$

where θ is the dispersion parameter and u_j follows the gamma distribution with a parameter w. This corresponds to model (11.4) for the standard regression model. Given $\lambda = \phi/\theta$, the estimate of v_1 can be found using the IWLS by solving (11.6), which in this case is given by

$$(z_1^t z_1 I_p + W_\lambda)\hat{v}_1 = X^t z_1, \qquad (11.16)$$

using $W_\lambda = \text{diag}\{\lambda/\hat{u}_j\}$, and $\hat{u}$ is given by (11.5).

As for the standard regression problems, the SPCA solution from the IWLS algorithm is the optimizer of the PLS criterion

$$Q_\lambda(v_1, X) = \frac{1}{2}\text{trace}[(X - z_1 v_1^t)^t(X - z_1 v_1^t)] + \sum_{j=1}^{p} p_\lambda(|v_{1j}|), \qquad (11.17)$$

where $p_\lambda(\cdot)$ is the corresponding penalty function as we have discussed previously. Other penalty functions such as LASSO or SCAD can also be tried; computationally these are equivalent to using different formulas for $\hat{u}$ in the IWLS algorithm above; see Section 11.4.

Example: NCI-60 cancer data

For illustration we consider the so-called NCI-60 microarray data downloaded from the CellMiner program package, National Cancer Institute (http://discover.nci.nih.gov/cellminer/). Only $n = 59$ of the 60 human cancer cell lines were used in the analysis, as one of the cell lines had missing microarray information. The cell lines consisted of nine different cancers, including $1 =$ breast, $2 =$ central nervous system, $3 =$ colorectal, $4 =$ lung, $5 =$ leukemia, $6 =$ melanoma, $7 =$ ovarian, $8 =$ prostate, $9 =$ renal. The number of variables or genes measured in each sample is $p = 21{,}225$. As PCA scores may capture some underlying biological processes, we do not expect every gene in the genome to be involved. Among approximately 21,000 genes we can expect only a small fraction, probably fewer than 1,000, to be involved in a cellular process. Hence sparsity constraint can help in reducing the number of loading parameters to estimate.

In the case of PCA, we found it is possible to produce an extremely sparse version without sacrificing the total variance captured by the components. This involves modifying the observed singular values in (11.14) to reduce its variability and recomputing the data matrix X. The super sparse PCA method is then applied to this new data; the details are given in Lee *et al.* (2010). This method gives extremely sparse results, with only 1,260, 681 and 375 non–zero loadings for the first three principal components, compared to 13,259, 4,086 and 15,362 for the SPCA method.

Figure 11.4 shows the scatterplot matrix of the first three SPCA scores. Except for breast cancer, the different cancer types appear in recogniz- able clusters in the plot. This means that the sparse vector loadings capture some underlying biological differences between the cancers. The breast cancers (coded as 1) are spread across the other cancer types; this may indicate that at gene-expression level breast cancer has more diverse underlying biological processes than other cancers. Further bio- logical analyses based on the sparse non-zero loadings are likely to be more informative than some arbitrary thresholding of the standard PCA loading vectors.

11.10 Structured variable selection

In many regression problems, the explanatory variables often possess a natural group structure. For example, (a) categorical factors are often represented by a group of indicator variables, and (b) to capture flexi- ble functional shapes, continuous factors can be represented by a linear combination of basis functions such as splines or polynomials. In these situations, the problem of selecting relevant variables involves selecting groups rather than selecting individual variables. Depending on the situ- ation, the individual variables in a group may or may not be meaningful scientifically. If they are not, we are typically not interested in selecting individual variables and the interest is limited to *group selection*. How- ever, if the individual variables are meaningful, we would be interested in selecting individual variables within each selected group; we refer to this as *bi-level selection* (Huang et al., 2012).

In this section we show how the group and bi-level selections can be achieved using the random effect model approach. Specific distributional assumptions reflecting a given structure on the random effects produce a flexible class of penalties that include, for example, the group LASSO (Yuan and Lin, 2006) as a special case.

Suppose that the explanatory variables can be divided into K groups, and the outcome $y = (y_1, \ldots, y_n)^t$ has mean $\mu = (\mu_1, \ldots, \mu_n)^t$ that follows a GLM with link function $\eta_i \equiv h(\mu_i)$, such that we have a linear predictor

$$\eta = X\beta \equiv X_1\beta_1 + \cdots + X_K\beta_K, \qquad (11.18)$$

where $\eta = (\eta_1, \ldots, \eta_n)^t$ is the vector of linear predictors; $X \equiv (X_1, \ldots, X_K)$

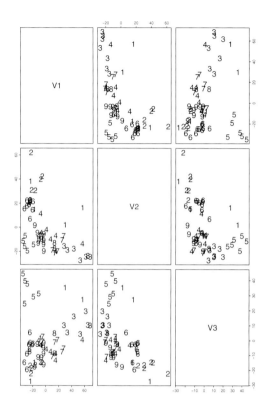

Figure 11.4 *The scatterplot matrix of the first three SPCA scores. 1 = breast, 2 = central nervous system, 3 = colorectal, 4 = lung, 5 = leukemia, 6 = melanoma, 7 = ovarian, 8 = prostate, 9 = renal. The different cancer types appear in recognizable clusters in the plot.*

is the collection of design matrices of the predictors and $\beta = (\beta_1, \ldots, \beta_K)^t$ is the vector of regression coefficients. Here, X_k is $n \times p_k$ matrix and β_k is a vector of length p_k; they are the design matrix and coefficients vector for the k–th group, respectively. Note that with this GLM specification our method immediately works with various types of data, including continuous, binary and count data.

Using the penalized approach, group variable selection can be achieved by maximizing

$$Q_\lambda(\beta) = \ell(\beta) - \sum_{k=1}^{K} J_{\lambda_k}(\|\beta_k\|_2), \qquad (11.19)$$

where $\ell(\beta) = \sum_{i=1}^{n} \log f_\phi(y_i|\beta)$ is the corresponding log-likelihood, with

$f_\phi(y_i|\beta)$ being the density function with the dispersion parameter ϕ, $\lambda_k > 0$ the regularization parameter of the k–th group and $||\cdot||_2$ the ℓ_2-norm. Yuan and Lin (2006) proposed the group LASSO, using penalty function $J_{\lambda_k}(t) = \lambda_k t$, $t > 0$. To adjust for different group sizes they chose $\lambda_k \equiv \lambda\sqrt{p_k}$, with $\lambda > 0$. Thus, the group LASSO can be constructed by applying the LASSO penalty to the ℓ_2-norm of sub-coefficients within each group. One conceptual problem with this approach is that how group selection is imposed is not obvious.

The basic random effect model in (11.4) does not consider the group structure, so it is not appropriate when we are interested in group selection. Lee, Pawitan and Lee (2015) considered a random effect model

$$\beta_{kj}|u_k \sim \mathrm{N}(0, u_k\theta), \quad k = 1, \ldots, K, \; j = 1, \ldots, p_k, \qquad (11.20)$$

and

$$u_k \sim \mathrm{gamma}(w_k), \quad k = 1, \ldots, K,$$

where, as before, θ and w_k are regularization parameters that control the degree of shrinkage and sparseness of the estimates. For a given θ, the sparsity among the groups increases as w_k's get larger, while for fixed w_k's the shrinkage becomes smaller as θ increases.

Thus the model assumes a single random effect u_k that corresponds to the k–th group. Group selection is achieved as follows: if $\hat{u}_k = 0$, then $\hat{\beta}_{kj} = 0$ for all $j = 1, \ldots, p_k$, whereas if $\hat{u}_k > 0$, then $\hat{\beta}_{kj} \neq 0$ for all $j = 1, \ldots, p_k$. This means that the model (11.20) is limited to group-only selection, as it does not impose sparsity within the selected groups.

Bi-level selection can be done by extending the model (11.20) as follows:

$$\beta_{kj}|u_k, v_{kj} \sim \mathrm{N}(0, u_k v_{kj}\theta), \quad k = 1, \ldots, K, \; j = 1, \ldots, p_k \qquad (11.21)$$

and

$$u_k \sim \mathrm{gamma}(w_k) \text{ and } v_{kj} \sim \mathrm{gamma}(\tau).$$

Here u_k is the random effect corresponding to the k–th group and v_{kj} is the random effect corresponding to the j–th variable in the k–th group. Hence this model selects variables at both the group level and the individual variable level within selected groups. For example, if $\hat{u}_k = 0$, then $\hat{\beta}_{kj} = 0$ for all $j = 1, \ldots, p_k$. However, even if $\hat{u}_k > 0$, $\hat{\beta}_{kj} = 0$ whenever $\hat{v}_{kj} = 0$.

These two random effect models illustrate one key advantage of the modeling approach, as they show how the group selection is achieved more transparently than that using other methods based directly on the penalized criterion (11.19).

Computations by the IWLS algorithm

We illustrate briefly the computations for the group-only selection; further details are given in Lee *et al.* (2015). First, the h-loglihood construction from the model (11.20) is similar as given in Section 11.2: $h = h_1 + h_2$, where

$$h_1 = \sum_{i=1}^{n} \log f_\phi(y_i|\beta)$$

$$h_2 = \sum_{k=1}^{K} \{\sum_{j=1}^{p_k} \log f_\theta(\beta_{kj}|u_k) + \log f_{w_k}(v_k)\},$$

using appropriate densities for $f_\phi(y_i|\beta)$, $f_\theta(\beta_{kj}|u_k)$ and $f_{w_k}(v_k)$, where as before $v_k = \log u_k$.

Given a fixed set of tuning parameters, the IWLS algorithm iterates between these two steps until convergence:

- Given β, let $u(\beta) \equiv (u_1(\beta_1), \ldots, u_K(\beta_K))^t$ be the solution of $\partial h/\partial u = 0$, which in this case

$$\hat{u}_k \equiv u_k(\beta_k) = \frac{1}{4}\Big\{(2 - p_k w_k) + \{(2 - p_k w_k)^2 + (8w_k/\sigma)\|\beta_k\|_2^2\}^{1/2}\Big\}.$$
(11.22)

- Given $\hat{u}_k$'s, update $\hat{\beta}$ using the WLS formula (11.6).

As for the tuning parameters, for simplicity we may set $w_k \equiv w$. However, to adjust for the different group sizes, we set $w_k \equiv w/p_k$, where w is a fixed constant. In practice the tuning parameter is then estimated using the cross-validation technique given in Section 11.5.

11.11 Interaction and hierarchy constraints

The number of interaction terms grows quadratically as a function of the number of predictors in the model, so sparse selection and estimation of the interaction terms are desirable. These terms in regression model form a natural hierarchy with the main effects, so their selection requires special considerations. For example, it is common practice that the presence of an interaction term requires both of the corresponding main effects in the model. This may be called a strong hierarchy constraint, while the weak version requires only one of the main effects to be present. This notion has also been called the heredity principle (Hamada, 1992), and is closely related to the functional marginality rule discussed in Chapter

2, e.g., if the second-order term X^2 is in the model, the first–order term X should be also. In these cases, it is sensible to select variables that respect the hierarchy. Accounting for hierarchy complicates the selection procedure, but it maintains the interpretability of the model. Our main goal now is to show a random effect model that imposes sparse estimation of interaction terms under the hierarchy constraints.

Consider a p-predictor regression model with both main and interaction terms. The outcome y has mean μ, and it follows a GLM with a link function $\eta \equiv h(\mu)$, such that we have linear predictor

$$\eta_i = \beta_0 + \sum_{j=1}^{p} x_{ij}\beta_j + \sum_{j<k} x_{ij}x_{ik}\delta_{jk}, \quad i = 1, \ldots, n,$$

which we write in matrix form as

$$\eta = X\beta + Z\delta,$$

where $\eta \equiv (\eta_1, \ldots, \eta_n)$ is the vector of linear predictors, $\beta = (\beta_1, \ldots, \beta_p)$ and $\delta \equiv (\delta_{12}, \ldots, \delta_{p-1,p})$ are the vectors of the corresponding regression coefficients for main and interaction terms, respectively. Similarly, X is the design matrix of the intercept and linear terms for the main effects, and Z is that of the cross product terms for the interactions.

Zhao et al. (2009) proposed a general LASSO-type penalty-based approach to deal with structured sparsity. Applied here, the parameters $(\beta_j, \beta_k, \delta_{jk})$ form a natural group, and their penalty would be

$$\sum_{j<k} \{|\delta_{jk}| + ||(\beta_j, \beta_k, \delta_{jk})||_\gamma\}, \quad \gamma > 1.$$

Bien et al. (2013) proposed a hierarchical LASSO with penalty

$$\sum_j \max\{|\beta_j|, ||\Delta_j||_1\} + 0.5||\Delta||_1,$$

where Δ is a $p \times p$ symmetric matrix of the interaction terms constructed from δ with zero on the diagonal, and Δ_j is the j-th column of Δ. As in the original LASSO method, sparsity is achieved because a number of variables gets zero estimated coefficients. Hierarchy is achieved since if a group is selected to be in the model, then both main effects and interaction terms are allowed to be non–zero; on the other hand, if a group is not selected, all main effects and interactions are set to zero. What is less obvious is how both main effects are free to be non-zero when the interaction is zero.

We now describe the random effect model approach for sparse estimation of interaction terms with hierarchy constraint. The key advantage is that it is obvious how the constraint is achieved; see Lee *et al.* (2015) for computational details. Recall that the *strong hierarchy constraint* means that if the model includes an interaction term, both of the corresponding main terms should also be included. This implies that if $\hat{\delta}_{kj} \neq 0$, then $\hat{\beta}_k \neq 0$ and $\hat{\beta}_j \neq 0$. For variable selection under the strong hierarchy constraint, Lee *et al.* (2015) considered the random effect model

$$\beta_j | u_j \sim N(0, u_j \theta), \quad \delta_{kj} | u_k, u_j, v_{kj} \sim N(0, u_k u_j v_{kj} \theta) \text{ for } k > j \quad (11.23)$$

and

$$u_j \sim \text{gamma}(w_1) \quad \text{and} \quad v_{kj} \sim \text{gamma}(w_2).$$

Under this model, if $\hat{u}_j = 0$ or $\hat{u}_k = 0$, then $\hat{\delta}_{jk} = 0$. Conversely, if $\hat{\delta}_{jk} \neq 0$, then $\hat{\beta}_j \neq 0$ and $\hat{\beta}_k \neq 0$ and strong hierarchy holds; see Figure 11.5. Furthermore, sparsity is achieved if these random effect estimates are zero with high probability. Thus, we have a conceptually transparent framework to impose the hierarchy constraint.

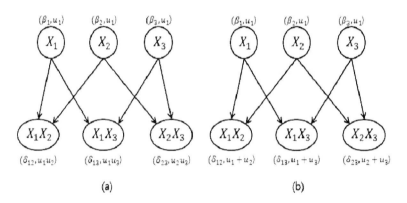

Figure 11.5 *A model with main effects and interaction terms under (a) strong hierarchy and (b) weak hierarchy constraints.*

Under the *weak hierarchy constraint* a model with an interaction term needs to include at least one of the corresponding main effects, i.e., if $\hat{\delta}_{kj} \neq 0$, then $\hat{\beta}_k \neq 0$ or $\hat{\beta}_j \neq 0$. To achieve this, consider the random effect model

$$\beta_j | u_j \sim N(0, u_j \theta),$$
$$\delta_{kj} | u_k, u_j, v_{kj} \sim N(0, (u_k + u_j) v_{kj} \theta) \quad (11.24)$$

and

$$u_j \sim \text{gamma}(w_1) \text{ and } v_{kj} \sim \text{gamma}(w_2).$$

With this model, the weak hierarchy constraint holds. If $\hat{u}_k = 0$ and

$\hat{u}_j = 0$, then $\hat{\delta}_{kj} = 0$. This means that if $\hat{\delta}_{kj} \neq 0$, then $\hat{\beta}_k \neq 0$ or $\hat{\beta}_j \neq 0$, i.e., an interaction can be included in the model if at least one corresponding main effect is included; see Figure 11.5.

Example: gene-gene interaction

As an illustration we analyse gene-gene interaction in a cohort study called ULSAM (Uppsala Longitudinal Study of Adult Men). This is an ongoing population-based study of all available men born between 1920 to 1924 in Uppsala County, Sweden. We shall analyse a subset of $n = 1 \sim 179$ subjects for which we have genetic data. These subjects were on average 71 years old, ranging from 70 to 74. The primary outcome is body-mass index (BMI), a major risk factor for many cardiovascular diseases; in this cohort, the BMI average was 26.3 and ranged from 16.7 to 46.3. Based on several criteria, we selected 10 single-nucleotide polymorphisms (SNPs) as the predictor variables; see Lee *et al.* (2015) for details and for the exact identities of the SNPs.

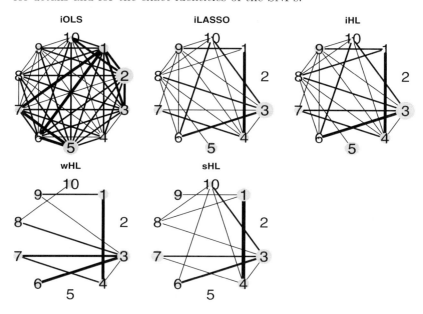

Figure 11.6 *Results from various methods applied to ULSAM data. Each node in each graph represents a SNP. The sizes of the main effects are represented by the circle size and the interactions by the thickness of the lines between two nodes.*

Figure 11.6 shows the results from various methods based on the interaction model; we used the prefix i to indicate that all pairwise interactions are included in the model. The ordinary least squares (iOLS) method estimates all the interaction terms, which are not informative; specifically, it cannot recognize the linkage disequilibrium between SNPs 1, 2 and 5. The largest interactions are between SNP pairs (1,6), (1,7) and (5,7). In contrast, all the sparse methods select (1,4) and (3,6) as the most interesting pairs. The hierarchy constrained wHL and sHL show comparable sparsity, and they both select only one of the linked SNPs 1, 2 and 5. As expected, the unconstrained iLASSO and iHL methods select interaction terms without main effects, e.g., (6,10) and (1,8). Including an interaction term without main effects can lead to misleading conclusions, since for example the interpretation depends on the coding of the variables, which is why it is not recommended in practice. If strong hierarchy is desired, the sHL method provides a sensibly sparse solution in this case.

Functional marginality and general graph structure

For completeness we describe here other statistical models in which the notion of hierarchy applies, and show how to model them using the random effects approach. Suppose we want to fit the second-order mixed polynomial model

$$\eta = X_1\beta_1 + \cdots + X_p\beta_p + X_1^2\delta_{11} + X_1X_2\delta_{12}\cdots + X_p^2\delta_{pp}, \qquad (11.25)$$

where X_kX_j denotes the component-wise product between two column vectors X_k and X_j of X. To maintain the functional marginality rule, we consider a random effect model

$$\beta_j|u_j \sim \mathrm{N}(0, u_j\theta),$$
$$\delta_{jj}|u_j, v_{jj} \sim \mathrm{N}(0, u_jv_{jj}\theta),$$
$$\delta_{kj}|u_k, u_j, v_{kj} \sim \mathrm{N}(0, u_ku_jv_{kj}\theta) \qquad (11.26)$$

and

$$u_j \sim \mathrm{gamma}(w_1) \quad \text{and} \quad v_{kj} \sim \mathrm{gamma}(w_2).$$

With this model the functional marginality rule holds. If $\hat{u}_k = 0$ then $\hat{\delta}_{kj} = 0$ for all j. For example if $\hat{\delta}_{kj} \neq 0$, then $\hat{\beta}_k \neq 0$ and $\hat{\beta}_j \neq 0$. This model is analogous to the strong hierarchy in the previous model, but now we include δ_{jj}. It is not difficult to extend the model (11.26) to general higher-order models.

Various hierarchical structures can be represented by a directed graph as in Figure 11.7. We could consider graphs with a strong or weak hierarchy constraint. In Figure 11.7(a) for strong hierarchy, X_5 can be included

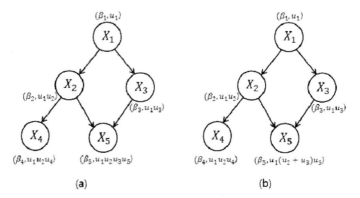

(a)   **(b)**

Figure 11.7 *The directed graph structures representing hierarchy of variables under (a) strong hierarchy and (b) weak hierarchy constraints.*

if (X_1, X_2, X_3) are included in the model. This directed graph can be modeled by the following random effect model:

$$\beta_1|u_1 \sim \mathrm{N}(0, u_1\theta), \quad \beta_2|u_1, u_2 \sim \mathrm{N}(0, u_1 u_2 \theta), \quad \beta_3|u_1, u_3 \sim \mathrm{N}(0, u_1 u_3 \theta),$$
$$\beta_4|u_1, u_2, u_4 \sim \mathrm{N}(0, u_1 u_2 u_4 \theta), \quad \beta_5|u_1, u_2, u_3, u_5 \sim \mathrm{N}(0, u_1 u_2 u_3 u_5 \theta)$$

and

$$u_j \sim \mathrm{gamma}(w), \quad j = 1, \ldots, 5.$$

In Figure 11.7(b) for weak hierarchy, X_5 can be included if the model includes, besides X_1, at least one of X_2 and X_3. This directed graph can be modeled by

$$\beta_5|u_1, u_2, u_3, u_5 \sim \mathrm{N}(0, u_1(u_2 + u_3)u_5\theta).$$

This illustrates how the random effect model can be adapted to describe various hierarchical structures in the covariates. Furthermore, it is straightforward to apply HL to various class of HGLM models via penalized h-loglihood; see Ha *et. al* (2014a) for general fraity models and Ha *et al.* (2014b) for competing risks models.

CHAPTER 12

Multivariate and missing data analysis

Up to now, we have developed univariate models having single response variables. Linear models (Chapter 2) have been extended to GLMs (Chapter 2) and linear mixed models (Chapter 5). Two extensions are combined in HGLMs (Chapter 6). GLMs can be further extended to joint GLMs, allowing structured dispersion models (Chapter 3). This means that a further extension of HGLMs can be made to allow structured dispersion models (Chapter 7) and include models for temporal and spatial correlations (Chapter 8) and for smoothing (Chapter 9). DHGLMs allow heavy-tailed distributions in various components of HGLMs, providing robust analysis with respect to violation of various model assumptions. DHGLMs allow many models to be unified and further extended, for example, financial models, genetic models and variable selections. All these models are useful for the analysis of data. DHGLMs can be further extended by adding more features in the model components. In this book, we present the DHGLM as the most general model for a single response.

In this chapter, we show that multivariate models are easily developed by assuming correlations among random effects in DHGLMs for different responses. Various methods can be used and developed to fit multivariate models. We show that the use of h-likelihood allows straightforward likelihood inferences from multivariate models, which is otherwise difficult because of intractable integration. The resulting algorithm is numerically efficient while giving statistically valid inferences. The GLM fits can be linked together (by augmentation or joint fit of the mean and dispersion) to fit DHGLMs, so that various inferential tools developed for GLMs can be used for inferences about multivariate models too. The h-likelihood leads to the decomposition of component GLMs for multivariate models, allowing us to gain insights, find inferences and check the model assumptions. Thus, h-likelihood leads to new kinds of likelihood inferences.

This multivariate approach also allows us to handle missing data by providing an additional response for missingness. To obtain the MLEs using the complete data likelihood (an extended likelihood), Dempster *et al.* (1977) developed the EM algorithm, while in the h-likelihood approach, the MLEs and REMLEs can be directly obtained by maximization of the adjusted profile h-likelihoods. Furthermore, it gives a straightforward way of computing the standard error estimates of MLEs and REMLEs using the Hessian matrix from adjusted profile h-likelihoods, while the EM algorithm needs additional procedures to compute them.

For such likelihood inferences from these broad classes of multivariate models to be feasible, classical likelihood (Chapter 1) has to be extended to hierarchical likelihood (Chapter 4). We have dealt with likelihood inferences for the GLM class of models. In this chapter we show how these likelihood inferences can be extended to more general classes of multivariate models in a straightforward way.

12.1 Multivariate models

12.1.1 Fitting multivariate models via h-likelihood

First consider a bivariate response $y_i = (y_{1i}, y_{2i})^t$ with *continuous* data y_{1i} and *count* data y_{2i}. Given a random component v_i, suppose that y_{1i} and y_{2i} are independent as follows:

(a) $y_{1i}|v_i \sim N(\mu_{1i}, \phi)$, where $\mu_{1i} = x_{1i}^t \beta + v_i$.

(b) $y_{2i}|v_i \sim \text{Poisson}(\mu_{2i})$, where $\log(\mu_{2i}) = x_{2i}^t \gamma + \delta v_i$.

(c) $v_i \sim N(0, \lambda)$.

This is a shared random effect model: If $\delta > 0$, the two responses are positively correlated, and vice versa for $\delta < 0$; if $\delta = 0$ they are independent. Here, an immediate extension of the h-likelihood is

$$h = \sum_{i=1}^{n} \{\log f(y_i|v_i) + \log f(v_i)\},$$

where

$$\log f(y_i|v_i) = -\frac{(y_{1i} - \mu_{1i})^2}{2\phi} - \frac{1}{2}\log(2\pi\phi) + y_{2i}\log\mu_{2i} - \mu_{2i} - \log y_{2i}!$$

and

$$\log f(v_i) = -\frac{v_i^2}{2\lambda} - \frac{1}{2}\log(2\pi\lambda).$$

Let $\xi = (\beta^t, \gamma^t, v^t)^t$ be fixed and random effect parameters and $\tau =$

(ϕ, δ, λ) be dispersion parameters. Then, we can show that the maximum h-likelihood solution $\hat{\xi}$ of $\partial h/\partial \xi = 0$ can be obtained via a GLM procedure with the augmented response variables $(y^t, \psi^t)^t$, assuming

$$\mu_{1i} = E(y_{1i}|v_i), \quad \mu_{2i} = E(y_{2i}|v_i), \quad v_i = E(\psi_i),$$
$$\text{var}(y_{1i}|v_i) = \phi, \quad \text{var}(y_{2i}|v_i) = \mu_{2i}, \quad \text{var}(\psi_i) = \lambda,$$

and the linear predictor

$$\eta = (\eta_{01}^t, \eta_{02}^t, \eta_1^t)^t = T\xi,$$

where

$$\eta_{01i} = \mu_{1i} = x_{1i}^t \beta + v_i, \quad \eta_{02i} = \log \mu_{2i} = x_{2it}^t \gamma + \delta v_i, \quad \text{and} \quad \eta_{1i} = v_i,$$

$$T = \begin{pmatrix} X_1 & 0 & Z_1 \\ 0 & X_2 & \delta Z_2 \\ 0 & 0 & I_n \end{pmatrix}.$$

In T, $(X_1, 0, Z_1)$ corresponds to the data y_{1i}, $(0, X_2, \delta Z_2)$ to the data y_{2i}, and $(0, 0, I_n)$ to the quasi–data ψ_i. The IWLS equation can be written as

$$T^t \Sigma_a^{-1} T \xi = T^t \Sigma_a^{-1} z, \tag{12.1}$$

where $z = (z_{01}^t, z_{02}^t, z_1^t)^t$ are the GLM augmented dependent variables defined as

$$z_{01i} = \eta_{01i} + (y_{1i} - \mu_{1i})(\partial \eta_{01i}/\partial \mu_{1i}) = y_{1i}$$

corresponding to the data y_{1i},

$$z_{02i} = \eta_{02i} + (y_{2i} - \mu_{1i})(\partial \eta_{02i}/\partial \mu_{2i})$$

corresponding to the data y_{2i}, and $z_{1i} = 0$ corresponding to the quasi–data ψ_i. In addition, $\Sigma_a = \Gamma W^{-1}$ where $\Gamma = \text{diag}(\phi I, I, \lambda I)$, and $W = \text{diag}(W_{01}, W_{02}, W_1)$ are the GLM weight functions defined as $W_{01i} = 1$ for the data y_{1i}, $W_{02i} = (\partial \mu_{2i}/\partial \eta_{02i})^2/\mu_{2i} = \mu_{2i}$ for the data y_{2i}, and $W_{1i} = 1$ for the quasi–data ψ_i. The dispersion parameter δ in the augmented model matrix T is updated via the Newton-Raphson method after estimating the other dispersion parameters (ϕ, λ) iteratively.

To estimate the dispersion parameters $\tau = (\phi, \delta, \lambda)$, we use the adjusted profile h-likelihood

$$p_\xi(h) = \left[h + \frac{1}{2} \log\{|2\pi(T^t \Sigma_a^{-1} T)^{-1}|\} \right]_{\xi = \hat{\xi}}.$$

Following Lee *et al.* (2005) we can show that the dispersion estimators for (ϕ, λ) can be obtained via IWLS equation for a gamma response and δ takes the place of the parameter ρ for correlated random effects in Chapter 8. This method can also be extended to multivariate responses.

Instead of a shared random effect model, it is straightforward to allow multivariate normal random effects for each response variable.

This shows that multivariate models can be easily built by allowing correlations among random effects in different DHGLMs. Furthermore, the use of h-likelihood indicates that interlinked GLM fitting methods for HGLMs can be easily extended to fit multivariate HGLMs. Lee *et al.* (2005) showed that the resulting h-likelihood method provides satisfactory estimators for all the cases they considered.

12.1.2 Joint models for continuous and binary data

Price *et al.* (1985) presented data from a study on the developmental toxicity of ethylene glycol (EG) in mice. Table 12.1 summarizes the data on malformation (binary response) and foetal weight (continuous response) and shows clear dose-related trends with respect to both responses. The rates of foetal malformation increase with dose, ranging from 0.3% in the control group to 57% in the highest dose (3 g/kg/day) group. Foetal weight decreases with increasing dose, with the average weight ranging from 0.972 g in the control group to 0.704 g in the highest dose group.

For analysis of this data set, Gueorguieva (2001) proposed the following joint HGLM. Let y_{1ij} be foetal weights and y_{2ij} an indicator for malformation obtained from the i–th dam. Let $y_{ij} = (y_{1ij}, y_{2ij})^t$ be the bivariate responses and $v_i = (w_i, u_i)^t$ be the unobserved random effects for the i–th cluster. We assume that y_{1ij} and y_{2ij} are conditionally independent given v_i, and propose the following bivariate HGLM:

(a) $y_{1ij}|w_i \sim N(\mu_{ij}, \phi)$ where $\mu_{ij} = x_{1ij}\beta_1 + w_i$,

(b) $y_{2ij}|u_i \sim \text{Bernoulli}(p_{ij})$ where $\text{logit}(p_{ij}) = x_{2ij}\beta_2 + u_i$, and

(c) $v_i \sim N(0, \Sigma)$, where $\Sigma = \begin{pmatrix} \sigma_1^2 & \rho\sigma_1\sigma_2 \\ \rho\sigma_1\sigma_2 & \sigma_2^2 \end{pmatrix}$.

Gueorguieva used the GHQ and MCEM methods. All the covariates (Dose and Dose^2) in this study are between-subject covariates. We present the results from the h-likelihood method by Yun and Lee (2004) which gives almost identical results to those from the GHQ method, but the h-likelihood method is computationally more efficient.

Gueorguieva (2001) ignored the quadratic trend of dose in the HGLM for binary outcomes because of its insignificance. He considered the quadratic trend only for the HGLM for continuous outcomes. However, we found that it is necessary for both HGLMs, as shown in Table 12.2, i.e., the quadratic trend becomes significant if it appears in both HGLMs.

Table 12.1 *Descriptive statistics for ethylene glycol data.*

Dose(g/kg)	Dams	Live	Malformations No.	Malformations %	Weight(g) Mean	Weight(g) (S.D)[a]
0.00	25	297	1	(0.34)	0.972	(0.0976)
0.75	24	276	26	(9.42)	0.877	(0.1041)
1.50	22	229	89	(38.86)	0.764	(0.1066)
3.00	23	226	129	(57.08)	0.704	(0.1238)

[a] Ignoring clustering.

Table 12.2 *Parameter estimates and standard errors for ethylene glycol data.*

Model Parameter	Responses Foetal Malformation Estimate	SE	t	Foetal Weight Estimate	SE	t
Intercept	−5.855	0.749	−7.817	0.978	0.017	58.740
Dose	4.739	0.979	4.841	−0.163	0.029	−5.611
Dose^2	−0.884	0.260	−3.398	0.025	0.009	2.709
$\hat{\sigma}_2$	1.356					
$\sqrt{\hat{\phi}}$				0.075		
$\hat{\sigma}_1$				0.084		
$\hat{\rho}$	−0.619					

SE = standard error; t = t-test value.

Table 12.2 shows a large negative correlation ρ. For testing $\rho = 0$ we can use the deviance based upon the restricted likelihood $p_{v,\beta}(h)$. It has a deviance difference of 11.5 with one degree of freedom, supporting non-null correlation. This negative correlation between the bivariate random effects indicates that high foetal malformation frequencies are associated with lower foetal weights.

This model can be again fitted with the augmented response variables $(y^t, \psi^t)^t$, assuming

$$\mu_{1i} = E(y_{1i}|v_i), \quad \mu_{2i} = E(y_{2i}|v_i), \quad w_i = E(\psi_{1i}), \quad u_i = E(\psi_{2i})$$
$$\text{var}(y_{1i}|v_i) = \phi, \quad \text{var}(y_{2i}|v_i) = \mu_{2i}(1 - \mu_{2i}), \quad \text{cov}(\psi_{1i}, \psi_{2i}) = \Sigma,$$

and the linear predictor

$$\eta = (\eta_{01}^t, \eta_{02}^t, \eta_1^t, \eta_2^t)^t = T\xi,$$

where

$$\eta_{01i} = \mu_{1i} = x_{1i}^t \beta + w_i$$
$$\eta_{02i} = \text{logit}(\mu_{2i}) = x_{2i}^t \gamma + u_i$$
$$\eta_{1i} = w_i$$
$$\eta_{2i} = u_i$$
$$\xi = (\beta^t, \gamma^t, w^t, u^t)^t,$$

$$T = \begin{pmatrix} X_1 & 0 & Z_1 & 0 \\ 0 & X_2 & 0 & Z_2 \\ 0 & 0 & I & 0 \\ 0 & 0 & 0 & I \end{pmatrix}.$$

In T, $(X_1, 0, Z_1, 0)$ corresponds to the data y_{1i}, $(0, X_2, 0, Z_2)$ to the data y_{2i}, $(0, 0, I, 0)$ to the quasi–data ψ_{1i} and $(0, 0, 0, I)$ to the quasi–data ψ_{2i}. Again, the corresponding IWLS equation can be written as

$$T^t \Sigma_a^{-1} T \xi = T^t \Sigma_a^{-1} z,$$

where $z = (z_{01}^t, z_{02}^t, z_1^t, z_2^t)^t$ are the GLM augmented dependent variables defined as

$$z_{01i} = \eta_{01i} + (y_{1i} - \mu_{1i})(\partial \eta_{01i})/(\partial \mu_{1i}) = y_{1i}$$

corresponding to the data y_{1i},

$$z_{02i} = \eta_{02i} + (y_{2i} - \mu_{1i})(\partial \eta_{02i})/(\partial \mu_{2i})$$

corresponding to the data y_{2i}, and

$$z_{1i} = z_{2i} = 0$$

corresponding to the quasi–data ψ_i. In addition, $\Sigma_a = \Gamma W^{-1}$ where $\Gamma = \text{diag}(\phi I, I, \Sigma \otimes I)$, and $W = \text{diag}(W_{01}, W_{02}, W_1, W_2)$ are the GLM weight functions defined as $W_{01i} = 1$ for the data y_{1i},

$$W_{02i} = (\partial \mu_{2i}/\partial \eta_{02i})^2/\mu_{2i} = \mu_{2i}$$

for the data y_{2i}, and $W_{1i} = W_{2i} = 1$ for the quasi–data ψ_i.

To estimate the dispersion parameters $\tau = (\phi, \rho, \sigma_1, \sigma_2)$, we use the adjusted profile likelihood

$$p_\xi(h) = \left[h + \frac{1}{2} \log\{|2\pi(T^t \Sigma_a^{-1} T)^{-1}|\} \right]_{\xi = \hat{\xi}}.$$

12.1.3 Model for vascular cognitive impairment data

As the elderly population increases, the proportion of stroke, one of the most common geriatric diseases, also rises. The increase of the number of patients with stroke results in the rise of the number of patients with vascular cognitive impairment (VCI) in which cognitive function is declined. Through an early intervention for VCI, the cognitive function can be improved. The purpose of the study was to examine the effects of demographic and acute neuroimaging variables on the cognitive functions in ischemic stroke patients.

The study was a retrospective analysis based on the Clinical Research Center for Stroke 5th Division (CRCS-5) Registry database. The CRCS-5 registry is a prospective web-based database of ischemic stroke patients admitted to participating centers scattered throughout a major part of South Korea. Using the database, the 375 patients investigated were

- Admitted during May 2007 and September 2014 to the Seoul National University Bundang Hospital.
- Diagnosed with ischemic stroke.
- Hospitalized within 7 days of onset.
- Discharged within 30 days of admission.

For the evaluation of cognitive function in VCI, the Korean Vascular Cognitive Impairment Harmonization Standards Neuropsychology Protocol (K-VCIHS-NP) was first conducted for all subjects 3 months after stroke onset to assess comprehensively their cognitive functions. The results are standardized as z-scores composed of four domains: executive ($y1$), memory ($y2$), visuospatial ($y3$) and language ($y4$). The K-VCIHS-NP was examined repeatedly (average of three repetitions) for each hospital visit. Twenty covariates were considered as shown in Table 12.3.

To fit this data set, Lee, Rönnegård and Noh (2016) proposed the following multivariate model for four response variables y_{kit} for the t–th visit of the i–th subject; for $k = 1, \cdots, 4$,

$$y_{kit}|v_{ki} \sim N(X_{it}\beta_k + v_{ki}, \phi_{kit}),$$

where the residual variance ϕ_{kit} is given by

$$\log(\phi_{kit}) = \gamma_k + b_{ki}, \ b_{ki} \sim N(0, \alpha_k).$$

The random effects follow multivariate normal distribution:

$$\begin{pmatrix} v_{1i} \\ v_{2i} \\ v_{3i} \\ v_{4i} \end{pmatrix} \sim MVN \left[\begin{pmatrix} 0 \\ 0 \\ 0 \\ 0 \end{pmatrix}, \begin{pmatrix} \lambda_{1i} & \rho_1\lambda^*_{1i,2i} & \rho_2\lambda^*_{1i,3i} & \rho_3\lambda^*_{1i,4i} \\ \rho_1\lambda^*_{1i,2i} & \lambda_{2i} & \rho_4\lambda^*_{2i,3i} & \rho_5\lambda^*_{2i,4i} \\ \rho_2\lambda^*_{1i,3i} & \rho_4\lambda^*_{2i,3i} & \lambda_{3i} & \rho_6\lambda^*_{3i,4i} \\ \rho_3\lambda^*_{1i,4i} & \rho_5\lambda^*_{2i,4i} & \rho_6\lambda^*_{3i,4i} & \lambda_{4i} \end{pmatrix} \right],$$

with $\lambda^*_{ji,ki} = \sqrt{\lambda_{ji}\lambda_{ki}}$, where variances of random effects are modelled as

$$\log(\lambda_{ki}) = \delta_k + c_{ki}, \ c_{ki} \sim N(0, \tau_k).$$

Table 12.3 *Covariates for vascular cognitive impairment data*

Variables	x Code	Definition
	Demographic variable:	
Age	1	int(age/10)
Gender	2	1 = male, 0 = female
Education	3	0 = none, 1 = elementary, 2 = middle, 3 = high, 4 = college
Hypertension	4	1 = hypertension, 0 = none
Diabetes mellitus	5	1 = diabetes mellitus, 0 = none
Atrial Fibrillationf	6	1 = atrial fibrillation, 0 = none
Stroke history	7	1 = history of stroke, 0 = none
NIHSS	8	Score at admission
VCINP	9	Time interval from stroke onset to first K-VCIHS-NP
PCI	10	1 = IQCODE≥3.6, 0 = otherwise
	Neuroimaging variable:	
AcuteLeft	11	Left or bilateral involvement
AcuteMulti	12	Lesion multiplicity on acute DWI
AcuteCS	13	Cortical involvement of acute lesions
ChrCS	14	Cortical involvement of chronic territorial infarction
PVWM	15	0 = 0 or 1, 1 = 2 or 3
SCWM	16	0 = 0 or 1, 1 = 2 or 3
LAC	17	Lacunae
CMB	18	Cerebral microbleed
MTA1	19	1 = MTA 2, 0 = not MTA 2
MTA2	20	1 = MTA 3 or 4, 0 = not MTA 3 or 4

int(age/10) = quotient of age/10.
NIHSS = National Institutes of Health Stroke Scale score.
K-VCIHS-NP = Korean Vascular Cognitive Impairment Harmonization Standards Neuropsychology Protocol.
PCI = prophylactic cranial irradiation
IQCODE = informant questionnaire on cognitive decline in the elderly
DWI = diffusion-weighted magnetic resonance imaging
PVWM = periventicular white matter lesion.
SCWM = subcortical white matter lesion.
MTA = medial temporal lobe atrophy.

The fitting results are shown in Tables 12.4. Random effects of executive and visuospatial domains show the strongest correlation, 0.583. This indicates that higher scores in the executive domain are associated with higher scores in the visuospatial domain. Correlations between visuospatial and memory domains have the lowest value, 0.29. For all domains, NIHSS (x8) scores of demographic variables and higher MTA incidence (3 or 4; MTA2) of nueroimaging variables revealed significant negative effects. These indicate higher NIHSS scores and lower MTA2 scores of cognitive functions.

12.2 Missing data problems

Analysis of missing data requires handling of unobserved random variables, and this should be a natural area of application for extended likelihood. Unfortunately, due to problems similar to those examples discussed in Chapter 4, early efforts using the extended likelihood approach did not lead to an acceptable methodology. Suppose some values in Y are missing (unobserved). We write $Y = (Y_{obs}, Y_{mis})$, where Y_{obs} denotes the observed components and Y_{mis} denotes the missing components. Let

$$f_\theta(Y) \equiv f_\theta(Y_{obs}, Y_{mis})$$

denote the joint density of Y, composed of observed Y_{obs} and unobserved Y_{mis}. The marginal density of observed Y_{obs} is obtained by integrating out the missing data Y_{mis},

$$f_\theta(Y_{obs}) = \int f_\theta(Y_{obs}, Y_{mis}) dY_{mis}.$$

Given $Y_{obs} = y_{obs}$, classical likelihood inference about θ may be based on the marginal likelihood

$$L(\theta) = L(\theta; y_{obs}) = f_\theta(y_{obs}).$$

Yates (1933) was the first to use the technique whereby estimates of θ are found by treating the missing values Y_{mis} as parameters and maximizing an extended likelihood

$$L_e(\theta, Y_{mis}) = f_\theta(y_{obs}, Y_{mis})$$

with respect to (θ, Y_{mis}). As we have shown in Example 4.2, this joint estimation is generally not justified and can lead to wrong estimates for some parameters. The classic example of this approach is in analysing missing plots in the analysis of variance where missing outcomes Y_{mis} are treated as parameters and then filled in to allow computationally efficient methods to be used for analysis (Yates 1933; Bartlett 1937). Box *et al.* (1970) apply the same approach in a more general setting, where a multivariate normal mean vector has a non–linear regression on covariates. DeGroot and Goel (1980) described the joint maximization of an extended likelihood L_e for their problem as a maximum likelihood (ML) procedure. As argued by Little and Rubin (1983, 2002),

Table 12.4 MDHGLM Results for vascular cognitive impairment data.

y_k Covariate	y_1 (executive) Estimate	SE	p-value	y_2 (memory) Estimate	SE	p-value	y_3 (visuospatial) Estimate	SE	p-value	y_4 (language) Estimate	SE	p-value
Intercept	-0.995	0.351	0.005	-0.679	0.301	0.024	-1.091	0.530	0.040	-0.193	0.331	0.560
Age	0.018	0.046	0.702	0.113	0.041	0.006	0.080	0.072	0.268	0.027	0.043	0.532
Male	-0.214	0.150	0.154	-0.879	0.122	0.000	-0.662	0.209	0.002	-0.441	0.149	0.003
Education	0.048	0.013	0.000	-0.01	0.011	0.358	0.000	0.018	0.987	0.020	0.013	0.116
HTN	-0.116	0.141	0.408	-0.095	0.115	0.412	-0.277	0.195	0.157	-0.378	0.141	0.007
DM	-0.107	0.129	0.407	-0.093	0.104	0.371	-0.168	0.180	0.351	-0.079	0.129	0.538
Af	0.002	0.180	0.990	-0.333	0.145	0.021	-0.032	0.252	0.898	0.233	0.178	0.190
Stroke history	-0.271	0.161	0.093	-0.265	0.129	0.039	-0.335	0.225	0.136	0.135	0.161	0.404
NIHSS	-0.069	0.012	0.000	-0.030	0.010	0.002	-0.081	0.018	0.000	-0.042	0.012	0.001
VCINP	-0.026	0.013	0.040	0.013	0.012	0.286	0.000	0.022	1.000	0.015	0.012	0.196
PCI	-1.122	0.143	0.000	-0.674	0.114	0.000	-1.049	0.201	0.000	-1.006	0.143	0.000
AcuteLeft	-0.110	0.123	0.371	-0.288	0.100	0.004	0.605	0.171	0.000	-0.080	0.123	0.514
AcuteMulti	-0.168	0.144	0.243	-0.007	0.116	0.951	0.002	0.198	0.993	-0.113	0.143	0.431
AcuteCS	-0.107	0.145	0.462	0.054	0.117	0.646	-0.051	0.201	0.801	0.209	0.145	0.147
ChrCS	-0.077	0.187	0.681	0.239	0.150	0.110	0.348	0.259	0.179	-0.152	0.188	0.420
PVWM	0.028	0.143	0.845	-0.118	0.115	0.304	0.273	0.197	0.166	-0.254	0.142	0.073
SCWM	-0.306	0.167	0.067	0.070	0.133	0.599	-0.219	0.231	0.344	0.027	0.165	0.871
LAC	-0.222	0.139	0.111	-0.094	0.112	0.401	-0.446	0.193	0.021	-0.246	0.139	0.078
CMB	-0.102	0.139	0.464	-0.171	0.112	0.126	0.063	0.194	0.743	0.148	0.139	0.287
MTA1	-0.134	0.164	0.417	-0.170	0.131	0.196	-0.158	0.226	0.486	-0.300	0.163	0.065
MTA2	-0.723	0.204	0.000	-0.479	0.157	0.002	-0.877	0.288	0.002	-0.681	0.203	0.001
$\log \delta_k$	0.136	0.078		-0.360	0.075		0.812	0.077		0.191	0.077	
$\log \tau_k$	-1.401	0.809		-1.350	0.791		-1.285	0.317		-1.290	0.563	
$\log \gamma_k$	-1.139	0.045		-1.129	0.045		0.058	0.045		-1.347	0.045	
$\log \alpha_k$	-0.982	0.226		-1.073	0.191		-0.929	0.434		-1.040	0.256	

$\hat{\rho}_1=0.420$ $\hat{\rho}_2=0.584$ $\hat{\rho}_3=0.487$ $\hat{\rho}_4=0.210$ $\hat{\rho}_5=0.400$ $\hat{\rho}_6=0.301$

HTN = hypertension. DM = diabetes mellitus. Af = atrial fibrillation. NIHSS = National Institutes of Health Stroke Scale score. VCINP = time interval from onset of stroke to first K-VCIHS-NP. PCI = prophylactic cranial irradiation. ChrCS = cortical involvement of chronic territorial infraction. PVWM = periventicular white matter lesion. SCWM = subcortical white matter lesion. LAC = lacunae. CMB = cerebral microbleed. MTA = medial temporal lobe atrophy. MCAR = missing completely at random.
s.e. = standard error.

one problem in this joint maximization is that it is statistically efficient only when the fraction of missing values tends to zero as the sample size increases. Press and Scott (1976) also point this out in the context of their problem. Box *et al.*(1970) and Press and Scott (1976) argued for maximizing L_e on grounds of computational simplicity. This simplicity, however, usually does not apply unless the number of missing values is small.

Stewart and Sorenson (1981) discuss maximization of L_e and L for the problem considered by Box *et al.* (1970) and reject maximization of L_e. Little and Rubin (1983, 2002) state correctly that L is the true likelihood of θ and illustrate by various examples that joint maximiza- tion of L_e yields incorrect parameter estimators. However, the marginal likelihood L is generally hard to obtain because it involves intractable integration. Thus, in missing data problems, various methods such as factored likelihood (Anderson 1957; Rubin 1974), the EM algorithm (Dempster *et al.*, 1977) and the stochastic EM algorithm (Diebolt and Ip, 1996) have been proposed.

Suppose unobserved data are missing at random (MAR, Rubin 1976). This means that the probability of being missing does not depend on the values of missing data Y_{mis}, although it may depend on values of observed data y_{obs}. Under the MAR assumption, statistical inferences about θ can be based on the marginal loglihood of the observed responses $Y_{obs} = y_{obs}$ only, *ignoring the missing data mechanism*

$$m_{ign} = \log f_\theta(y_{obs}).$$

More generally, we include in the model the distribution of a variable indicating whether each component of Y is observed or missing. For $Y = (Y_{ij})$,

$$R_{ij} = 1, \text{ if } Y_{ij} \text{ is missing},$$
$$= 0, \text{ if } Y_{ij} \text{ is observed}.$$

This leads to a bivariate model with the probability function

$$f_{\theta,\lambda}(Y, R) \equiv f_\theta(Y) f_\lambda(R|Y).$$

The actual observed data consist of the values of observed (Y_{obs}, R). The distribution of all the observed data is obtained by integrating out the unobservables Y_{mis}

$$f_{\theta,\lambda}(Y_{obs}, R) = \int f_\theta(Y_{obs}, Y_{mis}) f_\lambda(R|Y_{obs}, Y_{mis}) dY_{mis}.$$

Thus, having observed bivariate responses $(Y_{obs} = y_{obs}, R = r)$, the full loglihood for the fixed parameters (θ, λ) is

$$m_{full} = \log f_{\theta,\lambda}(y_{obs}, r). \qquad (12.2)$$

Under MAR, i.e., $f_\lambda(R|Y_{obs}, Y_{mis}) = f_\lambda(R|Y_{obs})$ for all Y_{mis}, we have

$$f_{\theta,\lambda}(Y_{obs}, R) = f_\theta(Y_{obs})f_\lambda(R|Y_{obs}).$$

Thus, when θ and λ are distinct, loglihood inferences for θ from m_{full} and m_{ign} are the same.

To use the extended likelihood framework, given $(Y_{obs} = y_{obs}, R = r)$, we can define the h-loglihoods as

$$h_{ign} = \log f_\theta(y_{obs}, v(Y_{mis}))$$

and

$$h_{full} = \log f_\theta(y_{obs}, v(Y_{mis}); \theta) + \log f_\lambda(r|y_{obs}, Y_{mis}; \lambda),$$

where $v = v(Y_{mis})$ is an appropriate monotonic function that puts Y_{mis} on the canonical scale. When the canonical scale does not exist we use the scale $v()$, whose range is a whole real line. This allows us to avoid the boundary problem of the Laplace approximation, which is the adjusted profile loglihood $p_v(h)$.

12.2.1 Missing plot analysis of variance

This is a more general version of Example 4.2 and was considered by Yates (1933). Suppose $y = (y_{obs}, y_{mis})$ consists of n independent normal variates with mean $\mu = (\mu_1, \ldots, \mu_n)$ and common variance σ^2. The subset $y_{obs} = (y_1, \ldots, y_k)$ consists of k observed values and $y_{mis} = (y_{k+1}, \ldots, y_n)$ represents $(n - k)$ missing values. Let

$$\mu_i = E(Y_i) = x_i^t \beta,$$

where β represents $p \times 1$ regression coefficients. The h-loglihood is similarly defined as in Example 4.2:

$$h(\mu, \sigma^2, y_{mis}) = -\frac{n}{2}\log(2\pi\sigma^2) - \frac{1}{2\sigma^2}\sum_{i=1}^{k}(y_i - \mu_i)^2 - \frac{1}{2\sigma^2}\sum_{i=k+1}^{n}(y_i - \mu_i)^2.$$

Here for $i = k + 1, \ldots, n$, $\partial h/\partial y_i = -(y_i - \mu_i)/\sigma^2$ gives $\hat{y}_i = x_i^t\beta$ and

$$-\partial^2 h/\partial y_i^2 = 1/\sigma^2,$$

so $v(y_{mis}) = y_{mis}$ is a canonical scale for β, but not for σ^2. Because

$$h(\mu, \sigma^2, \hat{y}_{mis}) = -\frac{n}{2}\log(2\pi\sigma^2) - \frac{1}{2\sigma^2}\sum_{i=1}^{k}(y_i - x_i\beta)^2 - \frac{1}{2\sigma^2}\sum_{i=k+1}^{n}(\hat{Y}_i - x_i\beta)^2,$$

ordinary least squares estimates maximize the h-likelihood; they maximize the first summation and give a null second summation. Yates (1933) noted that if the missing values are replaced by their least squares estimates, correct estimates can be obtained by least squares applied to the filled-in data.

However, as in Example 4.2, the analysis of the filled-in data with missing y_i by $\widehat{y}_i$ gives an incorrect dispersion estimation

$$\widehat{\sigma}^2 = \frac{1}{n}\sum_{i=1}^{k}(y_i - \bar{y})^2$$

because $\widehat{y}_{mis}$ is not information-free for σ^2. With the use of adjusted profile loglihood

$$p_{y_{mis}}(h) = -\frac{n}{2}\log(2\pi\sigma^2) - \frac{1}{2\sigma^2}\sum_{i=1}^{k}(y_i - \bar{y})^2 + \frac{1}{2}\log|\sigma^2(X_{mis}^T X_{mis})^{-1}|,$$

where X_{mis} is the model matrix corresponding to missing data, we have the MLE

$$\widehat{\sigma}^2 = \frac{1}{k}\sum_{i=1}^{k}(y_i - \bar{y})^2.$$

Thus, proper profiling gives the correct dispersion estimate. Here y_{mis} represents unobservable random variables and $p_{y_{mis}}(h)$ is the Laplace approximation integrating out the unobservable random variables y_{mis}. This shows that missing data can be handled with the h-likelihood exactly as we handle random effects in HGLMs.

12.2.2 Regression with missing predictor

Suppose (X_i, Y_i), $i = 1, ..., n$ are n observations from a bivariate normal distribution with mean (μ_x, μ_y), variance (σ_x^2, σ_y^2) and correlation ρ, where responses $Y_i = y_i$ are observed for all n observations, and $X_{obs} = (x_1, ..., x_k)$ are observed, but some regressors $X_{mis} = (X_{k+1}, ..., X_n)$ are MAR. Assume that interest is focussed on the regression coefficient of Y_i on X_i, $\beta_{y \cdot x} = \rho \sigma_y / \sigma_x = \beta_{x \cdot y} \sigma_y^2 / \sigma_x^2$. Note that

$$E(X_i | Y_i = y_i) = \mu_x + \beta_{x \cdot y}(y_i - \mu_y)$$
$$\text{var}(X_i | Y_i = y_i) = \sigma_x^2(1 - \rho^2).$$

Thus, with $\theta = (\mu_x, \mu_y, \sigma_x^2, \sigma_y^2, \rho)$ we have

$$
\begin{aligned}
m_{ign} &= \log f(y; \theta) + \log f(x_{obs}|y; \theta) \\
&= -\frac{n}{2}\log(2\pi\sigma_y^2) - \frac{1}{2\sigma_y^2}\sum_{i=1}^{n}(y_i - \mu_y)^2 - \frac{k}{2}\log\{2\pi\sigma_x^2(1-\rho^2)\} \\
&\quad - \sum_{i=1}^{k}(x_i - \mu_x - \beta_{x\cdot y}(y_i - \mu_y))^2/\{2\sigma_x^2(1-\rho^2)\}.
\end{aligned}
$$

The marginal MLE of $\beta_{y\cdot x}$ is

$$
\hat{\beta}_{y\cdot x} = \hat{\beta}_{x\cdot y}\hat{\sigma}_y^2/\hat{\sigma}_x^2,
$$

where

$$
\hat{\beta}_{x\cdot y} = \sum_{i=1}^{k}(y_i - \bar{y})x_i / \sum_{i=1}^{n}(y_i - \bar{y})^2
$$
$$
\bar{y} = \sum_{i=1}^{k} y_i/k, \quad \bar{x} = \sum_{i=1}^{k} x_i/k
$$

$$
\hat{\sigma}_y^2 = \sum_{i=1}^{n}(y_i - \hat{\mu}_y)^2/n, \quad \hat{\mu}_y = \sum_{i=1}^{n} y_i/n,
$$
$$
\hat{\sigma}_x^2 = \hat{\beta}_{x\cdot y}^2\hat{\sigma}_y^2 + \sum_{i=1}^{k}\{x_i - \bar{x} - \hat{\beta}_{x\cdot y}(y_i - \bar{y})\}^2/k.
$$

Here

$$
\begin{aligned}
h_{ign} &= \log f(y; \theta) + \log f(x_{obs}|y; \theta) + \log f(X_{mis}|y; \theta) \\
&= -\frac{n}{2}\log(2\pi\sigma_y^2) - \frac{1}{2\sigma_y^2}\sum_{i=1}^{n}(y_i - \mu_y)^2 - \frac{n}{2}\log\{2\pi\sigma_x^2(1-\rho^2)\} \\
&\quad - \left\{\sum_{i=1}^{k}(x_i - \mu_x - \beta_{x\cdot y}(y_i - \mu_y))^2\right. \\
&\quad \left. + \sum_{i=k+1}^{n}(X_i - \mu_x - \beta_{x\cdot y}(y_i - \mu_y))^2\right\}/\{2\sigma_x^2(1-\rho^2)\}.
\end{aligned}
$$

For $i = k+1, ..., n$,

$$
\tilde{X}_i = E(X_i|Y_i = y_i) = \mu_x + \beta_{x\cdot y}(y_i - \mu_y)
$$

is the target of estimation (TOE). Thus, given $(\sigma_x, \sigma_y, \rho)$, joint maximization of (X_{mis}, μ_x, μ_y) gives the MLEs for location parameters

$$
\hat{\mu}_x = k^{-1}\sum_{i=1}^{k}\{x_i - \hat{\beta}_{x\cdot y}(y_i - \hat{\mu}_y)\} = \bar{x} - \hat{\beta}_{x\cdot y}(\bar{y} - \hat{\mu}_y) \text{ and } \hat{\mu}_y = \sum_{i=1}^{n} y_i/n
$$

and the estimators

$$
\hat{X}_i = \hat{\mu}_x + \hat{\beta}_{x\cdot y}(y_i - \hat{\mu}_y) = \bar{x} + \hat{\beta}_{x\cdot y}(y_i - \bar{y})
$$

for the TOE, $E(X_i|Y_i = y_i)$. The MLE $\hat{\mu}_x$ is of particular interest and can be obtained as a simple sample mean

$$\hat{\mu}_x = (\sum_{i=1}^{k} x_i + \sum_{i=k+1}^{n} \hat{X}_i)/n$$

after effectively imputing the predicted values $\hat{X}_i$ for missing X_i from linear regression of observed X_i on observed y_i. This shows that the h-likelihood procedure implicitly implements the factored likelihood method of Anderson (1957).

Little and Rubin (1983) showed that the joint maximization of h_{ign} does not give a proper estimate of $\beta_{y \cdot x}$. Here X_{mis} are canonical for location parameters, but not for $\beta_{y \cdot x} = \rho \sigma_y / \sigma_x$, so that we should use the adjusted profile loglihood

$$
\begin{aligned}
p_{X_{mis}}(h_{ign}) =\ & -\frac{n}{2}\log(2\pi\sigma_y^2) - \sum_{i=1}^{n} \frac{(y_i - \mu_y)^2}{2\sigma_y^2} - \frac{n}{2}\log\{\sigma_x^2(1-\rho^2)\} \\
& - \sum_{i=1}^{k}(x_i - \mu_x - \beta_{x \cdot y}(y_i - \mu_y))^2/\{2\sigma_x^2(1-\rho^2)\} \\
& + \{(n-k)/2\}\log\{\sigma_x^2(1-\rho^2)\},
\end{aligned}
$$

which is identical to marginal likelihood m_{ign}. Thus the marginal MLE of $\beta_{y \cdot x}$ is obtained by maximizing $p_{X_{mis}}(h_{ign})$. In summary, the use of h-likelihood is a convenient and correct way of handling missing data. We believe that all the existing counter–examples are caused by the misuse of the h-likelihood.

12.3 Missing data in longitudinal studies

The two types of missing data patterns in longitudinal studies are (a) *monotone* missingness (or dropout), in which once an observation is missing all subsequent observations on that individual are also missing and (b) *non-monotone* missingness, in which some observations of a repeated measure are missing but are followed by later observations. Diggle and Kenward (1994) proposed a logistic model for dropout. Troxel *et al.* (1998) extended Diggle and Kenward's method to non-monotone missingness. Little (1995) provided an excellent review of various modelling approaches. A difficulty in missing data problems is the integration necessary to obtain the marginal likelihood of the observed data after eliminating missing data. The use of h-likelihood enables us to overcome such difficulties.

Let $Y_i^* = (Y_{i1}, \cdots, Y_{iJ})^t$ be the complete measurements on the i–th subject if they are fully observed. The observed and missing components are denoted by Y_i^O and Y_i^M respectively. Let $R_i = (R_{i1}, \cdots, R_{iJ})^t$ be a vector of indicators of missing data, so that $R_{ij} = 0$ when the j–th measurement of the i–th subject is observed. Let v_i be an unobserved random effect and $f(Y_i^*, R_i, v_i)$ be the joint density function of Y_i^*, R_i and v_i. To model the missing mechanism we can use the selection model (Diggle and Kenward, 1994),

$$f(Y_i^*, R_i, v_i) = f(Y_i^*|v_i)f(R_i|Y_i^*, v_i)f(v_i),$$

or the pattern mixture model (Little, 1995),

$$f(Y_i^*, R_i, v_i) = f(R_i|v_i)f(Y_i^*|R_i, v_i)f(v_i).$$

In the selection model, we may assume

(a) $f(R_i|Y_i^*, v_i) = f(R_i|v_i)$ or

(b) $f(R_i|Y_i^*, v_i) = f(R_i|Y_i^*).$

Under assumption (a) the joint density becomes

$$f(Y_i^*, R_i, v_i) = f(Y_i^*|v_i)f(R_i|v_i)f(v_i),$$

leading to shared random effect models (Ten Have et $al.$, 1998); see Lee et $al.$ (2005) for the h-likelihood approach.

Suppose we have a selection model satisfying (b). For the i–th subject, the joint density function of (Y_i^O, R_i, v_i) can be written as

$$f(Y_i^O, R_i, v_i) = \int \{ \prod_{j \in obs} (1-p_{ij})f_i(Y_{ij}^*|v_i)\}\{ \prod_{j \in miss} p_{ij}f_i(Y_{ij}^*|v_i)\}f(v_i)dY_i^M,$$

where $j \in obs$ and $j \in miss$ indicate j–th observed and missing measurements of the i–th subject, respectively.

If $Y_{ij}^*|v_i$ follow linear mixed models, $f(Y_i^O, R_i, v_i)$ can be further simplified. Then, we define

$$\sum \log f(y_i^O, r_i, Y_i^M, v_i)$$

as the h-likelihood with unobservable random variables $w_i = (Y_i^M, v_i)$. We use the criteria h for w, $p_w(h)$ for mean parameters $\psi = (\beta, \delta, \rho)$ and $p_{w,\psi}(h)$ for the remaining dispersion parameters.

If the responses R_{ij} follow a Bernoulli GLM with probit link (Diggle and Kenward, 1994),

$$\eta = \Phi^{-1}(p_{ij}) = x_{ij}^t\delta + \rho Y_{ij}^*.$$

We can allow y_{ij} in the covariate x_{ij}. If $\rho = 0$, the data are missing at random and otherwise are non-ignorable . This leads to the joint density

$$f(Y_i^O, R_i, v_i) = \left\{ \prod_{j \in obs} (1 - p_{ij}) f_i(Y_{ij}^O | v_i) \right\}$$

$$\times \left\{ \prod_{j \in miss} \Phi \left(\frac{x_{ij}^t \delta + \rho E(Y_{ij}^* | v_i)}{\sqrt{1 + \rho^2 \phi^2}} | v_i \right) \right\} f(v_i).$$

Thus, with the probit link we can eliminate nuisance unobservables Y^M, so that we can use

$$\sum \log f(y_i^O, r_i, v_i)$$

as the h-likelihood for necessary inferences. For the h-likelihood approach in general for missing data problems see Yun *et al.* (2005).

12.3.1 Antidepressant data with dropout

Heyting *et al.* (1990) presented a longitudinal multicentre trial of antidepressants. These data were used by Diggle and Kenward (1994) to illustrate the use of a selection model with non-random dropout. Here we show how the same type of model can be fitted and appropriate measures of precision estimated in a much less computationally demanding way using an appropriate h-likelihood. A total of 367 subjects were randomized to one of three treatments in each of six centres. Each subject was rated on the Hamilton depression score, a sum of 16 test items producing responses on a 0 to 50 scale. Measurements were made on each of five weekly visits, the first made before treatment, the remaining four during treatment. We number these weeks 0 through 4. Subjects dropped out of the trial from week 2 onward and by the end of the trial 119 (32%) had left.

We fit a selection model that has the same non-random dropout mechanism and a similar response model to the one used originally by Diggle and Kenward (1994). The use of such non-random dropout models for primary analyses in such settings has drawn considerable and justified criticism because of the strong dependence of the resulting inferences on untestable modelling assumptions, particularly, the sensitivity of inference to the assumed shape of the distribution of the unobserved data: see the discussion in Diggle and Kenward. This point is made very clearly in Little and Rubin (2002, Section 15.4) and a simple illustration is given in Kenward (1998). If such models are to have a role, it should be more properly as part of a sensitivity analysis and a variety of alternatives needs to be considered. This implies in turn the need for relatively efficient estimation methods. The original analysis of Diggle and Kenward used the Nelder-Mead simplex search algorithm for optimization; this proved to be very slow. Other users of similar models used computer–intensive Markov chain Monte Carlo methods, both fully Bayesian (Best *et al.*, 1996) and hybrid frequentist (Gad, 1999). One advantage of the h-likelihood approach is that it can in principle provide a much less computationally demanding route for fitting these models.

Let $y_{ijk} = (y_{ijk0}, \cdots, y_{ijk4})^t$ and $r_{ijk} = (r_{ijk0}, \cdots, r_{ijk4})^t$ be respectively the complete (some possibly unobserved) responses and corresponding missing value indicators for the k–th subject in treatment group j and centre i. Dropout implies that if someone is missing at time l, i.e., $r_{ijkl} = 1$, he or she will be missing subsequently, i.e., $r_{ijkR} = 1$ for all $R > l$. If $d_{ijk} = \sum_{l=0}^{4} I(r_{ijkl} = 0)$, $y_{ijk0}, \cdots, y_{ijk(d_{ijk}-1)}$ are observed and $y_{ijkd_{ijk}}, \cdots, y_{ijk4}$ are missing. If $d_{ijk} = 5$, i.e., $r_{ijk4} = 0$, there is no dropout.

Diggle and Kenward (1994) proposed the following missing-not-at-random model for the dropout mechanism:

$$\text{logit}(p_{ijkl}) = \delta_0 + \delta_1 y_{ijk(l-1)} + \rho y_{ijkl}, \ l = 2, 3, 4, \tag{12.3}$$

where $p_{ijkl} = \Pr(r_{ijkl} = 1 \mid y_{ijk0}, \cdots, y_{ijk(l-1)}, y_{ijkl}, r_{ijk(l-1)} = 0)$. The underlying dropout rate is set to 0 for weeks 0 and 1 because there were no drop-outs in those weeks.

For a complete response y_{ijk}, Yun *et al.* (2005) considered two covariance models, namely a compound symmetric model using random subject effects and a saturated covariance model. Consider a model with random subject effects

$$y_{ijkl} = \gamma_i + \eta_j l + \xi_j l^2 + \varepsilon_{ijkl}^* \tag{12.4}$$

where $\varepsilon_{ijkl}^* = v_{ijk} + \varepsilon_{ijkl}$, $v_{ijk} \sim N(0, \lambda)$ are the random subject effects and $\varepsilon_{ijkl} \sim N(0, \phi)$ are the residual terms. This model has the same mean (fixed effects) structure as the original Diggle and Kenward model, but implies a different, more constrained covariance structure than the antedependence model of order 2 used by Diggle and Kenward.

Data from the antidepressant trial data are reported in Table 12.5. Results from missing-completely-at-random (MCAR) and missing-at-random (MAR) models (respectively $\delta_1 = \rho = 0$ and $\rho = 0$) are almost identical. However, from the fit of the full model, there is a suggestion that ρ seems not to be null, and thus the missing mechanism may not be ignorable. This conclusion must be treated with caution, however. First, the interpretation of this parameter as one governing the non-ignorability of the missing value process depends on the assumed (normal) distributional form for the missing data. Second, the usual asymptotic behaviour of likelihood-based tests for settings of this type has been called into question (Jansen *et al.*, 2005).

The random effect model above assumes a compound symmetric correlation in which the correlation among repeated measures remains constant. However, variances may change over time and correlations may differ with the time differences between pairs of measurements. The antedependence covariance structure can accommodate such patterns, and a second-order example (12 parameters in this setting) was used originally by Diggle and Kenward (1994) as a compromise between flexibility and parsimony. The efficiency of the simplex numerical maximization method used by these authors was highly dependent on the number of parameters, so it was important to restrict these as much as possible. With the current approach this is less important, and little efficiency is lost if the antedependence structure is replaced by an unstructured covariance matrix.

Table 12.5 shows results from a saturated covariance model with a missing-not-at-random (MNAR) mechanism. The magnitude of the regression coefficients of the two covariance models are similar. However, their standard error estimates can be quite different. The estimated covariance matrix is as follows:

$$
\begin{pmatrix}
12.72 & & & & \\
9.99 & 33.14 & & \text{symmetric} & \\
7.00 & 20.67 & 44.73 & & \\
6.93 & 18.24 & 31.21 & 51.14 & \\
6.28 & 13.45 & 20.71 & 29.56 & 52.23
\end{pmatrix},
$$

which shows that the variance increases with time while correlations may decrease as the time difference increases. The inverse of this matrix (precision matrix) is

$$
\begin{pmatrix}
0.10 & & & & \\
-0.03 & 0.05 & & \text{symmetric} & \\
0.00 & -0.02 & 0.05 & & \\
0.00 & 0.00 & -0.02 & 0.04 & \\
0.00 & 0.00 & 0.00 & -0.01 & 0.03
\end{pmatrix}.
$$

The very small elements off the main two diagonals strongly suggest a first-order antedependence structure and unsurprisingly the results are very close to those originally obtained by Diggle and Kenward (1994). Note that this structure is inconsistent with the compound symmetry structure imposed earlier.

Table 12.5 *Analysis of antidepressant trial data using h-likelihood methods.*

	Random effect Model						SC	
	MCAR		MAR		MNAR		MNAR	
Parameter	Estimate	s.e.	Estimate	s.e.	Estimate	s.e.	Estimate	s.e.
γ_1	22.30	0.64	22.23	0.63	22.34	0.61	21.34	0.47
γ_2	21.72	0.60	21.67	0.53	21.83	0.53	22.57	0.44
γ_3	19.04	0.58	19.00	0.63	19.30	0.59	19.47	0.42
γ_4	23.91	0.60	23.86	0.58	24.07	0.55	23.97	0.44
γ_5	20.56	0.59	20.51	0.58	20.82	0.57	20.92	0.43
γ_6	19.90	0.61	19.82	0.54	19.81	0.56	21.01	0.45
η_1	-2.74	0.27	-2.74	0.22	-2.94	0.23	-3.60	0.49
η_2	-4.47	0.28	-4.47	0.27	-5.01	0.27	-5.95	0.50
η_3	-2.79	0.26	-2.79	0.26	-3.01	0.27	-3.98	0.47
ξ_1	0.03	0.02	0.03	0.02	0.02	0.02	0.22	0.13
ξ_2	0.11	0.02	0.11	0.02	0.13	0.02	0.71	0.13
ξ_3	0.08	0.02	0.08	0.02	0.07	0.02	0.50	0.12
δ_0			-3.11	0.23	-3.28	0.37	-3.41	0.36
δ_1			0.11	0.02	0.32	0.02	0.29	0.04
ρ					-0.35	0.03	-0.29	0.06
σ_1^2	16.75		16.75		18.40			
σ_2^2	16.16		16.15		15.33			

SC = saturated covariance model.
MNAR = missing not at random.
MAR = missing at random.
MCAR = missing completely at random.
s.e. = standard error.

12.3.2 *Schizophrenic behaviour data with non-monotone missingness*

Consider the schizophrenic behaviour data in Chapter 10 again. Because abrupt changes among repeated responses were peculiar to schizophrenics, we proposed to use a DHGLM with

$$Y_{ij} = \beta_0 + x_{1ij}\beta_1 + x_{2ij}\beta_2 + t_j\beta_3 + sch_i\beta_4 + sch_i \cdot x_{1ij}\beta_5 + sch_i \cdot x_{2ij}\beta_6 + v_i + e_{ij}$$

where $v_i \sim N(0, \lambda_1)$ is the random effect; $e_{ij} \sim N(0, \phi_i)$; sch_i is equal to 1 if a subject is schizophrenic and 0 otherwise; t_j is the measurement time; x_{1ij} is the effect of PS versus CS; x_{2ij} is the effect of TR versus the average of CS and PS, and

$$\log(\phi_i) = \gamma_0 + sch_i\gamma_1 + sch_i b_i,$$

where $b_i \sim N(0, \lambda_2)$ are the dispersion random effects. We call this model DI (DHGLM with ignorable missingness).

Rubin and Wu (1997) ignored missingness after discussion with the psychologists. However, according to the physicians we consulted, missingness could be caused by eye blinks related to eye movements (responses) (Goosens *et al.*, 2000); this leads to a selection model:

$$\eta = \Phi^{-1}(p_{ij}) = \delta_0 + x_{1ij}\delta_1 + x_{2ij}\delta_2 + sex_i\delta_3 + sch_i\delta_4 + sex_i \cdot x_{1ij}\delta_5$$
$$+ sex_i \cdot x_{2ij}\delta_6 + sex_i \cdot sch_i\delta_7 + \rho Y_{ij}^*.$$

We can combine the model DI with the probit model (DN: DHGLM with non-ignorable missingness). We found that the time effect is not significant in the probit model. The analyses from DHGLMs with and without a model for missingness are in Table 12.6. The negative value of $\hat{\rho}$ supports the physicians' opinions that lower values of the outcome are more likely to be missing at each cycle. However, the conclusions concerning the non–ignorable missingness depend on untestable distributional assumptions so that sensitivity analysis has been recommended. Fortunately, the analysis of responses in these data is not sensitive to the assumption about the heavy tails or the missing mechanism.

12.4 Denoising signals by imputation

Wavelet shrinkage is a popular method for denoising signals corrupted by noise (Donoho and Johnstone, 1994). This problem can be typically expressed as the estimation of an unknown function $f(x)$ from noisy observations of the model

$$y_i = f(x_i) + \epsilon_i, \quad i = 1, 2, \ldots, \quad n = 2^J,$$

where $x_i = i/n$ and the errors ϵ_i are assumed independently $N(0, \sigma^2)$ distributed.

If some data are missing, most wavelet shrinkage methods cannot be directly applied because of two main restrictions: (a) the observations must be equally spaced, and (b) the sample size, say n, must be dyadic, i.e., $n = 2^J$ for some integer J.

Figure 12.1 illustrates an example of a signal with missing values, where we generate a degraded version of a John Lennon image with 70% missing pixels. Panel (b) shows an example with 70% of the image randomly chosen as missing, while in panel (e) 4×4 clusters are treated as missing. The white pixels represent the locations of the missing observations.

Table 12.6 Analyses from DHGLMs for the schizophrenic behaviour data.

Part	Parameter	DI est.	DI s.e.	DI t-value	DN est.	DN s.e.	DN t-value
Response	Int	0.811	0.014	59.29	0.802	0.014	55.89
	x_1	0.006	0.005	1.42	0.004	0.005	0.96
	x_2	-0.121	0.005	-24.63	-0.121	0.005	-24.40
	$time$	-0.002	0.000	-5.23	-0.002	0.000	-5.97
	sch	-0.036	0.019	-1.85	-0.051	0.020	-2.51
	$sch \cdot x_1$	-0.023	0.006	-3.54	-0.022	0.007	-3.33
	$sch \cdot x_2$	-0.004	0.007	-0.59	-0.005	0.007	-0.63
Missing	Int				2.148	0.231	9.30
	x_1				0.065	0.062	1.05
	x_2				-0.276	0.071	-3.86
	sex				-0.085	0.072	-1.18
	sch				-0.072	0.054	-1.30
	$sex \cdot x_1$				-0.171	0.123	-1.39
	$sex \cdot x_2$				-0.379	0.128	-2.97
	$sex \cdot sch$				-0.284	0.103	-2.76
	Y^*				-3.704	0.296	-12.53
Dispersion	λ_1	0.089			0.093		
	γ_0	-5.320			-5.287		
	γ_1	0.149			0.241		
	λ_2	0.583			0.738		

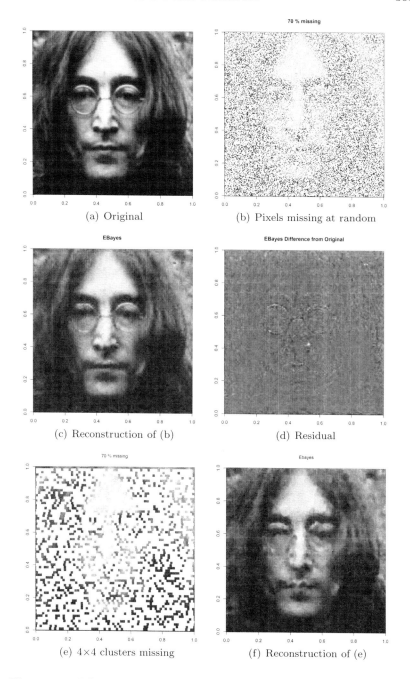

Figure 12.1 *John Lennon image. (a) The original image. (b) The image with 70% missing pixels. (c) The reconstructed image from (b). (d) The residual image (original reconstruction). (e) The image with missing clusters of 4×4 pixels. (f) Reconstruction of (e).*

Kovac and Silverman (2000) suggested a wavelet regression method for missing values based on coefficient-dependent thresholding. However, this method requires an efficient algorithm to compute the covariance structure, so that it may be hard to extend to image data like Figure 12.1. As shown in panels (c) and (f), the h-likelihood method described below recovers the image very effectively.

Suppose the complete data $y_{com} = \{y_{obs}, y_{mis}\}$ follow a normal distribution $N(f, \sigma^2 I_n)$, where y_{obs} denotes the subset of observed data and y_{mis} denotes the subset of missing data. Note that $f = (f_1, f_2, \ldots, f_n)$, where $f_i = (W^T \theta)_i$, W is the orthogonal wavelet operator and θ denotes wavelet coefficients. Suppose that

$$\ell_{com}(\theta, \sigma^2; y_{com}) = \log f_{\theta,\sigma^2}(y_{com})$$

is the complete-data loglihood. We compare the EM algorithm with the h-likelihood approach.

Since the wavelet transform requires equally spaced observations and the dyadic sample size, the wavelet algorithm cannot be directly applied in the presence of missing data. Thus, the imputation of missing data is necessary. Lee and Meng (2005) developed EM-based methods for the imputation by solving equations given by

$$E[\hat{f}_{com}|y_{obs}, f = \hat{f}_{obs}] = \hat{f}_{obs}, \tag{12.5}$$

where $\hat{f}_{com}$ and $\hat{f}_{obs}$ are estimates of f obtained from the complete data y_{com} and the observed data y_{obs}, respectively. To solve (12.5), we need the conditional expectation analogous to the E-step of the EM algorithm. Since the exact E-step is analytically infeasible in a wavelet application, Lee and Meng proposed a Monte Carlo simulation or an approximation that solves the equations (12.5). However, their methods suffer from slow convergence and distorted results when the proportion missing is large (over 30 %).

Kim et al., (2006) showed how the h-likelihood method handles this problem. As in Chapter 4.8 we impute random parameters (here missing data) by maximum h-likelihood estimates and then estimate fixed parameters (θ, σ^2) by maximizing appropriately modified criteria. Missing values y_{mis} are treated as unobserved random parameters and independence of y_{obs} and y_{mis} is assumed. Suppose $y_{obs} = (y_1, \ldots, y_k)$ consists of k observed values and $y_{mis} = (y_{k+1}, \ldots, y_n)$ represents $(n-k)$ missing values. Let $M_{obs} = -\frac{1}{2\sigma^2} \sum_{i=1}^{k}(y_i - f_i)^2$ and $M_{mis} = -\frac{1}{2\sigma^2} \sum_{i=k+1}^{n}(y_i - f_i)^2$. Then we define the h-log-likelihood as

$$\begin{aligned} h &= h(\theta, \sigma^2, y_{mis}; y_{obs}) = \ell_{com}(\theta, \sigma^2; y_{com}) \tag{12.6} \\ &= \ell_{obs}(\theta, \sigma^2; y_{obs}) + \ell_{mis}(\theta, \sigma^2; y_{mis}), \end{aligned}$$

where the log-likelihood for the observed data, $\ell_{obs}(\theta, \sigma^2; y_{obs}) = M_{obs} - \frac{k}{2}\log\sigma^2$, and the log-likelihood for the missing data, $\ell_{mis}(\theta, \sigma^2; y_{mis}) = M_{mis} - \frac{n-k}{2}\log\sigma^2$. In (12.6) we highlight a main philosophical difference between the complete-data likelihood and the h-likelihood. In the former, missing data y_{mis} are unobserved while in the latter y_{mis} are unobserved nuisance parameters. Therefore, instead of using the E-step, we use a profiling method to adjust for nuisance random parameters.

To estimate θ, if observed data are equally spaced and dyadic, we maximize the *penalized loglihood* with the observed data

$$P_{obs} = M_{obs} - \lambda q(\theta), \qquad (12.7)$$

where λ is the thresholding value for wavelet shrinkage and $q(\theta)$ is a penalty function. However, the above likelihood cannot be directly implemented for wavelet estimation because of missing values. By the h-likelihood approach, the missing data y_{mis} can be imputed by solving the score equations $\partial h/\partial y_i = -(y_i - f_i)$, $i = k+1, \ldots, n$. This results in the solution $\hat{y}_{mis,i} = f_i$ that maximizes the h-likelihood of (12.6).

Now consider the *penalized loglihood* of the complete data

$$P_{com} = M_{obs} + M_{mis} - \lambda q(\theta).$$

The profile log-likelihood is often used to eliminate nuisance parameters. Thus, on eliminating y_{mis} the profile h -loglihood becomes

$$P_{com}\big|_{y_{mis}=\hat{y}_{mis}} = (M_{obs} + M_{mis})\big|_{y_{mis}=\hat{y}_{mis}} - \lambda q(\theta) = P_{obs}. \qquad (12.8)$$

Thus, we can obtain the wavelet estimate of θ by maximizing (12.8) after missing values are imputed. This derivation does not use any Monte Carlo simulation or approximation so that the proposed method is indeed very fast.

In summary, we impute missing data by maximizing the h-likelihood (12.6) and then estimate θ by minimizing penalized least squares (12.8). Since the derivation minimizes the penalized log-likelihood of (12.8) for parameter estimation, the proposed approach provides a simple way to obtain good wavelet estimates (Kim *et al.*, 2006). Thus the h-likelihood provides a very effective imputation, which gives rise to an improved wavelet algorithm (Oh *et al.*, 2006).

CHAPTER 13

Multiple testing

Recently, technological developments in areas such as genomics and neuroimaging have led to situations where thousands or more hypotheses need to be tested simultaneously. The problem of such large-scale simultaneous testing is naturally expressed as a prediction problem for finding the true results in an optimal way (Lee and Bjørnstad, 2013). In this chapter, we formulate hypothesis testing as an HGLM with discrete random effects. Up to now, we have investigated models with continuous random effects, where the choice for the scale of random effects is crucial in defining the h-likelihood. From the transformation of continuous random effects in Chapter 4, we see that the Jacobian term appears in the extended likelihood, so that the mode of extended likelihood is not invariant with respect to the transformation of random effects. However, for the transformation of discrete random effects, such a Jacobian term does not appear in the extended likelihood. Thus, any extended likelihood defined on any scale of the discrete random effect becomes the h-likelihood. Until now, we have studied the estimation of parameters and prediction of unobservables by using h-likelihood. In this chapter we study how to use the h-likelihood for the testing of hypotheses by using discrete random effects. For this, we first review the classical single hypothesis testing.

13.1 Single hypothesis testing

Suppose we are interested in a simple hypothesis testing on the mean of a random variable $Y \sim N(\mu, 1)$,

$$H_0 : \mu = \mu_0 \text{ versus } H_1 : \mu = \mu_1. \qquad (13.1)$$

Let the random state $s = 0$ if H_0 is true, and $s = 1$ if H_1 is true. Then, given an observed data $Y = y$, a prediction of the true state s and the related inferential uncertainty are of interest. Here the h-likelihood becomes

$$L(s; y, s) = f(y, s) = f(y|s)P(s) = P(s|y)f(y),$$

where $p_0 = P(s = 0)$, $f(y|H_0) = f(y|s = 0)$, $f(y|H_1) = f(y|s = 1)$ and $f(y) = p_0 f(y|H_0) + (1 - p_0) f(y|H_1)$. In this chapter, we use the probability function $P()$ to highlight discrete random effects s. Note that $P(g(s)) = P(s)$ for any monotonic transformation of $u = g(s)$, so that $L(s; y, s) = L(u; y, u)$, i.e., the mode is invariant with respect to any monotonic transformation. This means that all extended likelihoods are h-likelihoods for discrete random effects.

Let $L(s) \equiv L(s; y, s) = f(y, s)$, so

$$L(s = 0) = f(y|H_0)P(s = 0) = p_0 N(\mu_0, 1)$$
$$L(s = 1) = f(y|H_1)P(s = 1) = (1 - p_0)N(\mu_1, 1).$$

The marginal density of y is

$$f(y) = f(y|H_0)P(s = 0) + f(y|H_1)P(s = 1),$$

so

$$P(s = i|y) = \frac{f(y|H_i)P(s = i)}{f(y)}$$

is the predictive probability in Chapter 4, satisfying

$$P(s = 0|y) + P(s = 1|y) = 1.$$

It is the unique predictive probability (Bjørnstad , 1990). The h-likelihood ratio for $s = 1$ versus $s = 0$ becomes the ratio of predictive probabilities because

$$R = \frac{L(s = 1)}{L(s = 0)} = \frac{P(s = 1|y)f(y)}{P(s = 0|y)f(y)} = \frac{P(s = 1|y)}{P(s = 0|y)}.$$

Let δ be a predictor (test) for the true state s, i.e. $\delta = 1$ (discovery) if we reject H_0 and $\delta = 0$ (non-discovery) if we do not reject H_0. We can now show that the optimal test is determined by the ratio of predictive probabilities R, which is equivalent to the h-likelihood ratio, but the latter is much easier to compute than the former.

Table 13.1 *Outcome of single hypothesis test.*

Random state	$\delta = 0$	$\delta = 1$
$s = 0$	No error	Type I error
$s = 1$	Type II error	No error

The outcome of the test can be summarized as in Table 13.1. Type I error is $P(\delta = 1|H_0)$, type II error is $P(\delta = 0|H_1)$ and the power is $P(\delta = 1|H_1)$. With the loss

$$s(1 - \delta) + \lambda(1 - s)\delta,$$

given the data, we have the risk,

$$E\{s(1-\delta)+\lambda(1-s)\delta|y\} = E\{s+\lambda(1-s)\delta-s\delta|y\}$$
$$= P(s=1|y)+P(s=0|y)(\lambda-R)\delta.$$

Thus, for each value of λ, the optimal test δ^λ minimizing the risk is determined by the h-likelihood ratio,

$$\delta^\lambda = I\{R>\lambda\}.$$

Note, however, that the h-likelihood ratio

$$R = \frac{(1-p_0)}{p_0}L$$

is the product of $(1-p_0)/p_0$, the odds for the hypotheses, and

$$L = \frac{N(\mu_1,1)}{N(\mu_0,1)} = \frac{f(y|H_1)}{f(y|H_0)},$$

the Neyman-Pearson likelihood ratio. With a single hypothesis testing, an estimation of p_0 (and therefore predictive probability) may not be possible, so that we need to define the optimal test δ^λ without estimating p_0. For example, if $\hat{p}_0 = 0$ or 1, δ^λ becomes 0 or 1 without depending upon λ. Thus, in the single hypothesis testing, we define the optimal δ^λ as

$$\delta^\lambda = I\{L>\lambda^*\},$$

by determining $\lambda^* = \lambda p_0/(1-p_0)$ for some given p_0 with $0 < p_0 < 1$, satisfying

$$P(\delta^\lambda = 1|H_0) \le \alpha$$

for a pre-specified significant level $\alpha = 0.05$ or 0.01. It is well established that this Neyman-Pearson likelihood-ratio test δ^λ is generally the most powerful (efficient). Under our random effect model, the optimal test δ^λ is a predictor for random effect s, and we can view the likelihood ratio test as a prediction of discrete random effect controlling a prediction error (here type I error) at a prespecified significant level α.

Consider a hypothesis testing

$$H_0 : \mu = 0 \text{ versus } H_1 : \mu \ne 0.$$

When the null hypothesis is rejected, it is common to state the sign of non-zero effect $\mu \ne 0$. For example, it is common to classify an abnormality as an abnormally positive or abnormally negative state. Thus, we should consider three states for hypotheses

$$H_0 : \mu = 0, \ H_1 : \mu < 0 \text{ and } H_2 : \mu > 0$$

and a test with three actions, $\delta = 0, 1$ and 2. A single hypothesis testing with three (directional) states was addressed by Cox and Hinkley (1974). Leventhal and Huynh (1996) pointed out that non–directional two-sided tests overestimate power because the standard power $P(\delta \neq 0 | s \neq 0)$ contains the type III errors

$$P(\delta = 2 | s = 1) \text{ and } P(\delta = 1 | s = 2).$$

They proposed the use of a revised power that excludes type III errors. We shall show how these directional errors can be controlled in multiple testing.

13.2 Multiple testing

Most literature on multiple testing focuses on the error control of the test without paying much attention to the power of the test. In this chapter, we model multiple testing as an HGLM with discrete random effects and show how the h-likelihood can lead to an extension of the most powerful Neyman-Pearson likelihood ratio test for single hypothesis testing to multiple testing.

13.2.1 False discovery rate

Suppose we have N null hypotheses $H_1, \cdots, H_N$ to test simultaneously. Table 13.2 summarizes the possible outcomes of a multiple testing, where V_{01} is the number of false discoveries (positives) and V_{10} is the number of false non-discoveries (nulls). In multiple testing, we observe M_0, M_1 and N while all remaining cells such as N_0, N_1, V_{00}, V_{01}, V_{10} and V_{11}, in Table 13.2 are unobserved random variables. In Chapter 4, we show that the predictive probability gives asymptotically valid inferences for the prediction of unobserved random variables. In this chapter, we show that it serves as the optimal test for given fixed parameters. Thus, it is the asymptotically optimal test where MLEs for fixed parameters are asymptotically consistent.

In multiple testing, a control of the family-wise error rate

$$P(V_{01} > 0 | \text{all } s_i = 0)$$

has been studied. Instead of controlling the family-wise error rate, Benjamini and Hochberg (1995) proposed to control the false discovery rate (FDR) among declared discoveries. They define FDR as the expected proportion of errors among rejected hypotheses (discoveries). The proportion of errors among the discoveries is V_{01}/M_1 and the authors suggested controlling the false discovery rate $E(V_{01}/M_1)$. Following Efron

Table 13.2 *Outcomes of multiple testing*

Random state	$\delta = 0$	$\delta = 1$	Total
$s = 0$	V_{00}	V_{01}(Type I error)	N_0
$s = 1$	V_{10} (Type II error)	V_{11}	N_1
Total	M_0	M_1	N

(2004) in this book, we use the (marginal) FDR as

$$\text{FDR} = E(V_{01})/E(M_1)$$

and the false non-discovery rate (FNDR),

$$\text{FNDR} = E(V_{10})/E(M_0).$$

Note that $E(V_{01})/E(N_0)$ and $E(V_{10})/E(N_1)$ are type I and type II error rates, respectively. The type I is the false discovery rate under the null. We can also develop an optimal multiple test controlling type I error rate too.

13.2.2 *Directional errors*

In two-sided multiple testing, there are two (directional) alternatives (states) to yield three types of errors as in Table 13.3. Let V_{jk} for $j \neq k$ be the number of observations in state j falsely declared to be the state k. Let $T_1 = V_{01} + V_{02}$, $T_2 = V_{10} + V_{20}$, and $T_3 = V_{12} + V_{21}$ be the total numbers of cases corresponding to the types I, II, and III errors, respectively. With $M = M_1 + M_2$ we can define FDR for the type I error rate as

$$\text{FDR}_I = E(T_1)/E(M),$$

and those for the sum of the types I and III error rates as

$$\text{FDR}_{I+III} = E(T_1 + T_3)/E(M).$$

The false non-discovery rate (FNDR) and the false assignment rate (FAR) under alternatives are defined by

$$\text{FNDR} = E(T_2)/E(M_0) \text{ and } \text{FAR} = E(T_2 + T_3)/E(N_1 + N_2).$$

It is important to note that the likelihood approach allows consistent prediction of all of these error rates, which is useful for understanding the performance of a test.

We may control FDR_{I+III} minimizing type II error rate or control FDR_I

minimizing sum of type II and type III error rates. If we control FDR$_I$ minimizing the type II error, the type III error is considered as power of the test, which is not desirable.

Table 13.3 *Outcomes of multiple testing with two alternatives.*

Hypothesis	Declared as null	Declared as alternative 1	Declared as alternative 2	Total
Null	V_{00}	V_{01}(Type I error)	V_{02}(Type I error)	N_0
Alternative 1	V_{10}(Type II error)	V_{11}	V_{12}(Type III error)	N_1
Alternative 2	V_{20}(Type II error)	V_{21}(Type III error)	V_{22}	N_2
Total	M_0	M_1	M_2	N

13.3 Multiple testing with two states

To give a clear context for the notations, consider the neuroimaging data example in Section 13.5.2. Suppose the response y_{ij1} for the i–th site of the j–th individual in the control group can be modeled for $i = 1, \ldots, N$ and $j = 1, \ldots, n_1$ as

$$y_{ij1} = \xi_i + \epsilon_{ij1},$$

while response y_{ij2} in the treatment group is modeled for $j = 1, \ldots, n_2$ as

$$y_{ij2} = \xi_i + w_i + \epsilon_{ij2},$$

where ξ_i is the site effect, w_i is the treatment effect for the i–th site, and ϵ_{ijm} is the error with $E(\epsilon_{ijm}) = 0$ and $Var(\epsilon_{ijm}) = \phi_{im}$, for $m = 1, 2$. We express the model in terms of the difference in means:

$$d_i = \bar{y}_{i2} - \bar{y}_{i1},$$

where $\bar{y}_{im} = \sum_j y_{ijm}/n_m$ and $\psi_i = \phi_{i1}/n_1 + \phi_{i2}/n_2$. Then we have

$$E(d_i|w_i) = w_i \quad \text{and} \quad Var(d_i|w_i) = \psi_i.$$

The formal null hypotheses are of the form H_{0i}: $w_i = 0$. The inferential problem of interest is to identify only those sites with large effect sizes w_i, not those with w_i close to 0. To achieve this, it is natural to assume that for the "Uninteresting" (null) sites, the w_is are independent with

$$E(w_i|H_{0i}) = 0 \quad \text{and} \quad Var(w_i|H_{0i}) = \sigma^2,$$

with typically $0 < \sigma \ll 1$. The "Interesting" (alternative) sites are assumed to be independent with

$$E(w_i|H_{1i}) = \mu \neq 0 \quad \text{and} \quad Var(w_i|H_{1i}) = \tau^2.$$

Since this multiple testing problem has two states, let $s_i = 1$ if site i is

Interesting (H_{1i}), and $s_i = 0$ if site i is Uninteresting (H_{0i}). Suppose $P(s_i = 0) = p_0$ and $P(s_i = 1) = 1 - p_0$. We first consider the case that s_i are independent. Suppose we have the following hierarchical model:

Conditional on w_i and s_i, $E(d_i|w_i, s_i) = w_i$ and $Var(d_i|w_i, s_i) = \psi_i$

Conditional on $s_i = 0$, $E(w_i|H_{0i}) = 0$ and $Var(w_i|H_{0i}) = \sigma^2$

Conditional on $s_i = 1$, $E(w_i|H_{1i}) = \mu$ and $Var(w_i|H_{1i}) = \tau^2$.

In this model we have parameters, $\theta = (p_0, \sigma, \mu, \tau, \psi_1, \cdots, \psi_N)$, and $2N$ unobservables, $v = (w, s)$, where w and s are respectively the vector of w_i and s_i. Let y be the set of all observations. The h-likelihood for the unknown quantities (v, θ) is defined to be

$$L(v, \theta; y, v) = f_\theta(y, v) = f_\theta(y)P_\theta(v|y).$$

Parameters (p_0, σ, μ, τ) can be estimated by maximizing the marginal likelihood. The error variances ψ_i are estimated by standard variance estimators $\hat{\psi}_i$ (Lee and Bjørnstad, 2013). We call them the MLE for $\hat{\theta}$.

If effect sizes w_i are not of interest, we can eliminate them by integration. This leads to a hierarchical model for $d = (d_1, \cdots, d_N)$ with independent d_i:

$$\text{Given } s_i \;=\; 0, \; E(d_i|H_{0i}) = 0 \text{ and } Var(d_i|H_{0i}) = \psi_i + \sigma^2,$$
$$\text{given } s_i \;=\; 1, \; E(d_i|H_{1i}) = \mu \text{ and } Var(d_i|H_{1i}) = \psi_i + \tau^2.$$

Without loss of generality, from now on we assume that the null and alternative distributions are normal. The h-likelihood is given by

$$L(s) \equiv L(s, \theta; d, s) = f_\theta(d, s) = \prod_{i=1}^{N} f_\theta(d_i, s_i) = \prod_{i=1}^{N} L(s_i),$$

where $L(s_i) \equiv L(s_i, \theta; d_i, s) = f_\theta(d_i, s_i)$, so we have

$$L(s_i = 1) = P(s_i = 1)f_\theta(d_i|s_i = 1) = (1 - p_0)N(\mu, \psi_i + \tau^2),$$
$$L(s_i = 0) = P(s_i = 0)f_\theta(d_i|s_i = 0) = p_0 N(0, \psi_i + \sigma^2).$$

In practice, θ is evaluated at the estimated value as described above.

13.3.1 Optimal test

The h-likelihood ratio of $s_i = 1$ versus $s_i = 0$ is

$$R_i = \frac{L(s_i = 1)}{L(s_i = 0)} = \frac{P(s_i = 1|d_i)f_\theta(d_i)}{P(s_i = 0|d_i)f_\theta(d_i)} = \frac{P(s_i = 1|d_i)}{P(s_i = 0|d_i)} \frac{(1 - p_0)}{p_0}L_i,$$

where

$$f_\theta(d_i) = f_\theta(d_i|H_{0i})p_0 + f_\theta(d_i|H_{1i})(1 - p_0)$$

is the marginal likelihood component from the i–th difference d_i and

$$L_i = f_\theta(d_i|H_{1i})/f_\theta(d_i|H_{0i}) = N(\mu, \psi_i + \tau^2)/N(0, \psi_i + \sigma^2)$$

is the Neyman-Pearson likelihood ratio test for the i–th hypothesis. Let δ_i be a test for the i–th hypothesis H_i, i.e., $\delta_i = 1$ (discovery) if H_i is rejected and 0 (non-discovery) if not. Consider the loss

$$\sum \{s_i(1 - \delta_i) + \lambda(1 - s_i)\delta_i\}.$$

Similarly to Section 13.1 it can be shown that given λ, the optimal rule $\delta^\lambda = \{\delta_1^\lambda, \cdots, \delta_N^\lambda\}$, becomes

$$\delta_i^\lambda = I(R_i > \lambda). \tag{13.2}$$

This shows that the likelihood ratio test in single hypothesis testing can be straightforwardly extended to multiple testing via the h-likelihood.

In multiple testing,

$$p_0 = \frac{E(N_0)}{N}$$

becomes an estimable quantity, so that the predictive probabilities can be properly estimated. Thus, R_i is directly used to control the error rate. In a single hypothesis testing, there is no parameter to estimate and δ^λ is optimal. However, in multiple testing, δ^λ is asymptotically optimal provided that the parameters are consistently estimated. If not, numerical studies would be necessary to check whether the proposed test maintains the declared error rate. For controlling the FDR, we need to find an appropriate λ. Note that

$$\text{FDR}(\lambda) = \frac{E(V_{01})}{E(M_1)} = \frac{\sum_i P(s_i = 0, \delta_i^\lambda = 1)}{E(\sum_i \delta_i^\lambda)},$$

where $M_1 = \sum_i \delta_i^\lambda$ and $V_{01} = \sum_i \delta_i^\lambda(1 - s_i)$. It follows that an estimated FDR is given by

$$\widehat{\text{FDR}}(\lambda) = \frac{\hat{p}_0 \sum_i P(R_i > \lambda|H_{0i})}{\sum_i I(R_i > \lambda)}.$$

FDR can be controlled by calculating $\widehat{\text{FDR}}(\lambda)$ at various values of λ and choosing λ according to $\widehat{\text{FDR}}(\lambda) \leq \alpha$ for pre-specified significant level α.

13.4 Multiple testing with three states

In most applications, a discovery is always made with a statement whether the effect is positive or negative. Let $s_i = 0$ if the i–th null hypothesis is true, and $s_i = 1$ or 2 if the sign of the effect is negative or positive, respectively. The probability of type III error should never be considered as part of the power of the test. Conditional on $s_i = k$, suppose $w_i \sim N(\mu_k, \sigma_k^2)$ for $k = 0, 1, 2$. The inferential interest is to identify cases with large effect sizes w_i and not those with w_i close to 0, so that we let $\mu_0 \equiv 0$ with a small $\sigma_0 << 1$. Here, s_i is of inferential focus, whereas w_i is not. We consider the following model:

$$\text{Given } s_i = k, \ d_i \sim N(\mu_k, \psi_i + \sigma_k^2), \qquad (13.3)$$

with the h-likelihood

$$L(s) \equiv L(\theta, s; y, s) = f_\theta(d, s) = f_\theta(d|s)P(s) = \prod_i f_\theta(d_i|s_i)P(s_i).$$

Lee and Lee (2016) showed that the optimal test $\delta^{(I+III)} = \{\delta_i^{(I+III)}; i = 1, \cdots, N\}$, which controls FDR_{I+III} with the smallest expected frequency of type II errors becomes

$$\delta_i^{(I+III)} = 2 \text{ if } \frac{P(s_i \neq 0|d_i)}{P(s_i \neq 2|d_i)} = \frac{L(s_i \neq 0)}{L(s_i \neq 2)} > \lambda \text{ and } \frac{P(s_i = 2|d_i)}{P(s_i = 1|d_i)} = \frac{L(s_i = 2)}{L(s_i = 1)} > 1,$$

$$= 1 \text{ if } \frac{P(s_i \neq 0|d_i)}{P(s_i \neq 1|d_i)} = \frac{L(s_i \neq 0)}{L(s_i \neq 1)} > \lambda \text{ and } \frac{P(s_i = 2|d_i)}{P(s_i = 1|d_i)} = \frac{L(s_i = 2)}{L(s_i = 1)} \leq 1,$$

$$= 0 \text{ otherwise.}$$

Similarly, the optimal test $\delta^{(I)} = \{\delta_i^{(I)}; i = 1, \cdots, N\}$ that controls FDR_I with the smallest expected frequency of sum of the type II and type III errors becomes

$$\delta_i^{(I)} = 2 \text{ if } \frac{P(s_i = 2|d_i)}{P(s_i = 0|d_i)} = \frac{L(s_i = 2)}{L(s_i = 0)} > \lambda \text{ and } \frac{P(s_i = 2|d_i)}{P(s_i = 1|d_i)} = \frac{L(s_i = 2)}{L(s_i = 1)} > 1,$$

$$= 1 \text{ if } \frac{P(s_i = 1|d_i)}{P(s_i = 0|d_i)} = \frac{L(s_i = 1)}{L(s_i = 0)} > \lambda \text{ and } \frac{P(s_i = 2|d_i)}{P(s_i = 1|d_i)} = \frac{L(s_i = 2)}{L(s_i = 1)} \leq 1,$$

$$= 0 \text{ otherwise.}$$

This shows that all the optimal tests controlling various error rates are represented by h-likelihood ratios of components (ratios of predictive probabilities).

Let r be the total number of discoveries. For a given λ and θ, FDR_I and FNDR can be estimated by

$$\widehat{\text{FDR}}_I(\lambda) = r^{-1} \sum_i P_{\hat{\theta}}(s_i = 0|d_i)I(\delta_i \neq 0) \text{ and}$$

$$\widehat{\text{FNDR}}(\lambda) = (N - r)^{-1} \sum_i P_{\hat{\theta}}(s_i \neq 0|d_i)I(\delta_i = 0).$$

The step-down test procedure for controlling FDR$_I$ (Lee and Lee , 2016) is

$$
\begin{aligned}
\text{let } \lambda^* &= \inf\{\lambda : \widehat{\text{FDR}}_I(\lambda) \leq \alpha\}; \text{then} \\
\delta_i^{(I)} &= 2 \text{ if } \frac{L(s_i = 2)}{L(s_i = 0)} > \lambda^* \text{ and } \frac{L(s_i = 2)}{L(s_i = 1)} > 1, \quad (13.4) \\
&= 1 \text{ if } \frac{L(s_i = 1)}{L(s_i = 0)} > \lambda^* \text{ and } \frac{L(s_i = 2)}{L(s_i = 1)} \leq 1, \\
&= 0 \text{ otherwise.}
\end{aligned}
$$

Similarly, by using

$$
\widehat{\text{FDR}}_{I+III} = r^{-1} \sum_i \sum_{k=1}^{2} \{P_{\hat{\theta}}(s_i = k|d_i) I(\delta_i = k)\},
$$

FDR$_{I+III}$ can be controlled at a specific level.

13.4.1 Multiple testings for correlated tests

Most multiple testing procedures assume independence among tests. However, in microarray experiments, different genes may cluster into groups along biochemical pathways, and in neuroimaging data, adjacent voxels are correlated. Efron (2007) noted that dependency may result in overly liberal or overly conservative testing procedures. Sabatti *et al.* (2003) found that Benjamini and Hochberg's procedure (1995) suffers from power loss in genetic data under dependency among markers. Benjamini and Yekutieli (2001) and Efron (2007) proposed robust tests against general correlation structures in neuroimaging and genetic data, respectively, but these procedures often encounter severe loss of power, thus distorting conclusions (Lee and Lee , 2016).

To account for the correlation structure, Lee and Lee (2016) use the hidden Markov random field models (HMRFMs) for s. Suppose that $s = \{s_i; i \in S\}$ with $S = \{1, \ldots, N\}$ satisfies the Markovian property

$$
P(s_i|s_{S\setminus\{i\}}) = P(s_i|s_j, j \in \mathcal{N}_i),
$$

where $S \setminus \{i\} = \{j; j \in S \text{ and } j \neq i\}$ and $\mathcal{N}_i$ is the set of neighbors of the ith site. Among various models with the Markovian property, the multilevel logistic model for s is commonly used (Elliot *et al.*, 1984):

$$
P(s) \propto \exp\left[-\sum_{i \in S}\sum_{k=0}^{2}\{I(s_i = k)(\beta_k + \gamma \sum_{j \in \mathcal{N}_i} I(s_j = k))\}\right], \quad (13.5)
$$

where $\beta = (\beta_0 \equiv 0, \beta_1, \beta_2)$ and γ denote the fixed unknown parameters.

In this model, β_k has the effect of controlling the relative percentage of the hypotheses belonging to the k–th state, and the scalar γ plays the role of the interaction parameter because $\sum_{j \in \mathcal{N}_i} I(s_i = k) I(s_j = k)$ is the number of neighbors in $\mathcal{N}_i$ that belong to the same state as s_i. When γ is negative, neighbors tend to have the same effect direction, and vice versa for positive γ. When $\gamma = 0$, this model reduces to the independent model of Lee and Bjørnstad (2013).

Given s, we assume d_i are independent to give the h-likelihood

$$L(\theta, s; d, s) = f_\theta(d, s) = f_\theta(d|s) P(s) = P(s) \prod_{i \in S} f_\theta(d_i|s) \qquad (13.6)$$

where $\theta = (\mu_1, \mu_2, \sigma_0^2, \sigma_1^2, \sigma_2^2, \beta_1, \beta_2, \gamma)$ are the parameters, $d = (d_1, \cdots, d_N)$ and $s = (s_1, \cdots, s_N)$. It is straightforward to show that the optimal test can be obtained by replacing $P(s_i = k|d_i)$ for independent case by

$$P(s_i = k|d) = \sum_{s_{S \setminus \{i\}}} P(s_1, \cdots, s_i = k, \cdots, s_N | d),$$

which is computationally demanding because it requires summation over all possible realizations of $s_{S \setminus \{i\}}$ if N is large. Lee and Lee (2016) developed a method (HM) to estimate θ by using the mean-field approximation and compute $P(s_i = k|d)$ by the Gibbs sampling (Geman and Geman , 1984).

13.5 Examples

Via numerical studies, Lee and Bjørnstad (2013) and Lee and Lee (2016) showed that existing conventional methods control the FDR well at the nominal level. However, the methods often have low power. The h-likelihood ratio test maintains an error rate well at the nominal level, while giving more power compared to the existing methods.

13.5.1 Simulated data example

For 100 simulated data sets, we set $n_1 = n_2 = 10$ and generated lattice images of size 32×32 ($N = 1024$). Benjamini and Yekutieli (2001) showed that Benajmini and Hochberg's (1995) method (BH) can control FDR under so-called positive dependence and proposed a modified BH (which we call BY) for control FDR under general dependency. For BH and BY, we followed the directional FDR procedure of Benjamini and Yekutieli (2005). We also considered Lee and Bjørnstad's method (LB) under the assumption of independence (Lee and Bjørnstad , 2013).

Hidden fields s were generated from model (13.5) at $\beta_1 = \beta_2 = 0$ and $\gamma = -1$. We considered a first-order neighborhood with four neighbors for each pixel (Besag, 1986). Given $s_i = k$, Lee and Lee (2016) generated $w_i \sim N(\mu_k, \sigma_k^2)$ with $\mu_1 = \mu > 0$, $\mu_2 = -\mu$, $\sigma_0^2 = 0.05$, and $\sigma_1^2 = \sigma_2^2 = 0.1$. Finally, given w_i, responses y_{ij1} and y_{ij2} were generated from $N(0,1)$ and $N(w_i, 1)$, respectively. Figure. 13.1 shows a simulated data set and test results when $\mu = 1$. Figure. 13.1 (a) and (b) show the observed fields d and true states s, respectively. Because γ is negative, the same states tend to be clustered.

The tests δ are predictions of the true states s. Figures 13.1 (c) through (f) show predictions of true states from BH, BY, LB and HM when controlling FDR_{I+III} at 0.1. We found that all tests maintain FDR_{I+III} very well at the nominal level 0.1. From Figure 13.1, HM has the best prediction with FNDR around 17%. FNDRs of BH and LB, based on the independence assumption, are 45% and 42%, respectively. The FNDR of BY is 61%. Thus, the BY designed for general correlation structures has the least power, worse than those under an independence assumption. BH and BY do not provide the estimation of error rates, while likelihood methods LB and HM provide estimations of error rates. By comparing LB and HM, we see that the proper modeling of correlation is very important to enhance the power of a test.

13.5.2 Neuroimaging data example

Little is known about gender-related differences in behavioral performance or the functional activation of brain regions during the resting state. During specific cognitive tasks such as language use or visuospatial tasks, there is clearly a gender difference (Bell et al., 2006). Males have significantly greater glucose metabolism than females in bilateral motor cortices and the right temporal region, including the hippocampus, which may explain males' typically superior visuospatial and motor functions. In contrast, females tend to have greater glucose metabolism in the left frontal cortex, including the inferior and orbitofrontal cortices, that may be associated with better language and emotional processing. Lee and Lee (2016) consider positron emission tomography (PET) data from the study of the Korean standard template by Lee et al. (2003). The data consist of scans of 28 healthy males and 22 healthy females. Each image has $N = 189{,}201$ voxels. For detailed descriptions of the data, refer to Lee et al. (2003).

The goal was to identify the significantly different voxels of the brains of males and females. Lee and Bjørnstad's (2013) established that the

Figure 13.1 *Examples of image data. (a) Observed fields d. (b) True hidden fields z. Decision rules of (c) BH, (d) BY, (e) LB and (f) HM controlling* FDR_{I+III} *at 0.1.*

White = null. Gray = Alternative with a negative sign. Black = Alternative with a positive sign.

LB test is more powerful than existing methods. However, Figure 13.2 shows that the HM test is much more powerful than the independent model (LB). For example, in Figure. 13.2 (a) and (b), when controlling FDR_{I+III} at level 0.01, LB under independence modeling gives 525 negatively activated and 503 positively activated voxels, while HM yields 38,239 negatively activated and 67,081 positively activated voxels. We can see that by finding a better fitted model for a correlation structure of tests, the power of the tests can be enhanced.

For simplicity we sometimes assume normal distribution, but this is unnecessary. All our results are applicable if a statistical model can be built for multiple testing to allow the h-likelihood. This h-likelihood ratio test is optimal for certain models; see Lee and Bjørnstad (2013) for optimality under independent models and Lee and Lee (2016) for correlated models. We see that searching for a better fitted correlation model among tests is important to improve the power of the tests. In this book, we have shown various likelihood-based tests, model selection criteria and model checking tools for model selections that can be used to find better fitted models to achieve power gains in multiple tests. Finally, a random effect model provides insight and thorough understanding for hypothesis testing.

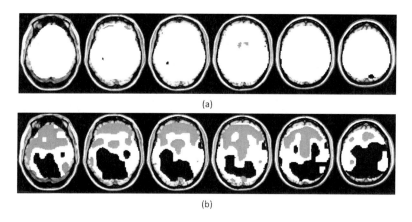

(a)

(b)

Figure 13.2 *Multiple testing for neuroimage data. (a) LB when controlling FDR_{I+III} at 0.01. (b) HM when controlling FDR_{I+III} at 0.01. The gray (black) regions are negatively (positively) activated.*

CHAPTER 14

Random effect models for survival data

In this chapter we study how the GLM class of models can be applied for the analysis of data in which the response variable is the lifetime of a component or the survival time of a patient. Survival data usually refers to medical trials, but the ideas are useful also in industrial reliability experiments. In industrial studies interest often attaches to the average durations of products: when we buy new tyres we may ask how long they will last. However, in medical studies such a question may not be relevant. For example some patients may have already outlived the average lifetime. So a more relevant question would be 'now that the patient has survived to the present age, what will his or her remaining lifetime be if he or she takes a certain medical treatment?' Thus, hazard modelling is often more natural in medical studies, while in industrial reliability studies modelling the average duration time is more common.

In survival-data analysis censoring can occur when the outcome for some patients is unknown at the end of the study. We may know only that a patient was still alive at certain time, but the exact failure time is unknown, either because the patient withdrew from the study or because the study ended while the patient was still alive. Censoring is so common in medical experiments that statistical methods must allow for it if they are to be generally useful. In this chapter we assume that censoring occurs missing at random (MAR) as described in Section 12.2. We show how to handle more general types of missingness in Chapter 12.

14.1 Proportional-hazard model

Suppose that the survival time T for individuals in a population has a density function $f(t)$ with the corresponding distribution function $F(t) = \int_0^t f(s)ds$, which is the fraction of the population dying by time t. The *survivor function*

$$S(t) = 1 - F(t)$$

measures the fraction still alive at time t. The *hazard function* $\alpha(t)$ represents the instantaneous risk, in that $\alpha(t)\delta t$ is the probability of dying in the next small time interval δt given survival to time t. Because

$$\text{pr(survival to } t + \delta t) = \text{pr(survival to } t)\text{pr(survival for } \delta t | \text{survival to } t)$$

we have

$$S(t + \delta t) = S(t)\{1 - \alpha(t)\delta t\}.$$

Thus,

$$S(t) - S(t + \delta t) = \int_{t}^{t+\delta t} f(s)ds = f(t)\delta t = S(t)\alpha(t)\delta t,$$

so that we have

$$\alpha(t) = \frac{f(t)}{S(t)}.$$

The cumulative hazard function is given by

$$\Lambda(t) = \int_{0}^{t} \alpha(s)ds = \int_{0}^{t} \frac{f(s)}{S(s)}ds = -\log S(t),$$

so that

$$f(t) = \alpha(t)S(t) = \alpha(t)\exp\{-\Lambda(t)\}.$$

Thus if we know the hazard function we know the survival density for likelihood inferences.

Consider the exponential distribution with

$$\alpha(t) = \alpha;$$

this gives

$$\Lambda(t) = \alpha t$$

$$f(t) = \alpha e^{-\alpha t}, \quad t \geq 0.$$

In the early stage of life of either a human or machine the hazard tends to decrease, while in old age it increases, so that beyond a certain point of life the chance of death or breakdown increases with time. In the stage of adolescence the hazard would be flat, corresponding to the exponential distribution.

Suppose that the hazard function depends on time t and on a set of covariates x, some of which could be time-dependent. The *proportional-hazard model* separates these components by specifying that the hazard at time t for an individual whose covariate vector is x is given by

$$\alpha(t; x) = \alpha_0(t)\exp(x^t \beta),$$

where $\alpha_0(t)$ is a hazard function and is specifically the baseline hazard function for an individual with $x = 0$. For identifiability purposes the linear predictor, $\eta = x^t\beta$, does not include the intercept. In this model the ratio of hazards for two individuals depends on the difference between their linear predictors at any time, and so is a constant independent of time. This proportional hazards assumption is a strong assumption that clearly needs checking in applications. To allow the ratio of hazards for two individuals to change during follow-up, we can introduce a time-dependent covariate that changes value with time. This corresponds to introducing time·covariate interactions in the regression model.

Various assumptions may be made about the $\alpha_0(t)$ function. If a continuous survival function is assumed, $\alpha_0(t)$ is a smooth function of t, defined for all $t \geq 0$. Various parametric models can be generated in this way, e.g. $\alpha_0(t) = \kappa t^{\kappa-1}$ for $\kappa > 0$. Cox's model (Cox, 1972) assumes non-parametric baseline hazards by treating $\alpha_0(t)$ as analogous to the block factor in a blocked experiment, defined only at points where death occurs, thus making no parametric assumptions about its shape. This is equivalent to assuming that the baseline cumulative hazard $\Lambda_0(t)$ is a step function with jumps at the points where deaths occur. In practice it often makes little difference to estimates and inferences whether we make a parametric assumption about the baseline hazard function $\alpha_0(t)$ or not.

Cox's proportional-hazard models with parametric or non–parametric baseline hazard are used for analyzing univariate survival data. Aitkin and Clayton (1980) showed that Poisson GLMs could be used to fit proportional-hazards models with parametric baseline hazards. Laird and Olivier (1981) extended this approach to fit parametric models having piecewise exponential baseline hazards. Such models can be fitted by a Poisson GLM allowing a step function for an intercept, and these give very similar fits to the Cox model with non-parametric baselines. Whitehead (1980) and Clayton and Cuzick (1985) extended this approach to Cox's models with a non–parametric baseline hazard. Further extensions have been made for multivariate survival data, for example, for frailty models with parametric baseline hazards (Xue, 1998, Ha and Lee, 2003) and with non-parametric baseline hazards (Ma *et al.*, 2003, Ha and Lee, 2005a).

14.2 Frailty models and the associated h-likelihood

14.2.1 Frailty models

A frailty model is an extension of Cox's proportional-hazards model to allow for random effects, and has been widely used for the analysis of multivariate survival data in the form of recurrent or multiple-event times. The hazard function for each patient may depend on observed risk variables but usually not all such variables are known or measurable. The unknown component in the hazard function is usually termed the individual random effect or frailty. When the recurrence times of a particular type of event may be obtained for a patient, frailty is an unobserved common factor for each patient and is thus responsible for creating the dependence between recurrence times. This frailty may be regarded as a random quantity from some suitably defined distribution of frailties.

Let T_{ij} for $i = 1, \ldots, q$, $j = 1, \ldots, n_i$, be the survival time for the jth observation of the ith individual, $n = \sum_i n_i$, and C_{ij} be the corresponding censoring time. Let the observed quantities be $Y_{ij} = \min(T_{ij}, C_{ij})$ and $\delta_{ij} = I(T_{ij} \leq C_{ij})$, where $I(\cdot)$ is the indicator function. Denote by U_i the unobserved frailty random variable for the ith individual. We make the following two assumptions:

Assumption 1. Given $U_i = u_i$, the pairs $\{(T_{ij}, C_{ij}), j = 1, \ldots, n_i\}$ are conditionally independent, and T_{ij} and C_{ij} are also conditionally independent for $j = 1, \ldots, n_i$.

Assumption 2. Given $U_i = u_i$, the set $\{C_{ij}, j = 1, \ldots, n_i\}$ is noninformative about u_i. Thus, given $U_i = u_i$ the conditional hazard function of T_{ij} is of the form

$$\alpha_{ij}(t|u_i) = \alpha_0(t) \exp(x_{ij}^t \beta + v_i), \tag{14.1}$$

where $x_{ij} = (x_{ij1}, \ldots, x_{ijp})^t$ and $v_i = \log(u_i)$. When u_i is lognormal v_i is normal on the linear predictor scale. The frailties U_i are from some distribution with frailty parameter λ, which is the scale parameter of the random effects. The gamma or lognormal distribution is usually assumed for the distribution of U_i.

14.2.2 H-likelihood

We now construct the h-likelihood for random effect survival models such as frailty models. We define the $n_i \times 1$ observed random vectors related to the ith individual as $Y_i = (Y_{i1}, \ldots, Y_{in_i})^t$ and $\delta_i = (\delta_{i1}, \ldots, \delta_{in_i})^t$. The contribution h_i of the ith individual to the h-likelihood is given by the logarithm of the joint density of (Y_i, δ_i, V_i), where $V_i = \log(U_i)$:

$$h_i = h_i(\beta, \Lambda_0, \lambda; y_i, \delta_i, v_i) = \log\{f_{1i}(\beta, \Lambda_0; y_i, \delta_i|u_i) f_{2i}(\lambda; v_i)\}, \tag{14.2}$$

where f_{1i} is the conditional density of (Y_i, δ_i) given $U_i = u_i$, f_{2i} is the density of V_i and $\Lambda_0(\cdot)$ is the baseline cumulative hazard function. By the conditional independence of $\{(T_{ij}, C_{ij}), j = 1, \ldots, n_i\}$ in Assumption 1 we have

$$f_{1i}(\beta, \Lambda_0; y_i, \delta_i | u_i) = \prod_j f_{1ij}(\beta, \Lambda_0; y_{ij}, \delta_{ij} | u_i), \qquad (14.3)$$

where f_{1ij} is the conditional density of (Y_{ij}, δ_{ij}) given $U_i = u_i$. By the conditional independence of both T_{ij} and C_{ij} in Assumption 1 and the non-informativeness of Assumption 2, f_{1ij} in equation (14.3) becomes the ordinary censored-data likelihood given $U_i = u_i$:

$$f_{1ij} = \{\alpha(y_{ij} | u_i)\}^{\delta_{ij}} \exp\{-\Lambda(y_{ij} | u_i)\},$$

where $\Lambda(\cdot | u_i)$ is the conditional cumulative hazard function of T_{ij} given $U_i = u_i$. Thus, its contribution for all individuals is given, as required, by

$$h = h(\beta, \Lambda_0, \lambda) = \sum_i h_i = \sum_{ij} \ell_{1ij} + \sum_i \ell_{2i},$$

where $\ell_{1ij} = \ell_{1ij}(\beta, \Lambda_0; y_{ij}, \delta_{ij} | u_i) = \log f_{1ij}$, $\ell_{2i} = \ell_{2i}(\lambda; v_i) = \log f_{2i}$, and $\eta'_{ij} = \eta_{ij} + v_i$ with $\eta_{ij} = x^t_{ij}\beta$ and $v_i = \log(u_i)$.

14.2.3 Parametric baseline hazard models

Following Ha and Lee (2003), we show how to extend Aitkin and Clayton's (1980) results for parametric proportional-hazards models to frailty models. The first term ℓ_{1ij} in h can be decomposed as follows:

$$\begin{aligned} \ell_{1ij} &= \delta_{ij}\{\log \alpha_0(y_{ij}) + \eta'_{ij}\} - \{\Lambda_0(y_{ij}) \exp(\eta'_{ij})\} \\ &= \delta_{ij}\{\log \Lambda_0(y_{ij}) + \eta'_{ij}\} - \{\Lambda_0(y_{ij}) \exp(\eta'_{ij})\} \\ &\quad + \delta_{ij} \log\{\alpha_0(y_{ij})/\Lambda_0(y_{ij})\} \\ &= \ell_{10ij} + \ell_{11ij}, \end{aligned}$$

where $\ell_{10ij} = \delta_{ij} \log \mu'_{ij} - \mu'_{ij}$, $\ell_{11ij} = \delta_{ij} \log\{\alpha_0(y_{ij})/\Lambda_0(y_{ij})\}$ and $\mu'_{ij} = \mu_{ij} u_i$ with $\mu_{ij} = \Lambda_0(y_{ij}) \exp(x^t_{ij}\beta)$. The first term ℓ_{10ij} is identical to the kernel of a conditional Poisson likelihood for δ_{ij} given $U_i = u_i$ with mean μ'_{ij}, whereas the second term ℓ_{11ij} depends neither on β nor v_i.

By treating $\delta_{ij} | u_i$ as the conditional Poisson response variable with mean μ'_{ij}, frailty models can be fitted by a Poisson HGLM with log-link:

$$\log \mu'_{ij} = \log \mu_{ij} + v_i = \log \Lambda_0(y_{ij}) + x^t_{ij}\beta + v_i.$$

We now give three examples, with θ_0 and φ representing the parameters specifying the baseline hazard distribution.

Example 14.1: *Exponential distribution.* If $\alpha_0(t)$ is a constant hazard rate

θ_0, $\Lambda_0(t) = \theta_0 t$ becomes the baseline cumulative hazard for an exponential distribution with parameter θ_0. Thus, $\alpha_0(t)/\Lambda_0(t) = 1/t$ and no extra parameters are involved. It follows that

$$\log \mu'_{ij} = \log y_{ij} + \log \theta_0 + x^t_{ij}\beta + v_i.$$

By defining $\beta_0 \equiv \log \theta_0$, we can rewrite

$$\log \mu'_{ij} = \log y_{ij} + x^{*t}_{ij}\beta^* + v_i,$$

where $x^*_{ij} = (x_{ij0}, x^t_{ij})^t$ with $x_{ij0} = 1$ and $\beta^* = (\beta_0, \beta^t)^t$. The exponential parametric frailty models, where frailty may have various parametric distributions including the gamma and log-normal, can be directly fitted using a Poisson HGLM with the offset $\log y_{ij}$. $\square$

Example 14.2: *Weibull distribution.* Setting $\Lambda_0(t) = \theta_0 t^\varphi$ produces a Weibull distribution with scale parameter θ_0 and shape parameter φ; this gives the exponential distribution when $\varphi = 1$. Now $\alpha_0(t)/\Lambda_0(t) = \varphi/t$ depends on the unknown parameter φ. Similarly, we have

$$\log \mu'_{ij} = \varphi \log y_{ij} + x^{*t}_{ij}\beta^* + v_i.$$

Given the frailty parameter λ, the maximum h-likelihood estimators for β^*, φ and v are obtained by solving

$$\frac{\partial h}{\partial \beta_r} = \frac{\partial(\sum_{ij} \ell_{10ij})}{\partial \beta_r} = \sum_{ij}(\delta_{ij} - \mu'_{ij})x_{ijr} = 0 \ (r = 0, 1, \ldots, p),$$

$$\frac{\partial h}{\partial \varphi} = \sum_{ij}(\delta_{ij} - \mu'_{ij})z_{ij0} + \frac{\partial(\sum_{ij} \ell_{11ij})}{\partial \varphi} = 0, \qquad (14.4)$$

$$\frac{\partial h}{\partial v_i} = \sum_{j}(\delta_{ij} - \mu'_{ij}) + \frac{\partial(\ell_{2i})}{\partial v_i} = 0 \ (i = 1, \ldots, q), \qquad (14.5)$$

where $z_{ij0} = \partial\{\log \Lambda_0(y_{ij})\}/\partial\varphi = \log y_{ij}$.

Although the nuisance term ℓ_{11ij} reappears in the estimating equation (14.4) due to φ, the Weibull-parametric models can still be fitted using Poisson HGLMs, with the following trick. Let $v^* = (v_0, v^t)^t$, where $v_0 = \varphi$ and $v = (v_1, \ldots, v_q)^t$. By treating φ as random with the log-likelihood $\sum_{ij} \ell_{11ij}$, these estimating equations are those for Poisson HGLMs with random effects v^* whose log-likelihood is $\ell_{12} = \sum_{ij} \ell_{11ij} + \sum_i \ell_{2i}$. We then follow h-likelihood procedures for the inferences. We can rewrite the equations (14.4) and (14.5) in the form

$$\frac{\partial h}{\partial v^*_s} = \sum_{ij}(\delta_{ij} - \mu'_{ij})z_{ijs} + \frac{\partial \ell_{12}}{\partial v^*_s} = 0 \ (s = 0, 1, \ldots, q),$$

where z_{ijs} is z_{ij0} for $s = 0$ and $\partial\eta'_{ij}/\partial v_s$ for $s = 1, \ldots, q$.

Let

$$\eta'^* = X^*\beta^* + Z^*v^*,$$

where X^* is the $n \times (p + 1)$ matrix whose ijth row vector is x^{*t}_{ij}, Z^* is the

$n \times (q+1)$ group indicator matrix whose ijth row vector is $z_{ij}^{*t} = (z_{ij0}, z_{ij}^{t})$ and $z_{ij} = (z_{ij1}, \ldots, z_{ijq})^{t}$ is the $q \times 1$ group indicator vector. Given λ this leads to the IWLS score equations:

$$\begin{pmatrix} X^{*t}WX^{*} & X^{*t}WZ^{*} \\ Z^{*t}WX^{*} & Z^{*t}WX^{*}+U^{*} \end{pmatrix} \begin{pmatrix} \widehat{\beta^{*}} \\ \widehat{v^{*}} \end{pmatrix} = \begin{pmatrix} X^{*t}Ww^{*} \\ Z^{*t}Ww^{*} + R^{*} \end{pmatrix}, \quad (14.6)$$

where W is the diagonal weight matrix whose ijth element is μ_{ij}', U^{*} is the $(q+1) \times (q+1)$ diagonal matrix whose ith element is $-\partial^{2}\ell_{12}/\partial v_{i}^{*2}$,

$$w^{*} = \eta'^{*} + W^{-1}(\delta - \mu')$$

and

$$R^{*} = U^{*}v^{*} + \partial\ell_{12}/\partial v^{*}.$$

The asymptotic covariance matrix for $\widehat{\tau^{*}}-\tau^{*}$ is given by $H^{-1} = (-\partial^{2}h/\partial\tau^{*2})^{-1}$ with H being the square matrix on the left-hand side of (14.6). So, the upper left-hand corner of H^{-1} gives the variance matrix of $\widehat{\beta^{*}}$:

$$\mathrm{var}(\widehat{\beta^{*}}) = (X^{*t}\Sigma^{-1}X^{*})^{-1},$$

where $\Sigma = W^{-1} + Z^{*}U^{*-1}Z^{*t}$. For non-log-normal frailties a second-order correction is necessary. See Lee and Nelder (2003c) for simulation studies. $\square$

Example 14.3: *Extreme-value distribution.* Imposing $\Lambda_{0}(t) = \theta_{0}\exp(\varphi t)$ produces an extreme-value distribution. Because the transformation $\exp(t)$ transforms this distribution to the Weibull distribution, for fitting the extreme-value frailty models we need only to replace y_{ij} by $\exp(y_{ij})$ in the estimating procedure for the Weibull frailty models. $\square$

The procedure can be extended to fitting parametric models with other baseline hazard distributions such as the Gompertz, the generalized extreme-value, discussed in Aitkin and Clayton (1980), and the piecewise exponential, studied by Holford (1980) and Laird and Olivier (1981). Here the baseline cumulative hazard of the Gompertz distribution is

$$\Lambda_{0}(t) = \theta_{0}\varphi^{-1}\{\exp(\varphi t) - 1\},$$

a truncated form of the extreme-value distribution.

14.2.4 Non–parametric baseline hazard models

Suppose that in the frailty models (14.1) the functional form of baseline hazard, $\alpha_{0}(t)$, is unknown. Following Breslow (1972), we consider the baseline cumulative hazard function $\Lambda_{0}(t)$ to be a step function with jumps at the observed death times, $\Lambda_{0}(t) = \sum_{k:y_{(k)}\leq t}\alpha_{0k}$, where $y_{(k)}$ is the kth $(k = 1, \ldots, s)$ smallest distinct death time among the observed event times or censored times t_{ij}, and $\alpha_{0k} = \alpha_0(y_{(k)})$. Ma *et al.* (2003) and Ha and Lee (2005a) noted that

$$
\begin{aligned}
\sum_{ij} \ell_{1ij} &= \sum_{ij} \delta_{ij} \{\log \alpha_0(y_{ij}) + \eta'_{ij}\} - \sum_{ij} \{\Lambda_0(y_{ij}) \exp(\eta'_{ij})\} \\
&= \sum_k d_{(k)} \log \alpha_{0k} + \sum_{ij} \delta_{ij} \eta'_{ij} - \sum_k \alpha_{0k} \{ \sum_{(i,j) \in R(y_{(k)})} \exp(\eta'_{ij})\},
\end{aligned}
$$

where $d_{(k)}$ is the number of deaths at $y_{(k)}$ and $R(y_{(k)}) = \{(i,j) : t_{ij} \geq y_{(k)}\}$ is the risk set at $y_{(k)}$.

Let $y_{ij,k}$ be 1 if the (i,j)th individual dies at $y_{(k)}$ and 0 otherwise and let $\kappa = (\kappa_1, \ldots, \kappa_s)^t$, where $\kappa_k = \log \alpha_{0k}$. Let y and v denote the vectors of $y_{ij,k}$'s and v_i's, respectively. Since $\mu_{ij,k} = \alpha_{0k} \exp(\eta'_{ij})$ and

$$
\sum_k \sum_{(i,j) \in R(y_{(k)})} y_{ij,k} \log(\mu_{ij,k}) = \sum_k d_{(k)} \log \alpha_{0k} + \sum_{ij} \delta_{ij} \eta'_{ij},
$$

$\sum_{ij} \ell_{1ij}$ becomes

$$
\ell_{P1}(\gamma; y|v) = \sum_k \sum_{(i,j) \in R(y_{(k)})} \{y_{ij,k} \log(\mu_{ij,k}) - \mu_{ij,k}\},
$$

which is the likelihood from Poisson model. Thus, a frailty model with non–parametric baseline hazards can be fitted by using the following Poisson HGLM.

Given frailty v_i, let $y_{ij,k}$ be conditionally independent with

$$
y_{ij,k}|v_i \sim \text{Poisson}(\mu_{ij,k}), \text{ for } (i,j) \in R(y_{(k)});
$$

here

$$
\log \mu_{ij,k} = \kappa_k + x_{ij}^t \beta + v_i = x_{ij,k}^t \gamma + v_i, \tag{14.7}
$$

where $x_{ij,k} = (e_k^t, x_{ij}^t)^t$, e_k is a vector of components 0 and 1 such that $e_k^t \kappa = \kappa_k$, and $\gamma = (\kappa^t, \beta^t)^t$.

Note that it is not necessary to assume that the binary responses $y_{ij,k}|v_i$ follow a Poisson distribution. Such an assumption would be unrealistic in this setting. Rather, it is the equivalence of the h-likelihoods for the frailty model with a non–parametric baseline hazard and the Poisson HGLM above. Thus, likelihood inferences can be based on Poisson HGLMs.

A difficulty that arises in fitting frailty models via Poisson HGLMs results from the number of nuisance parameters associated with the baseline hazards κ increasing with sample size. Thus, for the elimination of these nuisance parameters it is important to have a computationally efficient algorithm. By arguments similar to those in Johansen (1983), given $\tau = (\beta^t, v^t)^t$ the score equations $\partial h / \partial \alpha_{0k} = 0$ provide the non–parametric maximum h-likelihood estimator of $\Lambda_0(t)$:

$$\widehat{\Lambda}_0(t) = \sum_{k:y_{(k)} \leq t} \widehat{\alpha}_{0k}$$

with

$$\widehat{\alpha}_{0k} = \exp(\widehat{\kappa}_k) = \frac{d_{(k)}}{\sum_{ij \in R(y_{(k)})} \exp(\eta'_{ij})}.$$

Ha *et al.* (2001) showed that the maximum h-likelihood estimator for $\tau = (\beta^t, v^t)^t$ can be obtained by maximizing the profile h-likelihood h^*

$$h^* = h|_{\alpha_0 = \widehat{\alpha}_0},$$

after eliminating $\Lambda_0(t)$, equivalent to eliminating $\alpha_0 = (\alpha_{01}, \ldots, \alpha_{0s})^t$. Let $\eta'_{ij} = x^t_{ij}\beta + z^t_{ij}v$, where $z_{ij} = (z_{ij1}, \ldots, z_{ijq})^t$ is the $q \times 1$ group indicator vector whose rth element is $\partial \eta'_{ij} / \partial v_r$. The kernel of h^* becomes

$$\sum_k \left[s^T_{1(k)}\beta + s^T_{2(k)}v - d_{(k)} \log \left\{ \sum_{ij \in R(y_{(k)})} \exp(\eta'_{ij}) \right\} \right] + \sum_i \ell_{2i}(\lambda; v_i),$$

where $s^t_{1(k)} = \sum_{(i,j) \in D_{(k)}} x^t_{ij}$ and $s^t_{2(k)} = \sum_{(i,j) \in D_{(k)}} z^t_{ij}$ are the sums of the vectors x^t_{ij} and z^t_{ij} over the sets $D_{(k)}$ of individuals who die at $y_{(k)}$.

Note that the estimators $\widehat{\Lambda}_0(t)$ and the profile h-likelihood h^* are, respectively, extensions to frailty models of Breslow's (1974) estimator of the baseline cumulative hazard function and his partial likelihood for the Cox model, and also that h^* becomes the kernel of the penalized partial likelihood (Ripatti and Palmgren, 2000) for gamma or log-normal frailty models. In particular, the profile likelihood h^* can also be derived directly by using the properties of the Poisson HGLM above. It arises by considering the conditional distribution of $y_{ij,k} | (v_i, \sum_{(i,j) \in R(y_{(k)})} y_{ij,k} = d_{(k)})$ for $(i, j) \in R(y_{(k)})$, which becomes a multinomial likelihood with the κ_k's eliminated. This also shows that multinomial random effect models can be fitted using Poisson random effect models.

Although several authors have suggested ways of obtaining valid estimates of standard errors from the EM algorithm and also of accelerating its convergence, the h-likelihood procedures are faster and provide a direct estimate of var$(\widehat{\beta})$ from the observed information matrix used in the Newton-Raphson method.

For the estimation of the frailty parameter λ given estimates of τ, we use the adjusted profile likelihood

$$p_{v,\beta}(h^*),$$

which gives an equivalent inference using $p_{v,\kappa,\beta}(h)$. For the Cox model without frailties the score equations $\partial h^*/\partial \tau = 0$ become those of Breslow (1974), and for the log-normal frailty model without ties they become those of McGilchrist and Aisbett (1991), and McGilchrist (1993): for the detailed form see equation (3.9) of Ha and Lee (2003). In estimating frailty (dispersion) parameters the term $\partial \hat{v}/\partial \lambda$ should not be ignored. The second-order correction is useful for reducing the bias for non-log-normal frailty models, exactly the same conclusion as drawn in HGLMs: see the discussion in Section 6.2.

For gamma frailty models with non-parametric baseline hazards, Ha and Lee (2003) showed numerically that the second-order correction method reduces the bias of ML method of Nielsen *et al.* (1992). Recently, Rondeau *et al.* (2003), Baker and Henderson (2005) showed numerically that biases of the ML procedure are nonnegligible. Ha and Lee (2007) showed how the h-likelihood gives a way of modifying the ML estimation to reduce such biases. Score equations for the frailty parameter λ are in Section 14.5. We also show that this approach is equivalent to a Poisson HGLM procedure without profiling. Our procedure can be easily extended to various frailty models with multi-level frailties (Yau, 2001), correlated frailties (Yau and McGilchrist, 1998; Ripatti and Palmgren, 2000) or structured dispersion (Lee and Nelder, 2001a).

For model selection we may use the scaled deviance D in Section 6.4 for goodness of fit and deviances based upon $p_v(h^*)$ and $p_{v,\beta}(h^*)$. However, care is necessary in using the scaled deviance because in non–parametric baseline hazards models the number of fixed parameters increases with the sample size (Ha *et al.*, 2007).

In summary, we show that frailty models with non-parametric baseline hazards are Poisson HGLMs, so that we may expect that methods that work well in HGLMs will continue to work well here. A difficulty with these models is the increase in nuisance parameters for baseline hazards with sample size. We have seen that in the h-likelihood approach profiling is effective in eliminating them.

14.2.5 Interval-censored data

Suppose the event time T_i cannot be observed exactly, but we have information that the event time falls in a specific interval. Suppose that n individuals are observed for r (≥ 2) consecutive periods of time (say years or months). Some individuals may die without completing the total of r observations and other individuals may stay throughout the entire period (called right-censored). Divide time into r intervals $[0 \equiv a_1, a_2)$,

... , $[a_{r-1}, a_r)$, $[a_r, a_{r+1} \equiv \infty)$, where the time point a_t denotes the starting point of the tth time period $(t = 1, ..., r)$. For example, if the ith individual survives the first interval but not the second, we know that $a_2 \le T_i < a_3$. The binary random variable $d_{it} = 0$ if the ith individual survives for the tth interval, and is 1 otherwise. Suppose that T_i takes a value within $[a_{m_i}, a_{m_i+1})$, then m_i becomes the number of intervals for the ith individual. Thus, we have the binary responses, represented by $d_{i2} = 0$, ... , $d_{im_i} = 0$, and $d_{im_i+1} = 1$. If $m_i < r$, we say that T_i is right-censored at a_r.

Suppose that the event time T_i has a frailty model as follows. Given the latent variable v_i, the conditional hazard rate of T_i for $a_{t-1} \le s < a_t$ with $t = 2, ..., r$ is of the form

$$\alpha(s|v_i) = \alpha_0(s) \exp(x_{it}^t \beta + v_i),$$

where $\alpha_0(\cdot)$ is the baseline hazard function, x_{it} is a vector of fixed covariates affecting the hazard function and β are unknown fixed effects. For identifiability, because $\alpha_0(s)$ already contains the constant term, x_{it} does not include an intercept. The unobservable frailties v_i are assumed to be independent and identically distributed.

Given v_i, the distribution of d_{it}, $t = 1, ..., \min(m_i + 1, r)$, follows the Bernoulli distribution with conditional probability

$$p_{0it} = P(d_{it} = 1|v_i) = 1 - P(T_i \ge a_t | T_i \ge a_{t-1}, v_i).$$

Under the frailty model (14.1) we have

$$1 - p_{0it} = P(T_i \ge a_t | T_i \ge a_{t-1}, v_i) = \exp(-\exp(\gamma_t + x_{it}^t \beta + v_i)),$$

where $\gamma_t = \log \int_{a_{t-1}}^{a_t} \alpha_0(u) du$. Thus, given v_i, the complementary log-log link leads to the binomial GLM

$$\log(-\log(1 - p_{0it})) = \gamma_t + x_{it}^t \beta + v_i.$$

This model has been used for the analysis of hospital closures by Noh *et al.* (2006) and for modelling dropout by Lee *et al.* (2005) in the analysis of missing data.

In econometric data time intervals are often fixed, for example unemployment rates can be reported in every month. However, in biomedical data time intervals can be different for each subject. For example, in the pharmaceutical industry it is often of interest to find time to reach some threshold score on an index. However, the scores can only be measured at a clinic visit, and clinic visits are unequally spaced to fit in with the patients' lifestyles. An event occurs between the last visit (below threshold) and the current visit (above threshold), but the times of visit are different from subject to subject. The model above can be easily extended to analyse such data.

14.2.6 Example: kidney infection data

We now illustrate an extension of non–parametric baseline hazard models to structured dispersion (Chapter 7). McGilchrist and Aisbett (1991) reported kidney infection data for the first and second recurrence times (in days) of infections of 38 kidney patients following insertion of a catheter until it has to be removed owing to infection. The catheter may have to be removed for reasons other than infection, and we regard this as censoring. In the original data there are three covariates: Age, Sex and Type of disease. Sex is coded as 1 for males and 2 for females, and the type of disease is coded as 0 for GN, 1 for AN, 2 for PKD and 3 for other diseases. From the type of disease we generate the three indicator variables for GN, AN and PKD. The recurrence time of the jth ($j = 1, 2$) observation in the ith patient ($i = 1, \cdots, 38$) is denoted by t_{ij}.

For the analysis of these data McGilchrist and Aisbett (1991) considered a frailty model with homogeneous frailty, $v_i \sim N(0, \sigma^2)$. Hougaard (2000) considered various other distributions. Using a Bayesian non–parametric approach Walker and Mallick (1997) found that there might be a heterogeneity in the frailties, which could be attributed to the sex effect. To describe this Noh $et\ al.$ (2006) considered a structured dispersion model given by

$$\log \sigma_i^2 = w_i^t \theta, \tag{14.8}$$

where w_i are (individual specific) covariates and θ is a vector of parameters to be estimated. We can deal with a sex effect in this model and so check Walker and Mallick's conjecture by testing whether the corresponding regression coefficient θ is zero.

For model selection Ha $et\ al.$ (2007) proposed to use the following AIC:

$$\text{AIC} = -2p_{\gamma,v}(h) + 2d, \tag{14.9}$$

where d is the number of dispersion parameters θ, not the number of all fitted parameters. This is an extension of AIC based upon the restricted likelihood. Therneau and Grambsch (2000) considered a homogeneous frailty model with a linear predictor

$$\eta_{ij} = \text{Age}_{ij}\beta_1 + \text{Female}_{ij}\beta_2 + \text{GN}_{ij}\beta_3 + \text{AN}_{ij}\beta_4 + \text{PKD}_{ij}\beta_5 + v_i,$$

where $\text{Female}_{ij} = \text{Sex}_{ij} - 1$ is an indicator variable for female. The results of fitting this homogenous frailty model

$$\text{M1:} \quad \log \sigma_i^2 = \theta_0,$$

is in Table 14.1, and we see that the Sex effect is significant. Noh *et al.* (2006) found that female heterogeneity is negligible compared to that for males. So they proposed the following model

$$\text{M2:} \quad \log \sigma_{mi}^2 = \theta_0 + \text{PKD}_{mi}\theta_2$$

and $\sigma_f^2 = 0$, where σ_f^2 and σ_m^2 are frailty variances for female and male, respectively. Here $\sigma_f^2 = 0$ means that we assume $v_i = 0$ for females. The AIC for M1 is 370.7, while that for M2 is 366.7. Because M1 has only one additional parameter the AIC indicates that M2 fits the data better. In Table 14.1 we also show the analysis using M2.

The results show that the female patients have a significantly lower average infection rate. In M1 and M2 the estimates of the PKD effect are rather different, but not significantly so. This means that if heterogeneity is ignored the estimate of the hazard can be biased or vice versa. Heterogeneity among female patients is relatively much smaller, so that we assume heterogeneity only for male patients. Walker and Mallick (1997) noted that male and female heterogeneities were different in both hazard and variance. M2 supports their findings by showing significant sex effects for both the hazard (14.1) and between-patient variation (14.8), and provides an additional finding that the male PKD patients have relatively larger heterogeneity than the rest. However, there were only 10 males and most had early failures. Subject 21, who is a male PKD patient, had very late failures at 152 and 562 days. Under M2, the estimated male frailties $\hat{v}_i$ (for the subject $i = 1, 3, 5, 7, 10, 16, 21, 25, 29, 38$) are respectively as follows:

$$0.475, 0.068, 0.080, 0.638, -0.800,$$
$$-0.039, -3.841, -0.392, 0.256, -1.624.$$

The two largest negative values, -3.841 and -1.624, correspond to subjects 21 and 38 respectively; both are PKD patients. This indicates a possibility of larger heterogeneity among the male PKD patients. However, with only two patients, this is not enough for a firm conclusion. Therneau and Grambsch (2000) raised the possibility of the subject 21 being an outlier.

Because of the significant Sex effect, Noh *et al.* (2006) considered a hazard (14.1) model with the Female-PKD interaction

$$\eta_{ij} = \text{Age}_{ij}\beta_1 + \text{Female}_{ij}\beta_2 + \text{GN}_{ij}\beta_3 + \text{AN}_{ij}\beta_4 + \text{PKD}_{ij}\beta_5$$
$$+ \text{Female}_{ij} \cdot \text{PKD}_{ij}\beta_6 + v_i.$$

Table 14.1 *Estimation results for the kidney infection data.*

Model	Effect	estimate	s.e.	t	Model	estimate	s.e.	t
M1 for η_{ij}	Age	0.005	0.015	0.352	M2 for η_{ij}	0.003	0.012	0.216
	Female	-1.679	0.459	-3.661		-2.110	0.462	-4.566
	GN	0.181	0.537	0.338		0.226	0.435	0.520
	AN	0.394	0.537	0.732		0.550	0.435	1.266
	PKD	-1.138	0.811	-1.403		0.673	0.735	0.916
for $\log \sigma_i^2$	Constant	-0.716	0.910	-0.787	for $\log \sigma_i^2$	-0.654	1.303	-0.502
	PKD					2.954	1.680	1.758
M3 for η_{ij}	Age	0.004	0.013	0.271	M4 for η_{ij}	0.004	0.012	0.297
	Female	-2.088	0.436	-4.791		-2.267	0.498	-4.550
	GN	0.121	0.491	0.246		0.219	0.442	0.494
	AN	0.468	0.488	0.957		0.551	0.439	1.256
	PKD	-2.911	1.019	-2.857		-2.705	1.141	-2.371
	Female·PKD	3.700	1.226	3.017		3.584	1.262	2.841
for $\log \sigma_i^2$	Constant	-1.376	1.370	-1.004	for $\log \sigma_i^2$	-0.423	1.147	-0.369

They found that the PKD effect and Female-PKD interaction are no longer significant in the dispersion model Among all possible dispersion models they considered summaries of two models are as follows:

M3: $\log \sigma_i^2 = \theta_0$ has AIC $= 359.6$. This is the homogenous frailty model, having $\sigma_i^2 = \sigma^2 = \exp(\theta_0)$.

M4: $\log \sigma_{mi}^2 = \theta_0$ and $\sigma_f^2 = 0$ has AIC $= 358.5$, so M4 is better.

Because M2 and M4 have different hazard models we cannot compare them using AIC (14.9). From the results from M4, the relative risk of PKD patients over non-PKD patients for females is given by $\exp(-2.709 + 3.584) = 2.40$, whereas for males it is $\exp(-2.709) = 0.07$. M3 and M4 give similar estimates for models of hazards. Walker and Mallick (1996) considered only the Sex covariate and noted that male and female heterogeneities were different in both hazard and variance. The models support their findings by showing a significant Sex effect for between-patient variation (14.8). Heterogeneity among female patients is relatively much smaller, so that we assume heterogeneity only for male patients.

Now under the model M4 with a Female-PKD interaction in the hazard model, the corresponding male frailties $\hat{v}_i$ are respectively as follows:

$$0.515, 0.024, 0.045, 0.720, -0.952,$$
$$-0.087, -0.422, -0.504, 0.239, 0.422.$$

Now all the large negative values have vanished; both PKD and non-PKD patients have similar frailties. Noh *et al.* (2006) also considered models that allow stratification by sex for hazards model. It would be interesting to develop model-selection criteria to distinguish between M2 and M4, and between M4 and stratification models.

14.3 *Mixed linear models with censoring

14.3.1 Models

In (14.1) the frailty is modelled by random effects acting multiplicatively on the individual hazard rates. The mixed linear model (MLM) has been introduced as an alternative in which the random effect acts linearly on each individual's survival time, thus making interpretation of the fixed effects easier (as mean parameters) than in the frailty model (Klein *et al.*, 1999). Pettitt (1986) and Hughes (1999) proposed maximum-likelihood estimation procedures using, respectively, the EM algorithm and a Monte Carlo EM algorithm based on Gibbs sampling, both of which are computationally intensive. Klein *et al.* (1999) derived the Newton-Raphson method for models with one random component, but it is very complicated to obtain the marginal likelihood. Ha *et al.* (2002) showed that the use of the h-likelihood avoids such difficulties, providing a conceptually simple, numerically efficient and reliable inferential procedure for MLMs.

For T_{ij} we assume the normal HGLM as follows: for $i = 1, \ldots, q$ and $j = 1, \ldots, n_i$,

$$T_{ij} = x_{ij}^t \beta + U_i + \epsilon_{ij},$$

where $x_{ij} = (x_{ij1}, \ldots, x_{ijp})^t$ is a vector of fixed covariates, β is a $p \times 1$ vector of fixed effects including the intercept. And $U_i \sim N(0, \sigma_u^2)$ and $\epsilon_{ij} \sim N(0, \sigma_\epsilon^2)$ are independent. Here, the dispersion or variance components σ_ϵ^2 and σ_u^2 stand for variability within and between individuals, respectively. The T_{ij} could be expressed on some suitably transformed scale, e.g. $\log(T_{ij})$. If the log-transformation is used, the normal HGLM becomes an accelerated failure-time model with random effects.

14.3.2 H-likelihood and fitting method

We shall first present a simple method for estimating the parameters in the normal HGLM. Because the T_{ij} may be subject to censoring, only Y_{ij} are observed. Defining

$$\mu_{ij} \equiv E(T_{ij}|U_i = u_i) = x_{ij}^t \beta + u_i$$

we have

$$E(Y_{ij}|U_i = u_i) \neq \mu_{ij}.$$

Ha *et al.* (2002) extended the pseudo-response variable Y_{ij}^* of Buckley and James (1979) for the linear model with censored data as follows: Let

$$Y_{ij}^* = Y_{ij}\delta_{ij} + E(T_{ij}|T_{ij} > Y_{ij}, U_i = u_i)(1 - \delta_{ij}). \tag{14.10}$$

Then

$$
\begin{aligned}
E(Y_{ij}^*|U_i = u_i) &= E\{T_{ij}\ I(T_{ij} \leq C_{ij})|U_i = u_i\} \\
&+ E\{E(T_{ij}|T_{ij} > C_{ij}, U_i = u_i)\ I(T_{ij} > C_{ij})|U_i = u_i\}.
\end{aligned}
$$

By the conditional independence of T_{ij} and C_{ij} in Assumption 1, the first term on the right-hand side (RHS) of the above equation is

$$E[T_{ij}I(T_{ij} \leq C_{ij})|U_i = u_i] = E[E\{T_{ij}\ I(T_{ij} \leq C_{ij})|T_{ij}, U_i\}|U_i = u_i]$$

$$= E(T_{ij}|U_i = u_i) - \int_0^\infty t\ G_{ij}(t|u)\ dF_{ij}(t|u)$$

and the second term on the RHS is given by

$$E\{E(T_{ij}|T_{ij} > C_{ij}, u_i)\ I(T_{ij} > C_{ij})|U_i = u_i\} = \int_0^\infty t\ G_{ij}(t|u)\ dF_{ij}(t|u),$$

where $G_{ij}(\cdot|u)$ and $F_{ij}(\cdot|u)$ are arbitrary continuous conditional distribution functions of $C_{ij}|U_i = u_i$ and $T_{ij}|U_i = u_i$, respectively. Thus by combining the two equations we obtain the expectation identity

$$E(Y_{ij}^*|U_i = u_i) = E(T_{ij}|U_i = u_i)$$
$$= \mu_{ij}.$$

Let y_{ij} be the observed value for Y_{ij} and let $y_{ij}^* = Y_{ij}^*|_{Y_{ij}=y_{ij}}$ be the pseudo-response variables, computed based upon the observed data $Y_{ij} = y_{ij}$. Explicit formulae are possible under certain models. Suppose that $T_{ij}|(U_i = u_i) \sim N(\mu_{ij}, \sigma_\epsilon^2)$. Let $\alpha(\cdot) = \phi(\cdot)/\bar{\Phi}(\cdot)$ be the hazard function for $N(0,1)$, ϕ and $\bar{\Phi}(= 1 - \Phi)$ the density and cumulative distribution functions for $N(0,1)$, respectively, and

$$m_{ij} = (y_{ij} - \mu_{ij})/\sigma_\epsilon.$$

Then

$$E(T_{ij}|T_{ij} > y_{ij}, U_i = u_i) = \int_{y_{ij}}^{\infty} \{tf(t|U_i)\}/S(y_{ij})dt$$
$$= \int_{m_{ij}}^{\infty} \{(\mu_{ij} + \sigma_\epsilon z)\phi(z)\}/\bar{\Phi}(m)dz$$
$$= \mu_{ij} + \{\sigma_\epsilon/\bar{\Phi}(m)\} \int_{m_{ij}}^{\infty} z\phi(z)dz$$
$$= \mu_{ij} + \sigma_\epsilon\alpha(m_{ij}),$$

where at the last step we use $\phi'(z) = -z\phi(z)$. Thus, we have the pseudo-responses

$$y_{ij}^* = y_{ij}\delta_{ij} + \{\mu_{ij} + \sigma_\epsilon\alpha(m_{ij})\}(1 - \delta_{ij}).$$

Analogous to frailty models, the h-likelihood for the normal HGLM becomes

$$h = h(\beta, \sigma_\epsilon^2, \sigma_u^2) = \sum_{ij} \ell_{1ij} + \sum_i \ell_{2i},$$

where

$$\ell_{1ij} = \ell_{1ij}(\beta, \sigma_\epsilon^2; y_{ij}, \delta_{ij}|u_i)$$
$$= -\delta_{ij}\{\log(2\pi\sigma_\epsilon^2) + (m_{ij})^2\}/2 + (1 - \delta_{ij})\log\{\bar{\Phi}(m_{ij})\}.$$

and

$$\ell_{2i} = \ell_{2i}(\sigma_u^2; u_i) = -\{\log(2\pi\sigma_u^2) + (u_i^2/\sigma_u^2)\}/2.$$

Given $\theta = (\sigma_\epsilon^2, \sigma_u^2)$, the estimate $\hat{\tau} = (\hat{\beta}^t, \hat{u}^t)^t$ is obtained by IWLS with pseudo-responses y^*:

$$\begin{pmatrix} X^tX/\sigma_\epsilon^2 & X^tZ/\sigma_\epsilon^2 \\ Z^tX/\sigma_\epsilon^2 & Z^tZ/\sigma_\epsilon^2 + I_q/\sigma_u^2 \end{pmatrix} \begin{pmatrix} \hat{\beta} \\ \hat{u} \end{pmatrix} = \begin{pmatrix} X^ty^*/\sigma_\epsilon^2 \\ Z^ty^*/\sigma_\epsilon^2 \end{pmatrix}, \quad (14.11)$$

where X is the $n \times p$ matrix whose ijth row vector is x_{ij}^t, Z is the $n \times q$ group indicator matrix whose ijkth element z_{ijk} is $\partial \mu_{ij}/\partial u_k$, I_q is the $q \times q$ identity matrix, and y^* is the $n \times 1$ vector with ijth element y_{ij}^*.

The asymptotic covariance matrix for $\hat{\tau} - \tau$ is given by H^{-1} with

$$H = -\frac{\partial^2 h}{\partial \tau^2}, \tag{14.12}$$

giving

$$H = \begin{pmatrix} X^t W X/\sigma_\epsilon^2 & X^t W Z/\sigma_\epsilon^2 \\ Z^t W X/\sigma_\epsilon^2 & Z^t W Z/\sigma_\epsilon^2 + I_q/\sigma_u^2 \end{pmatrix}.$$

Here, $W = \mathrm{diag}(w_{ij})$ is the $n \times n$ diagonal matrix with the ijth element $w_{ij} = \delta_{ij} + (1 - \delta_{ij})\lambda(m_{ij})$ and $\lambda(m_{ij}) = \alpha(m_{ij})\{\alpha(m_{ij}) - m_{ij}\}$. So, the upper left-hand corner of H^{-1} in (14.12) gives the variance matrix of $\hat{\beta}$ in the form

$$\mathrm{var}(\hat{\beta}) = (X^t \Sigma^{-1} X)^{-1}, \tag{14.13}$$

where $\Sigma = \sigma_\epsilon^2 W^{-1} + \sigma_u^2 Z Z^t$. Note that both y^* in (14.11) and W in (14.12) depend on censoring patterns. The weight matrix W takes into account the loss of information due to censoring, so that $w_{ij} = 1$ if the ijth observation is uncensored.

When there is no censoring the IWLS equations above become the usual Henderson's mixed model equations using the data y_{ij} (Section 5.3). These two estimating equations are also extensions of those given by Wolynetz (1979), Schmee and Hahn (1979) and Aitkin (1981) for normal linear models without random effects.

For the estimation of the dispersion parameters θ given the estimates of τ, we use

$$p_{v,\beta}(h),$$

which gives McGilchrist's (1993) REML estimators (Section 5.4.3); he showed by simulation that the REML method gives a good estimate of the standard-errors of $\hat{\beta}$ for log-normal frailty models.

Since we cannot observe all the y_{ij}^*, they are imputed using estimates of other quantities

$$\widehat{y_{ij}^*} = y_{ij}\delta_{ij} + \{\hat{\mu}_{ij} + \hat{\sigma}_\epsilon \alpha(\hat{m}_{ij})\}(1 - \delta_{ij}),$$

where $\hat{m}_{ij} = (y_{ij} - \hat{\mu}_{ij})/\hat{\sigma}_\epsilon$ and $\hat{\mu}_{ij} = x_{ij}^t \hat{\beta} + \hat{u}_i$. Replacing y_{ij}^* by $\widehat{y_{ij}^*}$ increases the variance of $\hat{\beta}$. This variation inflation due to censoring is reflected in the estimation of θ, so that the variance estimator, $\widehat{\mathrm{var}}(\hat{\beta})$ in (14.13), works reasonably well (Ha *et al.*, 2002).

14.3.3 Advantages of the h-likelihood procedure

Ha and Lee (2005b) provided an interpretation for the pseudo responses

$$
\begin{aligned}
y_{ij}^* &= E(T_{ij}|Y_{ij} = y_{ij}, \delta_{ij}, U_i = u_i) \\
&= y_{ij}\delta_{ij} + E(T_{ij}|T_{ij} > y_{ij}, U_i = u_i)(1 - \delta_{ij}),
\end{aligned}
$$

which immediately shows that

$$
E(Y_{ij}^*|U_i = u_i) = \mu_{ij}.
$$

Thus, the h-likelihood method implicitly applies an EM-type algorithm to the h-likelihood procedure. Pettitt (1986) developed an EM algorithm for a marginal-likelihood procedure which uses the pseudo-responses

$$
E(T_{ij}|Y_{ij} = y_{ij}, \delta_{ij}) = y_{ij}\delta_{ij} + E(T_{ij}|T_{ij} > y_{ij})(1 - \delta_{ij}).
$$

However, due to the difficulty of integration in computing $E(T_{ij}|T_{ij} > y_{ij})$ the method was limited to single random effect models. Hughes (1999) avoided integration by using the Monte-Carlo method, which, however, requires heavy computation and extensive derivations for the E-step. In Chapter 5 we have shown that in normal HGLMs without censoring the h-likelihood method provides the ML estimators for fixed effects and the REML estimators for dispersion parameters. Now we see that for normal HGLMs with censoring it implicitly implements an EM-type algorithm. This method is easy to extend to models with many random components, for example imputing unobserved responses T_{ij} by $E(T_{ij}|Y_{ij} = y_{ij}, \delta_{ij}, U_i = u_i, U_j = u_j)$ in the estimating equations. With the use of h-likelihood the numerically difficult E-step or integration is avoided by automatically imputing the censored responses to y_{ijk}^*.

14.3.4 Example: chronic granulomatous disease

Chronic granulomatous disease (CGD) is an inherited disorder of phago-cytic cells, part of the immune system that normally kill bacteria, leading to recurrent infection by certain types of bacteria and fungi. We reanalyse the CGD data set in Fleming and Harrington (1991) from a placebo-controlled randomized trial of gamma interferon. The aim of the trial was to investigate the effectiveness of gamma interferon in preventing serious infections among CGD patients. In this study, 128 patients from 13 hospitals were followed for about 1 year. The number of patients per hospital ranged from 4 to 26. Of the 63 patients in the treatment group, 14(22%) patients experienced at least one infection and a total of 20 infections was recorded. In the placebo group, 30(46%) out of 65 patients experienced at least one infection, with a total of 56 infections being recorded.

The survival times are the recurrent infection times of each patient from the different hospitals. Censoring occurred at the last observation of all patients, except one, who experienced a serious infection on the date he left the study. In this study about 63% of the data were censored. The recurrent infection times for a given patient are likely to be correlated. However, since patients may come from any of 13 hospitals, the correlation may also be due to a hospital effect. This data set has previously been analyzed by Yau (2001) using multilevel log-normal frailty models with a single covariate x_{ijk} (0 for placebo and 1 for gamma interferon). The estimation of the variances of the random effects is also of interest.

Let T_{ijk} be the infection time for the kth observation of the jth patient in the ith hospital. Let U_i be the unobserved random effect for the ith hospital and let U_{ij} be that for the jth patient in the ith hospital. For the responses $\log T_{ijk}$, we consider a three-level MLM, in which observations, patients and hospitals are the units at levels 1, 2 and 3 respectively

$$\log T_{ijk} = \beta_0 + x_{ijk}^t \beta_1 + U_i + U_{ij} + \epsilon_{ijk}, \tag{14.14}$$

where $U_i \sim N(0, \ \sigma_1^2)$, $U_{ij} \sim N(0, \sigma_2^2)$ and $\epsilon_{ijk} \sim N(0, \sigma_\epsilon^2)$ are mutually independent error components. This model allows an explicit expression for correlations between recurrent infection times;

$$\text{cov}(\log T_{ijk}, \log T_{i'j'k'}) = \begin{cases} 0 & \text{if } i \neq i', \\ \sigma_1^2 & \text{if } i = i', j \neq j', \\ \sigma_1^2 + \sigma_2^2 & \text{if } i = i', j = j', k \neq k', \\ \sigma_1^2 + \sigma_2^2 + \sigma_\epsilon^2 & \text{if } i = i', j = j', k = k'. \end{cases}$$

Thus, the intra-hospital (ρ_1) and intra-patient (ρ_2) correlations are defined as

$$\rho_1 = \sigma_1^2/(\sigma_1^2 + \sigma_2^2 + \sigma_\epsilon^2) \quad \text{and} \quad \rho_2 = (\sigma_1^2 + \sigma_2^2)/(\sigma_1^2 + \sigma_2^2 + \sigma_\epsilon^2). \tag{14.15}$$

For analysis we use model (14.14), which allows the following four submodels:

M1: $(\sigma_1^2 = 0, \sigma_2^2 = 0)$ corresponds to a one-level regression model without random effects,

M2: $(\sigma_1^2 > 0, \sigma_2^2 = 0)$ to a two-level model without patient effects,

M3: $(\sigma_1^2 = 0, \sigma_2^2 > 0)$ to a two-level model without hospital effects, and

M4: $(\sigma_1^2 > 0, \sigma_2^2 > 0)$ to a three-level model, requiring both patient and hospital effects.

The results from these MLMs are given in Table 14.2. Estimated values of β_1 vary from 1.49 under M1 to 1.24 under M3, with similar standard errors of about 0.33, all indicating significant positive benefit of gamma interferon. For testing the need for a random component, we use the deviance $(-2p_{v,\beta}(h)$ in Table 14.2) based upon the restricted likelihood $p_{v,\beta}(h)$ (Chapter 5). Because such a hypothesis is on the boundary of the parameter space the critical value is $\chi^2_{2\kappa}$ for a size κ test. This value results from the fact that the asymptotic distribution of likelihood ratio test is a 50:50 mixture of χ^2_0 and χ^2_1 distributions (Chernoff, 1954; Self and Liang, 1987): for application to random effect models see Stram and Lee (1994), Vu et $al.$ (2001), Vu and Knuiman (2002) and Verbeke and Molenberghs (2003).

The deviance difference between M3 and M4 is 0.45 , which is not significant at a 5% level ($\chi^2_{1,0.10} = 2.71$), indicating the absence of the random hospital effects, i.e. $\sigma^2_1 = 0$. The deviance difference between M2 and M4 is 4.85, indicating that the random patient effects are necessary, i.e. $\sigma^2_2 > 0$. In addition, the deviance difference between M1 and M3 is 8.92, indicating that the random patient effects are indeed necessary with or without random hospital effects. Between the frailty models, corresponding to M3 and M4, Yau (2001) chose M3 by using a likelihood ratio test. AIC also chooses the M3 as the best model. In M3 the estimated intra-patient correlation in (14.15) is $\widehat{\rho}_2 = \widehat{\sigma^2_2}/(\widehat{\sigma^2_2 + \sigma^2_\epsilon}) = 0.250$.

14.4 Extensions

The h-likelihood procedure can be extended to random effect survival models allowing various structures. For example, frailty models allowing stratification and/or time-dependent covariates (Andersen et $al.$, 1997), mixed linear survival models with autoregressive random effect structures, joint models with repeated measures and survival time data (Ha et $al.$, 2003) and non-proportional hazard frailty models (MacKenzie et $al.$, 2003).

Survival data can be left-truncated (Lai and Ying, 1994) when not all subjects in the data are observed from the time origin of interest, yielding both left-truncation and right-censoring (LTRC). The current procedure can also be extended to random effect models with LTRC structure. In particular, as in Cox's proportional hazard models, the semiparametric frailty models for LTRC can be easily handled by replacing the risk set $R(y_{(k)}) = \{(i,j) : y_{(k)} \leq t_{ij}\}$ by $\{(i,j) : w_{ij} \leq y_{(k)} \leq t_{ij}\}$, where w_{ij} is the left truncation time. The development of multi-state frailty models based on LTRC would provide interesting future work.

In this chapter, the h-likelihood procedures assume the non-informative censoring defined in Assumption 2, but they can be extended to informa-

Table 14.2 Analyses using normal HGLMs for the CGD data.

Model	$\widehat{\beta}_0$	SE	$\widehat{\beta}_1$	SE	$\widehat{\sigma}_1^2$	$\widehat{\sigma}_2^2$	$\widehat{\sigma}_\epsilon^2$	$-2p_{v,\beta}(h)$	AIC
M1	5.428	0.185	1.494	0.322	—	—	3.160	426.52	428.52
M2	5.594	0.249	1.470	0.313	0.294	—	2.872	422.00	426.00
M3	5.661	0.202	1.237	0.331	—	0.722	2.163	417.60	421.60
M4	5.698	0.220	1.255	0.334	0.067	0.710	2.185	417.15	423.15

tive censoring (Huang and Wolfe, 2002) where censoring is informative for survival.

Recently, the h-likelihood approach has been greatly extended to various random effect survival models, including competing risks models allowing for several frailty terms (Ha et al., 2016; Christian et al., 2016), semi-competing risks frailty models, joint semi-parametric survival modelling with longitudinal and time-to-event 'outcomes, and variable selection of general frailty models (Ha et al., 2014a,b).

Epilogue

Since the first paper on h-likelihood in 1996, interest in this topic grew to produce the first edition of this book in 2006. This rather advanced mongrapgh has been developed in a second edition and two separate books on an applied data analysis with R (Lee, Rönnegård and Noh, 2017) and survival analysis (Ha, Jeong and Lee, 2017), which shows how wide and deep the subject is. We hope that more books are published for a wide audiance.

14.5 Proofs

Score equations for the frailty parameter in semiparametric frailty models

For the semiparametric frailty models (14.1) with $E(U_i) = 1$ and $\mathrm{var}(U_i) = \lambda = \sigma^2$, the adjusted profile h-likelihood (using the first-order Laplace approximation, Lee and Nelder, 2001a) for the frailty parameter λ is defined by

$$p_\tau(h^*) = [h^* - \frac{1}{2}\log\det\{A(h^*,\tau)/(2\pi)\}]|_{\tau=\widehat{\tau}(\lambda)}, \qquad (14.16)$$

where $\tau = (\beta^t, v^t)^t$, $\widehat{\tau}(\lambda) = (\widehat{\beta}^t(\lambda), \widehat{v}^t(\lambda))^t$ estimates of τ for given λ, $h^* = h|_{\alpha_0=\widehat{\alpha}_0(\tau)}$, $\widehat{\alpha}_0(\tau)$ estimates of α_0 for given τ, $h = \ell_1(\beta, \alpha_0; y, \delta|u) + \ell_2(\lambda; v)$, $v = \log u$, and

$$A(h^*, \tau) = -\partial^2 h^*/\partial\tau^2 = \begin{pmatrix} X^t W^* X & X^t W^* Z \\ Z^t W^* X & Z^t W^* Z + U \end{pmatrix}.$$

Here, X and Z are respectively the model matrices for β and v, $U = \mathrm{diag}(-\partial^2\ell_2/\partial v^2)$, and the weight matrix $W^* = W^*(\widehat{\alpha}_0(\tau), \tau)$ is given in Appendix 2 of Ha and Lee (2003). We first show how to find the score equation

$$\partial p_\tau(h^*)/\partial\lambda = 0.$$

Let $\omega = (\tau^t, \alpha_0^t)^t$, $\widehat{\omega}(\lambda) = (\widehat{\tau}^t(\lambda), \widehat{\alpha}_0^t(\lambda))^t$, $h\{\widehat{\omega}(\lambda), \lambda\} = h^*|_{\tau=\widehat{\tau}(\lambda)}$ and $H\{\widehat{\omega}(\lambda), \lambda\} = A(h^*, \tau)|_{\tau=\widehat{\tau}(\lambda)}$. Then $p_\tau(h^*)$ in (14.16) can be written as

$$p_\tau(h^*) = h\{(\widehat{\omega}(\lambda), \lambda\} - \frac{1}{2}\log\det[H\{\widehat{\omega}(\lambda), \lambda\}/(2\pi)].$$

Thus, we have

$$\frac{\partial p_\tau(h^*)}{\partial\lambda} = \frac{\partial h\{\widehat{\omega}(\lambda), \lambda\}}{\partial\lambda} - \frac{1}{2}\text{trace}\left[H\{\widehat{\omega}(\lambda), \lambda\}^{-1} \frac{\partial H\{\widehat{\omega}(\lambda), \lambda\}}{\partial\lambda}\right]$$

(14.17)

Here

$$\begin{aligned}\frac{\partial h\{\widehat{\omega}(\lambda), \lambda\}}{\partial\lambda} &= \frac{\partial h(\omega, \lambda)}{\partial\lambda}|_{\omega=\widehat{\omega}(\lambda)} + \left(\frac{\partial h(\omega, \lambda)}{\partial\omega}|_{\omega=\widehat{\omega}(\lambda)}\right)\left(\frac{\partial\widehat{\omega}(\lambda)}{\partial\lambda}\right) \\ &= \frac{\partial h(\omega, \lambda)}{\partial\lambda}|_{\omega=\widehat{\omega}(\lambda)},\end{aligned}$$

since the second term is equal to zero in the h-likelihood score equations. Note that $H\{\widehat{\omega}(\lambda), \lambda\}$ is function of $\widehat{\beta}(\lambda)$, $\widehat{v}(\lambda)$ and $\widehat{\alpha}_0(\lambda)$. Following Lee and Nelder (2001a), we ignore $\partial\widehat{\beta}(\lambda)/\partial\lambda$ in implementing $\partial H\{\widehat{\omega}(\lambda), \lambda\}/\partial\lambda$ in (14.17), but not $\partial\widehat{\alpha}_0(\lambda)/\partial\lambda$ and $\partial\widehat{v}(\lambda)/\partial\lambda$; this leads to

$$\begin{aligned}\frac{\partial H\{\widehat{\omega}(\lambda), \lambda\}}{\partial\lambda} &= \frac{\partial H(\omega, \lambda)}{\partial\lambda}|_{\omega=\widehat{\omega}(\lambda)} + \left(\frac{\partial H(\omega, \lambda)}{\partial\alpha_0}|_{\omega=\widehat{\omega}(\lambda)}\right)\left(\frac{\partial\widehat{\alpha}_0(\lambda)}{\partial\lambda}\right) \\ &+ \left(\frac{\partial H(\omega, \lambda)}{\partial v}|_{\omega=\widehat{\omega}(\lambda)}\right)\left(\frac{\partial\widehat{v}(\lambda)}{\partial\lambda}\right).\end{aligned}$$

Next, we show how to compute $\partial\widehat{\alpha}_0(\lambda)/\partial\lambda$ and $\partial\widehat{v}(\lambda)/\partial\lambda$. From the h, given λ, let $\widehat{\alpha}_0(\lambda)$ and $\widehat{v}(\lambda)$ be the solutions of $f_1(\lambda) = \partial h/\partial\alpha_0|_{\omega=\widehat{\omega}(\lambda)} = 0$ and $f_2(\lambda) = \partial h/\partial v|_{\omega=\widehat{\omega}(\lambda)} = 0$, respectively. From

$$\begin{aligned}\frac{\partial f_1(\lambda)}{\partial\lambda} &= \frac{\partial^2 h}{\partial\lambda\partial\alpha_0}|_{\omega=\widehat{\omega}(\lambda)} + \left(\frac{\partial^2 h}{\partial\alpha_0^2}|_{\omega=\widehat{\omega}(\lambda)}\right)\left(\frac{\partial\widehat{\alpha}_0(\lambda)}{\partial\lambda}\right) \\ &+ \left(\frac{\partial^2 h}{\partial v\partial\alpha_0}|_{\omega=\widehat{\omega}(\lambda)}\right)\left(\frac{\partial\widehat{v}(\lambda)}{\partial\lambda}\right) \\ &= 0\end{aligned}$$

and

$$\begin{aligned}\frac{\partial f_2(\lambda)}{\partial\lambda} &= \frac{\partial^2 h}{\partial\lambda\partial v}|_{\omega=\widehat{\omega}(\lambda)} + \left(\frac{\partial^2 h}{\partial\alpha_0\partial v}|_{\omega=\widehat{\omega}(\lambda)}\right)\left(\frac{\partial\widehat{\alpha}_0(\lambda)}{\partial\lambda}\right) \\ &+ \left(\frac{\partial^2 h}{\partial v^2}|_{\omega=\widehat{\omega}(\lambda)}\right)\left(\frac{\partial\widehat{v}(\lambda)}{\partial\lambda}\right) \\ &= 0\end{aligned}$$

we have

$$\frac{\partial \widehat{v}(\lambda)}{\partial \lambda} = \left[-\frac{\partial^2 h}{\partial v^2} + \left(\frac{\partial^2 h}{\partial v \partial \alpha_0} \right) \left(\frac{\partial^2 h}{\partial \alpha_0^2} \right)^{-1} \left(\frac{\partial^2 h}{\partial \alpha_0 \partial v} \right) \right]^{-1} \left(\frac{\partial^2 h}{\partial \lambda \partial v} \right) |_{\omega = \widehat{\omega}(\lambda)}$$

$$= \left[Z^T W^* Z + U \right]^{-1} \left(\frac{\partial^2 \ell_2}{\partial \lambda \partial v} \right) |_{\omega = \widehat{\omega}(\lambda)},$$

$$\frac{\partial \widehat{\alpha}_0(\lambda)}{\partial \lambda} = - \left(\frac{\partial^2 h}{\partial \alpha_0^2} |_{\omega = \widehat{\omega}(\lambda)} \right)^{-1} \left(\frac{\partial^2 h}{\partial v \partial \alpha_0} |_{\omega = \widehat{\omega}(\lambda)} \right) \left(\frac{\partial \widehat{v}(\lambda)}{\partial \lambda} \right).$$

For inference about λ in models with gamma frailty, where

$$\ell_{2i} = \ell_{2i}(\lambda; v_i) = (v_i - u_i)/\lambda + c(\lambda) \text{ with } c(\lambda) = -\log \Gamma(\lambda^{-1}) - \lambda^{-1} \log \lambda,$$

we use the second-order Laplace approximation (Lee and Nelder, 2001a), defined by

$$s_\tau(h^*) = p_\tau(h^*) - F/24,$$

where $F = \text{trace}(S)|_{\tau = \widehat{\tau}(\lambda)}$ with

$$S = - \{3(\partial^4 h^*/\partial v^4) + 5(\partial^3 h^*/\partial v^3) A(h^*, v)^{-1} (\partial^3 h^*/\partial v^3)\} A(h^*, v)^{-2}$$
$$= \text{diag}\{-2(\lambda^{-1} + \delta_{i+})^{-1}\}.$$

Here $\delta_{i+} = \Sigma_j \delta_{ij}$. The corresponding score equation is given by

$$\partial s_\tau(h^*)/\partial \lambda = \partial p_\tau(h^*)/\partial \lambda - \{\partial F/\partial \lambda\}/24 = 0,$$

where $\partial p_\tau(h^*)/\partial \lambda$ is given in (14.17) and $\partial F/\partial \lambda = \text{trace}(S')$ with $S' = \text{diag}\{-2(1 + \lambda \delta_{i+})^{-2}\}$.

Equivalence of the h-likelihood procedure for the Poisson HGLM and the profile-likelihood procedure for the semiparametric frailty model

Let h_1 and h_2 be respectively the h-likelihood for the semiparametric frailty model (14.1) and that for the auxiliary Poisson HGLM (14.7). They share common parameters and random effects, but h_1 involves quantities $(y_{ij}, \delta_{ij}, x_{ij})$ and h_2 quantities $(y_{ij,k}, x_{ij,k})$. In Section 10.2.4 we show that functionally

$$h = h_1 = h_2. \tag{14.18}$$

Using $\tau = (\beta^t, v^t)^t$ and $\omega = (\kappa^t, \tau^t)^t$ with $\kappa = \log \alpha_0$, in Section 10.2.4 we see that the h-likelihood h_1 and its profile likelihood, $h_1^* = h_1|_{\kappa = \widehat{\kappa}}$, provide common estimators for τ. Thus, (14.18) shows that the h-likelihood h_1 for the model (14.1) and the profile likelihood, $h_2^* = h_2|_{\kappa = \widehat{\kappa}}$, for the model (14.7) provide the common inferences for τ.

Consider adjusted profile likelihoods

$$p_\tau(h^*) = [h^* - \frac{1}{2}\log\det\{A(h^*,\tau)/(2\pi)\}]|_{\tau=\hat\tau}$$

and

$$p_\omega(h) = [h - \frac{1}{2}\log\det\{A(h,\omega)/(2\pi)\}]|_{\omega=\hat\omega}.$$

We first show that the difference between $p_\tau(h_1^*)$ and $p_\omega(h_1)$ is constant and thus that they give equivalent inferences for σ^2. Since

$$\partial h_1^*/\partial\tau = [\partial h_1/\partial\tau + (\partial h_1/\partial\kappa)(\partial\hat\kappa/\partial\tau)]|_{\kappa=\hat\kappa},$$
$$\partial\hat\kappa/\partial\tau = -(-\partial^2 h_1/\partial\kappa^2)^{-1}(-\partial^2 h_1/\partial\kappa\partial\tau)|_{\kappa=\hat\kappa}$$

we have

$$A(h_1^*,\tau) = P(h_1,\kappa,\tau)|_{\kappa=\hat\kappa},$$

where

$$P(h_1,\kappa,\tau) = (-\partial^2 h_1/\partial\tau^2) - (-\partial^2 h_1/\partial\tau\partial\kappa)(-\partial^2 h_1/\partial\kappa^2)^{-1}(-\partial^2 h_1/\partial\kappa\partial\tau).$$

Since

$$\det\{A(h_1,\omega)\} = \det\{A(h_1,\kappa)\}\cdot\det\{P(h_1,\kappa,\tau)\},$$

we have

$$p_\omega(h_1) = p_\tau(h_1^*) + c\ ,$$

where $c = -\frac{1}{2}\log\det\{A(h_1,\kappa)/(2\pi)\}|_{\kappa=\hat\kappa} = -\frac{1}{2}\sum_k\log\{d_{(k)}/(2\pi)\}$, which is constant. Thus, procedures based upon h_1 and h_1^* give identical inferences. In fact, Ha $et\ al.$'s profile-likelihood method based upon h_1^* is a numerically efficient way of implementing the h-likelihood procedure based upon h_1 for frailty models (14.1). Furthermore, from (14.18) this shows the equivalence of the h_1 and h_2^* procedures, which in turn implies that Lee and Nelder's (1996) procedure for the Poisson HGLM (14.4) is equivalent to Ha $et\ al.$'s (2001) profile-likelihood procedure for the frailty model (14.1). The equivalence of the h_2 and h_2^* procedures shows that Ha $et\ al.$'s profile-likelihood method can be applied to Ma $et\ al.$'s (2003) auxiliary Poisson models, effectively eliminating nuisance parameters. Finally, the extension of the proof to the second-order Laplace approximation, for example in gamma frailty models, can be similarly shown.

References

Agresti, A. (2002). *Categorical Data Analysis, 2nd ed.* New York: Wiley.

Airy, G.B. (1861). *On the Algebraic and Numerical Theory of Errors of Observations and the Combination of Observations.* London: Macmillan.

Aitkin, M.A. (1981). A note on the regression analysis of censored data. *Technometrics*, **23**, 161-163.

Aitkin, M. and Alfo, M. (1998). Regression models for binary longitudinal responses. *Statistics and Computing*, **8**, 289-307.

Aitkin, M., Anderson, D.A., Francis, B.J. and Hinde, J.P. (1989). *Statistical Modelling in GLIM.* Oxford: Clarendon Pres.

Aitkin, M. and Clayton, D. (1980). The fitting of exponential, Weibull and extreme value distributions to complex censored survival data using GLIM. *Journal of the Royal Statistical Society* C, **29**, 156-163.

Alam, M., Noh, M. and Lee, Y. (2013). Likelihood estimate of treatment effects under selection bias. *Statistics and Its Interface*, **6(3)**, 349-359.

Allison, D., Cui, X., Page, G. and Sabripour, M. (2006) Microarray data analysis: from disarray to consolidation and consensus, *Nature Reviews Genetics* **7**, 55-65.

Amos, C.I. (1994). Robust variance-components approach for assessing genetic linkage in pedigrees. *American Journal of Human Genetics*, **54**, 535-543.

Andersen, E.B. (1970). Asymptotic properties of conditional maximum-likelihood estimators. *Journal of the Royal Statistical Society* B, **32**, 283-301.

Andersen, P.K., Klein, J.P., Knudsen, K. and Palacios, R.T. (1997). Estimation of variance in Cox's regression model with shared gamma frailties. *Biometrics*, **53**, 1475-1484.

Anderson, T.W. (1957). Maximum likelihood estimates for the multivariate normal distribution when some observations are missing. *Journal of the American Statistical Association*, **52**, 200-203.

Azzalini, A., Bowman, A.W. and Hardle, W. (1989). On the use of nonparametric regression for model checking. *Biometrika*, **76**, 1-11.

Baltagi, B.H. (1995). *Economic Analysis of Panel Data.* New York: Wiley.

Barndorff-Nielsen, O.E. (1983). On a formulae for the distribution of the maximum likelihood estimator. *Biometrika*, **70**, 343-365.

Barndorff-Nielsen, O.E. (1997). Normal inverse Gaussian distributions and stochastic volatility modelling. *Scandinavian Journal of Statistics*, **24**, 1-13.

Barndorff-Nielsen, O.E. and Shephard, N. (2001). Non-Gaussian Ornstein-Uhlenbeck-based models and some of their uses in financial economics. *Journal of the Royal Statistical Society* B, **63**, 167-241.

Bartlett, M.S. (1937). Some examples of statistical methods of research in agriculture and applied biology. *Journal of the Royal Statistical Society Supplement*, **4**, 137-183.

Bayarri, M.J., DeGroot, M.H. and Kadane, J.B. (1988). What is the likelihood function? *Statistical Decision Theory and Related Topics IV, Vol. 1.* New York: Springer.

Bell, E. C., Willson, M. C., Wilman, A. H., Dave, S. and Silverstone, P. H. (2006). Males and females differ in brain activation during cognitive tasks. *Neuroimage*, **30(2)**, 529-538.

Benjamini, Y. and Hochberg, Y. (1995). Controlling the false discovery rate: a practical and powerful approach to multiple testing. *Journal of the Royal Statistical Society. Series B*, **57(1)**, 289-300.

Benjamini, Y. and Yekutieli, D. (2001). The control of the false discovery rate in multiple testing under dependency. *The Annals of Statistics*, **29(4)**, 1165-1188.

Benjamini, Y. and Yekutieli, D. (2005). False discovery rate: adjusted multiple confidence intervals for selected parameters. *The Annals of Statistics*, **100**, 71-80.

Berger, J.O. and Wolpert, R. (1988). *The Likelihood Principle.* Hayward: Institute of Mathematical Statistics Monograph Series.

Beseg, J. (1986). On the statistical analysis of dirty pictures. *Journal of the Royal Statistical Society. Series B*, **48**, 259-302.

Besag, J., Green, P., Higdon, D. and Mengersen, K. (1995). Bayesian computation and stochastic systems. *Statistical Science*, **10**, 3-66.

Besag, J. and Higdon, D. (1999). Bayesian analysis of agriculture field experiments. *Journal of the Royal Statistical Society* B, **61**, 691-746.

Best, N.G., Spiegelhalter, D.J., Thomas, A. and Brayne, C.E.G. (1996). Bayesian analysis of realistically complex models. *Journal of the Royal Statistical Society* A, **159**, 323-342.

Birnbaum, A. (1962). On the foundations of statistical inference. *Journal of the American Statistical Association*, **57**, 269-306.

Bissell, A.F. (1972). A negative binomial model with varying element sizes. *Biometrika*, **59**, 435-441.

Bjørnstad, J.F. (1990). Predictive likelihood principle: a review. *Statistical Science.* **5**, 242-265.

Bjørnstad, J.F. (1996). On the generalization of the likelihood function and likelihood principle. *Journal of the American Statistical Association*, **91**, 791-806.

Blangero, J., Williams, J.T. and Almasy, L. (2001). Variance component methods for detecting complex trait loci. *Advances in Genetics*, **42**, 151-181.

Bock, R.D. and Lieberman, M. (1970). Fitting a response model for dichotomously scored items. *Psychometrika*, **35**, 179-197.

Box, G.E.P. (1988). Signal-to-noise ratios, performance criteria and transformations. *Technometrics*, **30**, 1-17.

Box, M.J., Draper, N.R. and Hunter, W.G. (1970). Missing values in multi-response nonlinear data fitting. *Technometrics*, **12**, 613-620.

Breiman, L. (1995). Better subset regression using the nonnegative garrote. *Technometrics*, **37**, 373-384.

Breiman, L. (1996) Heuristics of instability and stabilization in model selection, *Annals of Statistics*, **24**, 2350-2383.

Breslow, N.E. (1972). Contribution to the discussion of the paper by D.R. Cox. *Journal of the Royal Statistical Society*, **34**, 216-217.

Breslow, N.E. (1974). Covariance analysis of censored survival data. *Biometrics*, **30**, 89-99.

Breslow, N.E. and Clayton, D. (1993). Approximate inference in generalized linear mixed models. *Journal of the American Statistical Association*, **88**, 9-25.

Breslow, N.E. and Lin, X. (1995). Bias correction in generalized linear mixed models with a single component of dispersion. *Biometrika*, **82**, 81-91.

Brinkley, P.A., Meyer, K.P. and Lu, J.C. (1996). Combined generalized linear modelling non-linear programming approach to robust process design: a case-study in circuit board quality improvement. *Applied Statistics*, **45**, 99-110.

Brownlee, K.A. (1960). *Statistical Theory and Methodology in Science and Engineering*. New York: Wiley.

Buckley, J. and James, I. (1979). Linear regression with censored data. *Biometrika*, **66**, 429-436.

Burdick, R.K. and Graybill, F.A. (1992). *Confidence Intervals on Variance Components*. New York: Marcel Dekker.

Burton, P.R., Palmer, L.J., Jacobs, K., Keen, K.J., Olson, J.M. and Elston, R.C. (2001). Ascertainment adjustment: where does it take us? *American Journal of Human Genetics*, **67**, 1505-1514.

Butler, R.W. (1986). Predictive likelihood inference with applications. *Journal of the Royal Statistical Society* B, **48**, 1-38.

Carlin, B.P. and Louis, T.A. (2000). *Bayesian and Empirical Bayesian Methods for Data Analysis*. London: Chapman & Hall.

Castillo, J. and Lee, Y. (2008). GLM methods for volatility models. *Statatistical Modelling*, **8**, 263-283.

Celeux, G., Forbes, F. and Peyrard, N. (2003). EM procedures using mean field-like approximations for Markov model-based images segmentation. *Pattern Recognition*, **36**, 131-144.

Chaganty, N.R. and Joe, H. (2004). Efficiency of generalized estimating equations for binary responses. *Journal of the Royal Statistical Society* B, **66**, 851-860.

Chambers, J.M., Cleveland, W.S., Kleiner, B. and Tukey, P.A. (1983). *Graphical Methods for Data Analysis*. Belmont, CA: Wadsworth.

Chernoff, H. (1954). On the distribution of the likelihood ratio. *Annals of Mathematical Statistics*, **25**, 573-578.

Clayton, D. and Cuzick, J. (1985). The EM algorithm for Cox's regression model using GLIM. *Applied Statistics*, **34**, 148-156.

Clayton, D. and Kaldor, J. (1987). Empirical Bayes estimates of age-standardized relative risks for use in disease mapping. *Biometrics*, **43**, 671-681.

Cliff, A.D. and Ord, J.K. (1981). *Spatial Processes: Models and Applications.* London: Pion.

Cochran, W.G. and Cox, G.M. (1957). *Experimental Designs, 2nd Ed.* New York: Wiley.

Cox, D.R. (1972). Regression models and Life-Tables. *Journal of the Royal Statistical Society* B, **34**, 187-220.

Cox, D.R. and Hinkley, D.V. (1974). *Theoretical Statistics.* London: Chapman & Hall.

Cox, D.R. and Reid, N. (1987). Parameter orthogonality and approximate conditional inference. *Journal of the Royal Statistical Society* B, **49**, 1-39.

Cressie, N. (1993). *Statistics for Spatial Data.* New York: Wiley.

Cressie, N., Kaiser, M., Daniels, M., Aldworth, J., Lee, J., Lahiri, S. and Cox, L. (1999). Spatial analysis of particulate matter in an urban environment. In: *Geostatistics for Enviromental Applications: Proceedings of Second European Conference on Geostatistics for Environmental Applications.*

Christian, N. J., Ha, I. D. and Jeong, J., H. (2016). Hierarchical likelihood inference on clustered competing risks data. *Statistics in Medicine*, **35**, 251-267.

Crouch, E.A.C. and Spiegelman, D. (1990). The evaluation of integrals of the form $\int_{-\infty}^{+\infty} f(t) \exp(-t^2) dt$: application to logistic-normal models. *Journal of the American Statistical Association*, **85**, 464-469.

Crowder, M.J. (1995). On the use of a working correlation matrix in using generalized linear models for repeated measurements. *Biometrika*, **82**, 407-410.

Curnow, R.N. and Smith, C. (1975). Multifactorial model for familial diseases in man. *Journal of the Royal Statistical Society* A, **137**, 131-169.

Daniels, M.J., Lee, Y.D., Kaiser, M. (2001). Assessing sources of variability in measurement of ambient particulate matter. *Environmetrics*, **12**, 547-558.

de Andrade, M. and Amos, C.I. (2000). Ascertainment issues in variance component models. *Genetic Epidemiology*, **19**, 333-344.

de Boor, C. (1978). *A Practical Guide to Splines.* Berlin: Springer.

DeGroot, M.H. and Goel, P.K. (1980). Estimation of the correlation coefficient from a broken random sample. *Annals of Statistics*, **8**, 264-278.

Dempster, A.P., Laird, N.M. and Rubin, D.B. (1977). Maximum likelihood from incomplete data via the EM algorithm. *Journal of the Royal Statistical Society* B, **39**, 1-38.

Diebolt, J. and Ip, E.H.S. (1996). A stochastic EM algorithm for approximating the maximum likelihood estimate. In: *Markov Chain Monte Carlo in Practice.* London: Chapman & Hall.

Diggle, P.J. and Kenward, M.G. (1994). Informative drop-out in longitudinal analysis. *Applied Statistics*, **43**, 49-93.

Diggle, P.J., Liang, K.Y. and Zeger, S.L. (1994). *Analysis of Longitudinal Data.* Oxford: Clarendon Press.

Diggle, P.J., Tawn, J.A. and Moyeed, R.A. (1998). Model-based geostatistics. *Applied Statistics*, **47**, 299-350.

Dodge, Y. (1997). LAD regression for detection of outliers in response and explanatory variables. *Journal of Multivariate Analysis*, **61**,144-158.

Donoho, D. L. and Johnstone, I. M.(1994). Ideal spatial adaptation by wavelet shrinkage. *Biometrika*, **81**, 425–455.

Cox, D.R. and Hinkley, D.V. (1974). *Theoretical Statistics*. London: Chapman & Hall.

Durbin, J. and Koopman, S.J. (2000). Time series analysis of non-Gaussian observations based on state space models from both classical and Bayesian perspectives. *Journal of the Royal Statistical Society* B, **62**, 3-56.

Eberlein, E. and Keller, U. (1995). Hyperbolic distributions in finance. *Bernoulli*, **3**, 281-299.

Edwards, A.W.F. (1972). *Likelihood*. Cambridge: Cambridge University Press.

Efron, B. (1986). Double exponential families and their use in generalized linear models. *Journal of the American Statistical Association*, **81**, 709-721.

Efron, B. (2004). Large scale simultaneous hypothesis testing: The choice of a null hypothesis. *Journal of the American Statistical Association* **99**, 96-104.

Efron, B. (2007). Correlation and large-scale simultaneous significance testing. *Journal of the American Statistical Association*, **102**, 93-103.

Efron, B. and Morris, C. (1975) Data analysis using Stein's estimator and its generalizations, *Journal of the American Statistical Association*, **70**, 311–319.

Eilers, P.H.C. and Marx, B.D. (1996). Flexible smoothing with B-splines and penalties. *Statistical Science*, **11**, 89-121.

Eilers, P.H.C. (2004). Fast computation of trends in scatterplots. Unpublished manuscript.

Eisenhart, C. (1947). The assumptions underlying the analysis of variance. *Biometrics*, **3**, 1-21.

Elliot, H., Derin, H., Christi, R. and Geman, D. (1983). Application of the Gibbs distribution to image segmentation. *DTIC Document*.

Elston, R.C. and Sobel, E. (1979). Sampling considerations in the gathering and analysis of pedigree data. *American Journal of Human Genetics*, **31**, 62-69.

Engel, J. (1992). Modelling variation in industrial experiments. *Applied Statistics*, **41**, 579-593.

Engel, J. and Huele, F.A. (1996). A generalized linear modelling approach to robust design. *Technometrics*, **38**, 365-373.

Engel, R.E. (1995). *ARCH*. Oxford: Oxford University Press.

Epstein, M.P., Lin, X. and Boehnke, M. (2002). Ascertainment-adjusted parameter estimates revisited. *American Journal of Human Genetics*, **70**, 886-895.

Eubank, R.L. (1988). *Spline Smoothing and Nonparametric Regression*. New York: Marcel Dekker.

Eubank, L., Lee, Y. and Seely, J. (2003). Unweighted mean squares for the general two variance component mixed models. *Proceedings of Graybill Conference*, Fort Collins: Colorado State University Press, 281-299.

Fan, J. (1997) Comments on "Wavelets in statistics: A review" by A. Antoniadis, *Journal of the International Statistical association*, **6**, 131-138.

Fan, J. and Li, R. (2001) Variable selection via nonconcave penalized likelihood and its oracle properties, *Journal of the American Statistical association*, **96**, 1348-1360.

Felleki, M., Lee, Y., Lee, D., Gilmour, A. and Rönnegård, L. (2012). Estimation of breeding values for mean and dispersion, their variance and correlation using double hierarchical generalized linear models, *Genetics Research*, **94**, 307-317.

Firth, D., Glosup, J. and Hinkley, D.V. (1991). Model checking with nonparametric curves. *Biometrika*, **78**, 245-252.

Fisher, R.A. (1918). The correlation between relatives and the supposition of Mendelian inheritance. *Transactions of the Royal Society of Edinburgh*, **52**, 399-433.

Fisher, R.A. (1921). On the probable error of a coefficient of correlation deduced from a small sample. *Metron*, **1**, 3-32.

Fisher, R.A. (1934). The effects of methods of ascertainment upon the estimation of frequencies. *Annals of Eugenics*, **6**, 13-25.

Fisher, R.A. (1935). *The Design of Experiments.* Edinburgh: Oliver and Boyd.

Fleming, T.R. and Harrington, D.R. (1991). *Counting processes and survival analysis.* New York: Wiley.

Fung, T. and Seneta, E. (2007) Tailweight, quantiles and kurtosis: a study of competing distribution. *Operations Research Letters*, **35**, 448–454.

Gabriel, K.R. (1962). Ante-dependence analysis of an ordered set of variables. *Annals of Mathematical Statistics*, **33**, 201-212.

Gad, A.M. (1999). Fitting selection models to longitudinal data with dropout using the stochastic EM algorithm. Unpublished PhD Thesis, University of Kent, Canterbury, UK.

Gardner, M. (1982) *Aha! Gotcha: Paradoxes to Puzzle and Delight.* W. H. Freeman & Co.

Gelfand, A.E. and Smith, A.F.M. (1990). Sampling-based approaches to calculating marginal densities. *Journal of the American Statistical Association*, **87**, 523-532.

Geman, S. and Geman, D. (1984). Stochastic relaxation, Gibbs distributions, and the Bayesian segmentation of images. *IEEE Transactions on Pattern Analysis and Machine Intelligence*, **6**, 721-741.

Glidden, D. and Liang, K.Y. (2002). Ascertainment adjustment in complex diseases. *Genetic Epidemiology*, **23**, 201-208.

Godambe, V.P. and Thompson, M.E. (1989). An extension of quasi-likelihood estimation. *Journal of Statistical Planning and Inference*, **22**, 137-152.

Golub, G. and Reinsch, C. (1971). Singular value decomposition and least squares solutions. In: *Handbook for Automatic Computation II: Linear Algebra*. New York: Springer Verlag.

Goossens, H.H.L.M. and Van Opstal, A.J. (2000). Blink-perturbed saccades in monkey. I. Behavioral analysis. *Journal of Neurophysiology*, **83** , 3411-3429.

Green, P.J. and Silverman, B.W. (1994). *Nonparametric Regression and Generalized Linear Models: A Roughness Penalty Approach.* London: Chapman & Hall.

Gueorguieva, R.V. (2001). A multivariate generalized linear mixed model for joint modelling of clustered outcomes in the exponential family. *Statistical Modelling*, **1**, 177-193.

Ha, I., Christian, N. J. Jeong, J., H., Park, J. and Lee Y. L. (2016). Analysis of clustered competing risks data using subdistribution hazard models with multivariate frailties. *Statistical Methods in Medical Research*, **25**, 2488-2505.

Ha, I., Jeong, J. and Lee, Y. (2017). *Statistical modelling for survival data with random effects: H-likelihood approach.* to appear in Springer.

Ha, I., Lee, M., Oh, S., Jeong, J., H., Sylvestere, R. and Lee, Y. (2014). Variable selection in subdistribution hazard frailty models with competing risks data. *Statistics in Medicine*, **33**, 4590-4604.

Ha, I. and Lee, Y. (2003). Estimating frailty models via Poission hierarchical generalized linear models. *Journal of Computational and Graphical Statistics*, **12**, 663-681.

Ha, I. and Lee, Y. (2005a). Comparison of hierarchical likelihood versus orthodox BLUP approach for frailty models. *Biometrika*, **92**, 717-723.

Ha, I. and Lee, Y. (2005b). Comparison of hierarchical likelihood versus orthodox BLUP approach for frailty models. *Biometrika*, **92**, 717-723.

Ha, I., Lee, Y. and McKenzie, G. (2007). Model selection for multi-component frailty models. *Statistics in Medicine*, **26**, 4790-4807.

Ha, I., Lee, Y. and Song, J.K. (2001). Hierarchical likelihood approach for frailty models. *Biometrika*, **88**, 233-243.

Ha, I., Lee, Y. and Song, J.K. (2002). Hierarchical likelihood approach for mixed linear models with censored data. *Lifetime Data Analysis*, **8**, 163-176.

Ha, I., Noh, M. and Lee, Y. (2010). bias reduction of likelihood estimators in semiparametric frailty models, *Scandinavian Journal of Statistics*, **37(2)**, 307-320.

Ha, I., Noh, M. and Lee, Y. (2012a). frailtyHL: a package for fitting models with h-likelihood. *The R Journal*, **4(2)**, 28-37.

Ha, I., Noh, M. and Lee, Y. (2012b). frailtyHL: frailty models via h-likelihood, URL http://CRAN.R-project.org/package=frailtyHL. R package version 1.1.

Ha I., Pan, J., Oh, S., Lee, Y. (2014). Variable selection in general frailty models using penalized h-likelihood. *Journal of Computational Graphical Statistics*, **23**, 1044-1060.

Ha, I., Lee, M., Oh, S., Jeong, J., Sylvestere, R. and Lee, Y. (2014). Variable selection in subdistribution hazard frailty models with competing risks data. *Statistics in Medicine*, **33**, 4590-4604.

Ha, I., Park, T.S. and Lee, Y. (2003). Joint modelling of repeated measures and survival time data. *Biometrical Journal*, **45**, 647-658.

Ha, I., Vaida, F. and Lee, Y. (2016). Interval estimation for random effects in proportional hazards models with frailties. *Statistical Methods in Medical Research*, **25**, 936-953.

Harvey, A.C. (1989). *Forecasting, Structural Time Series Models and the Kalman Filter*. Cambridge: Cambridge University Press.

Harvey, A.C., Ruiz, E. and Shephard, N. (1994). Multivariate stochastic variance models. *Review of Economic Studies*, **61**, 247-264.

Harville, D. (1974). Bayesian inference for variance components using only error contrasts. *Biometrika*, **61**, 383-385.

Hausman, J.A. (1978). Specification tests in econometrics. *Econometrica*, **46**, 1251-1271.

Heckman, J. and Singer, B. (1984). A method for minimizing the impact of distributional assumptions in econometric models for duration data. *Econometrica*, **52**, 271-320.

Henderson, C.R. (1973). Sire evaluation and genetic trends. In: *Proceedings of the Animal Breeding and Genetic Symposium in Honour of Dr. J. L. Lush*, pp 10-41. Champaign, IL: American Society of Animal Science.

Henderson, C.R, Kempthorne, O., Searle, S.R. and Von Krosigk, C.M. (1959). The estimation of genetic and environmental trends from records subject to culling. *Biometrics*, **15**, 192-218.

Heyting, A., Essers, J.G.A. and Tolboom, J.T.B.M. (1990). A practical application of the Patel-Kenward analysis of covariance to data from an anti-depressant trial with drop-outs. *Statistical Applications*, **2**, 695-307.

Hinkley, D.V. (1979). Predictive likelihood. *Annals of Statistics*, **7**, 718-728.

Hobert, J., and Casella, G. (1996). The effect of improper priors on Gibbs sampling in hierarchical linear mixed models, *Journal of American Statistical Association*, **91**, 1461-1473.

Hoerl, A. E. and Kennard, R. W. (1970) Ridge regression: biased estimation for nonorthogonal problems, *Technometrics*, **12**, 55-67.

Holford, T.R. (1980) The analysis of rates and survivorship using log-linear models, *Biometrics*, **36**, 299-305.

Hsiao, C. (1995). *Analysis of Panel Data*. Econometric Society Monograph, Cambridge: Cambridge University Press.

Huang, J., Breheny, P. and Ma, S. (2012). A selective review of group selection in high-dimensional models, *Statistical Science*, **27**, 481-499.

Huang, X. and Wolfe, R.A. (2002). A frailty model for informative censoring. *Biometrics*, **58**, 510-520.

Hudak, S.J., Saxena, A. Bucci, R.J. and Malcom, R.C. (1978). Development of standard methods of testing and analyzing fatigue crack growth rate data. Technical report. *AFML-TR-78-40*. Pittsburgh: Westinghouse Electric Corporation.

Hughes, J.P. (1999). Mixed effects models with censored data with application to HIV RNA levels. *Biometrics*, **55**, 625-629.

Hunter, D. and Li, R. (2005) Variable selection using MM algorithms, *Annals of Statistics*, **33**, 1617-1642.

Höskuldsson A. (1988). PLS regression methods. *Journal of Chemometrics*, **2**, 211-228.

Jackson, R.W.B. (1939). The reliability of mental tests. *British Journal of Psychology*, **29**, 267-287.

James, W. and Stein, C. (1961). Estimation with quadratic loss. In: *Proc. of Fourth Berkley Symp. Math. Statist. Probab. 1.* 361-380. Berkeley: University of California Press.

Johansen, S. (1983). An extension of Cox's regression model. *International Statistical Review*, **51**, 258-262.

Jørgensen, B. (1986). Some properties of exponential dispersion models. *Scandinavian Journal of Statistics*, **13**, 187-198.

Jørgensen, B. (1987). Exponential dispersion models. *Journal of the Royal Statistical Society* B, **49**, 127-162.

Kang, W., Lee, M. and Lee, Y. (2005). HGLM versus conditional estimators for the analysis of clustered binary data. *Statistics in Medicine*, **24**, 741-752.

Karim, M.R. and Zeger, S.L. (1992). Generalized linear models with random effects: salamander mating revisited. *Biometrics*, **48**, 681-694.

Kenward, M.G. (1998). Selection models for repeated measurements with non-random dropout: an illustration of sensitivity. *Statistics in Medicine*, **17**, 2723-2732.

Kenward, M.G. and Smith, D.M. (1995). Computing the generalized estimating equations with quadratic covariance estimation for repeated measurements. *Genstat Newsletter*, **32**, 50-62.

Kenward, M.G. and Roger, J.H. (1997). Small sample inference for fixed effects from restricted maximum likelihood. *Biometrics*, **53**, 983-997.

Kim, D., Lee, Y. and Oh, H. (2006). Hierarchical likelihood-based wavelet method for denoising signals with missing data. *IEEE Signal Processing Letters*, **13**, 361-364.

Kim, S., Shephard, N. and Chib, S. (1998). Stochastic volatility: likelihood inference and comparison with ARCH models. *Review of Economic Studies*, **98**, 361-393.

Klein, J.P., Pelz, C. and Zhang, M. (1999). Modelling random effects for censored data by a multivariate normal regression model. *Biometrics*, **55**, 497-506.

Koch, G.G., Landis, J.R., Freeman, J.L., Freeman, D.H. and Lehnen, R.G. (1977). A general method for analysis of experiments with repeated measurement of categorical data. *Biometrics*, **33**, 133-158.

Kooperberg, C. and Stone, C.J. (1992). Logspline density estimation for censored data. *Journal of Computational Graphical Statistics*, **1**, 301-328.

Kovac, A. and Silverman, B.W. (2000). Extending the scope of wavelet regression methods by coefficient-dependent thresholding. *Journal of the American Statistical Association*, **95**, 172-183.

Kwon, S., Oh, S. and Lee, Y. (2016) The use of random-effect models for high-dimensional variable selection problems, *Computational Statistics and Data Analysis*. In press.

Lai, T.Z. and Ying, Z. (1994). A missing information principle and M-estimators in regression analysis with censored and truncated data. *The Annals of Statistics*, **22**, 1222–1255.

Laird, N. (1978). Nonparametric maximum likelihood estimation of a mixing distribution. *Journal of the American Statistical Association*, **73**, 805-811.

Laird, N. and Olivier, D. (1981). Covariance analysis of censored survival data using log-linear analysis techniques. *Journal of the American Statistical Association*, **76**, 231-240.

Laird, N. and Ware, J.H. (1982). Random-effects models for longitudinal data. *Biometrics*, **38**, 963-974.

Lane, P.W. and Nelder, J.A. (1982). Analysis of covariance and standardization as instances of prediction. *Biometrics*, **38**, 613-621.

Lange, K.L., Little, R.J.A. and Taylor, J.M.G.(1989). Robust statistical modeling using the t distribution. *Journal of the American Statistical Association*, **84**, 881-896.

Lauritzen, S.L. (1974). Sufficiency, prediction and extreme models. *Scandinavian Journal of Statistics*, **1**, 128-134.

Lawless, J. and Crowder, M. (2004). Covariates and random effects in a gamma process model with application to degeneration and failure. *Lifetime Data Analysis*, **10**, 213-227.

Lee, D., Kang, H., Kim,E., Lee, H., Kim, H., Kim, Y.K., Lee, Y. and Lee, D.S. (2015). Optimal likelihood-ratio multiple testing with application to Alzheimer's disease and questionable dementia. *BMC Medical Research Methodology*, **15:9**.

Lee, D., Kang, H., Jang, M., Cho, S., Kang, W., Lee, J., Kang, E., Lee, K., Woo, J. and Lee, M. (2003). Application of false discovery rate control in the assessment of decrease of FDG uptake in early Alzheimer dementia. *Korean Journal of Nuclear Medicine*, **37**, 374-381.

Lee, D., Lee, W., Lee, Y. and Pawitan,Y. (2010) Super-sparse principal component analyses for high-throughput genomic data, *BMC Bioinformatics*, **11**:296.

Lee, D., Lee, W., Lee, Y. and Pawitan, Y. (2011a) Sparse partial least-squares regression and its applications to high-throughput data analysis, *Chemometrics and Intelligent Laboratory Systems*, **109**, 1-8.

Lee, D. and Lee,Y. (2016). Extended likelihood approach to multiple test with directional error control under hidden Markov random field model. *Journal of Multivariate Analysis*, Journal of Multivariate Analysis, **151**, 1-13.

Lee, D, Lee, Y, Pawitan, Y. and Lee, W. (2013) Sparse partial least-squares regression for high-throughput survival data analysis. *Statistics in Medicine*, **32** ,5340-5352.

Lee, W., Lee, D., Lee, Y. and Pawitan, Y. (2011b) Sparse canonical covariance analysis for high-throughput data, *Statistical Applications in Genetics and Molecular Biology*, **10(1)**, Article 30.

Lee, S, Pawitan, Y, Ingelsson, E. and Lee, Y. (2015) Sparse estimation of gene-gene interactions in prediction models. *Statistical Methods in Medical Research*. pii: 0962280215597261.

Lee, S, Pawitan, Y and Lee, Y. (2015) Random-effect model approach for group variable selection. *Computational Statistics and Data Analysis*, **89**, 147-157.

Lee, T.C.M. and Meng, X.L. (2005). A self-consistent wavelet method for denoising images with missing pixels. *Proceedings of 30th IEEE International Conference on Acoustics, Speech, and Signal Processing*, vol. II, 41–44.

Lee, Y. (1991). Jackknife variance estimators of the location estimator in the one-way random-effects model. *Annals of the Institute of Statistical Mathematics*, **43**, 707-714.

Lee, Y. (2000). Discussion of Durbin and Koopman's paper. *Journal of the Royal Statistical Society* B, **62**, 47-48.

Lee, Y. (2001). Can we recover information from concordant pairs in binary matched pairs? *Journal of Applied Statistics*, **28**, 239-246.

Lee, Y. (2002a). Robust variance estimators for fixed-effect estimates with hierarchical likelihood. *Statistics and Computing*, **12**, 201-207.

Lee, Y. (2002b). Fixed-effect versus random-effect models for evaluating therapeutic preferences. *Statistics in Medicine*, **21**, 2325-2330.

Lee, Y. (2004). Estimating intraclass correlation for binary data using extended quasi-likelihood. *Statistical Modelling*, **4**, 113-126.

Lee, Y., Alam, M., Noh, M., Rönnegård, L. and Skarin, A. (2016). Spatial model with excessive zero counts for analysis of the changes in reindeer distribution, *Ecology and Evolution*, published online.

Lee, Y., and Bjørnstad, J.F. (2013). Extended likelihood approach to large-scale multiple testing. *Journal of the Royal Statistical Society B*, **75**, 553-575.

Lee, Y. and Birkes, D. (1994). Shrinking toward submodels in regression. *Journal of Statistical Planning and Inference*, **41**, 95-111.

Lee, Y., Jang, M. and Lee, W. (2011). Prediction interval for diesease mapping using hierarchical likelihood. *Computational Statistics*, **26**, 159-179.

Lee, Y. and Kim, G. (2016). H-likelihood predictive intervals for unobservables. *International Statistical Review*. In press.

Lee, Y. and Nelder, J.A. (1996). Hierarchical generalised linear models (with discussion). *Journal of the Royal Statistical Society B*, **58**, 619-656.

Lee, Y. and Nelder, J.A. (1997). Extended quasi-likelihood and estimating equations approach. *IMS Notes Monograph Series*, 139-148.

Lee, Y. and Nelder, J.A. (1998). Generalized linear models for the analysis of quality-improvement experiments. *Canadian Journal of Statistics*, **26**, 95-105.

Lee, Y. and Nelder, J.A. (1999). The robustness of the quasi-likelihood estimator. *Canadian Journal of Statistics*, **27**, 321-327.

Lee, Y. and Nelder, J.A. (2000a). The relationship between double exponential families and extended quasi-likelihood families. *Applied Statistics*, **49**, 413-419.

Lee, Y. and Nelder, J.A. (2000b). Two ways of modelling overdispersion in non-normal data. *Applied Statistics*, **49**, 591-598.

Lee, Y. and Nelder, J.A. (2001a). Hierarchical generalised linear models: A synthesis of generalised linear models, random-effect model and structured dispersion. *Biometrika*, **88**, 987-1006.

Lee, Y. and Nelder, J.A. (2001b). Modelling and analysing correlated non-normal data. *Statistical Modelling*, **1**, 7-16.

Lee, Y. and Nelder, J.A. (2003c). Extended REML estimators. *Journal of Applied Statistics*, **30**, 845-856.

Lee, Y, and Nelder, J.A. (2004). Conditional and marginal models: another view (with discussion). *Statistical Science*, **19**, 219-238.

Lee, Y. and Nelder, J.A. (2005). Likelihood for random-effect models (with discussion). *Statistical and Operational Research Transactions*, **29**, 141-182.

Lee, Y. and Nelder, J.A. (2006a). Double hierarchical generalized linear models (with discussion). *Applied Statistics*, **55**, 139-185.

Lee, Y. and Nelder, J.A. (2006b). Fitting via alternative random effect models. *Statistics and Computing*, **16**, 69-75.

Lee, Y., Nelder, J.A. and Noh, M. (2007). H-likelihood: problems and solutions *Statistics and Computing*, **17**, 49-55.

Lee, Y. and Nelder, J.A. (2009). Likelihood Inference for Models with Unobservables: Another View (with discussion). *Statistical Science*, **24**, 255-302.

Lee, Y. and Noh, M. (2012). Modelling random effect variance with double hierarchical generalized linear models. *Statistical Modelling*, **12(6)**, 487-502.

Lee, Y. and Noh, M. (2015). dhglm: Double Hierarchical Generalized Linear Models, URL http://CRAN.R-project.org/package=dhglm. R package version 1.5.

Lee, Y. and Noh, M. (2016). mdhglm: Multivariate Double Hierarchical Generalized Linear Models, URL http://CRAN.R-project.org/package=mdhglm. R package version 1.2.

Lee, Y., Noh, M. and Ryu, K. (2005). HGLM modeling of dropout process using a frailty model. *Computational Statistics*, **20**, 295-309.

Lee, Y. and Oh, H. (2009). *Random-effect models for variable selection.* Department of Statistics, Stanford University.

Lee, Y. and Oh, H. (2014). A new sparse variable selection via random-effect model. *Journal of Multivariate Analysis*, **125**, 89-99.

Lee, Y., Rönnegård, L. and Noh,M. (2017).*Data analysis using hierarchical generalized linear models with R.* to appear in Chapman & Hall.

Lee, Y. and Seely, J. (1996). Computing the Wald interval for a variance ratio. *Biometrics*, **52**, 1486-1491.

Lee, Y., Yun, S. and Lee, Y. (2003). Analyzing weather effects on airborne particulate matter with HGLM. *Environmetrics*, **14**, 687-697.

Lehmann, E.L. (1983). *Theory of Point Estimation.* New York: Wiley.

Leon, R.V., Sheomaker, A.C. and Kackar, R.N. (1987). Performance measure independent of adjustment: an explanation and extension of Taguchi's signal to noise ratio. *Technometrics*, **29**, 253-285.

Leventhal, L. and Huynh, C. (1996). Directional decisions for two-tailed tests: power, error rates, and sample size. *Psychology Methods*, **1**, 278-292.

Liang, K.Y. and Zeger, S.L. (1986). Longitudinal data analysis using generalized linear models. *Biometrika*, **72**, 13-22.

Lichtenstein, P., Holm, N.V., Verkasalo, P.K., Iliadou, A., Kaprio, J., Koskenvuo, M., Pukkala, E., Skytthe, A. and Hemminki, K. (2000). Environmental and heritable factors in the causation of cancer: analyses of cohorts of twins from Sweden, Denmark and Finland. *New England Journal of Medicine*, **343**, 78-85.

Lin, X. and Breslow, N.E. (1996). Bias correction in generalized linear mixed models with multiple components of dispersion. *Journal of the American Statistical Association*, **91**, 1007-1016.

Lindsay, B. (1983). Efficiency of the conditional score in a mixture setting. *Annals of Statistics*, **11**, 486-497.

Lindsey, J.K. and Lambert, P. (1998). On the appropriateness of marginal models for repeated measurements in clinical trials. *Statistics in Medicine*, **17**, 447-469.

Lindström, L., Pawitan, Y., Reilly, M., Hemminki, K., Lichtenstein, P. and Czene, K. (2006). Estimation of genetic and environmental factors for melanoma onset using population-based family data. *Statistics in Medicine*, **25**, 3110-3123.

Little, R.J.A. (1995). Modeling the drop-out mechanism in repeated-measures studies. *Journal of the American Statistical Association*, **90**, 1112-1121.

Little, R.J.A. and Rubin, D.B. (1983). On jointly estimating parameters and missing data by maximizing the complete-data likelihood. *American Statistician*, **37**, 218-220.

Little, R.J.A. and Rubin, D.B. (2002). *Statistical Analysis with Missing Data*. New York: Wiley.

Lu, C.J. and Meeker, W.Q. (1993). Using degeneration measurements to estimate a time-to-failure distribution. *Technometrics*, **35**, 161-174.

Ma, R., Krewski, D. and Burnett, R.T. (2003). Random effects Cox models: a Poisson modelling approach. *Biometrika*, **90**, 157-169.

MacKenzie, G., Ha, I.D. and Lee, Y. (2003). Non-PH multivariate survival models based on the GTDL. In: *Proceedings of 18th International Workshop on Statistical Modelling*, Leuven, Belgium, pp. 273-277.

Madan, D.B. and Seneta, E. (1990). The variance gamma model for share marked returns. *Journal of Business*, **63**, 511-524.

McCullagh, P. (1980). Regression models for ordinal data. *Journal of the Royal Statistical Society B*, **42**, 109-142.

McCullagh, P. and Nelder, J.A. (1989). *Generalized Linear Models, 2nd Ed.* London: Chapman & Hall.

McCulloch, C.E. (1994). Maximum likelihood variance components estimation in binary data. *Journal of the American Statistical Association*, **89**, 330-335.

McCulloch, C.E. (1997). Maximum likelihood algorithms for generalized linear mixed models. *Journal of the American Statistical Association*, **92**, 162-170.

McGilchrist, C.A. (1993). REML estimation for survival models with frailty. *Biometrics*, **49**, 221-225.

McLachlan, G.J. (1987). On bootstrapping the likelihood ratio test statistic for the number of components in a normal mixture. *Applied Statistics*, **36**, 318-324.

McLachlan, G.J. and Krishnan, T. (1997). *The EM Algorithm and Extensions*. New York: Wiley.

McNemar, Q. (1947). Note on the sampling error of the difference between correlated proportions or percentages. *Psychometrika*, **12**, 153-157.

Molas, M., Noh, M., Lee, Y. and Lesaffre, E. (2013). Joint hierarchical generalized linear models with multivariate Gaussian random effects, *Computational Statistics and Data Analysis*, **68**, 239-250.

Milliken, G.A. and Johnson, D.E. (1984). *Analysis of Messy Data*. New York: Van Nostrand Reinhold.

Muralidharan, O. (2010). An empirical Bayes mixture method for effect size and false discovery rate estimation. *Annals of Applied Statistics*, **4**, 422-438.

Myers, P.H., Montgomery, D.C. and Vining, G.G. (2002). *Generalized Linear Models with Applications in Engineering and the Sciences*. New York: Wiley.

Nelder, J.A. (1965a) The analysis of randomized experiments with orthogonal block structure. I. Block structure and the null analysis of variance. *Proceedings of the Royal Society* A, **283**, 147-162.

Nelder, J.A. (1965b) The analysis of randomized experiments with orthogonal block structure. II. Treatment structure and the general analysis of variance. *Proceedings of the Royal Statistical Society A*, **283**, 163-178.

Nelder, J.A. (1968) The combination of information in generally balanced designs. *Journal of the Royal Statistical Society B*, **30**, 303-311.

Nelder, J.A. (1990). Nearly parallel lines in residual plots. *American Statistician*, **44**, 221-222.

Nelder, J.A. (1994). The statistics of linear models: back to basics. *Statistics and Computing*, **4**, 221-234.

Nelder, J.A. (1998). A large class of models derived from generalized linear models. *Statistics in Medicine*, **17**, 2747-2753.

Nelder, J.A. (2000). Functional marginality and response-surface fitting. *Journal of Applied Statistics*, **27**, 109-112.

Nelder, J.A. and Lee, Y. (1992). Likelihood, quasi-likelihood and pseudo-likelihood: some comparisons. *Journal of the Royal Statistical Society* B, **54**, 273-284.

Nelder, J.A. and Pregibon, D. (1987). An extended quasi-likelihood function. *Biometrika*, **74**, 221-232.

Ng, C., Oh, S. and Lee, Y. (2016). Going beyond oracle property: selection consistency of all local solutions of the generalized linear model, *Statistical Methodology*, **32**, 147-160.

Ng, T., Lee, W. and Lee Y. (2016) Change-point estimators with true identification property. *Bernoulli Journal*, In press.

Noh, M., Ha, I. and Lee, Y. (2006). Dispersion frailty models and HGLMs. *Statistics in Medicine*, **25**, 1341-1354.

Noh, M., Wu, L., and Lee, Y. (2012). Hierarchical likelihood methods for nonlinear and generalized linear mixed models with missing data and measurement errors in covariates. *Journal of Multivariate Analysis*, **109**, 42-51.

Noh, M. and Lee, Y. (2007a). Restricted maximum likelihood estimation for binary data in generalised linear mixed models. *Journal of Multivariate Analysis*, **98**, 896-915.

Noh, M. and Lee, Y. (2007b). Robust modelling for inference from GLM classes. *Journal of the American Statistical Association*, **102(479)**, 1059-1072.

Noh, M., Lee, Y., Yun, S., Lee, S., Lee, M., and Kang Y. (2006). Determinants of hospital closures in south Korea: use of HGLM. *Social Science and Medicine*, **63(9)**, 2320-2329.

Noh, M. and Lee, Y. (2008). Hierachical likelihood approach for nonlinear mixed effects models. *Computational Statistics and Data Analysis*, **52(7)**, 3517-3527.

Noh, M. and Lee, Y. (2007). REML estimation for binary data in GLMMs, *Journal of Multivariate Analysis*, **98(5)**, 896-915.

Noh, M., Lee, Y. and Kenward, M. G. (2011). Robust estimation of dropout models using hierarchical likelihood, *Journal of Statistical Computation and Simulation*, **81(6)**, 693-706.

Noh, M., Pawitan, Y. and Lee, Y. (2005). Robust ascertainment-adjusted parameter estimation. *Genetic Epidemiology*, **29**, 68-75.

Noh, M., Yip, B., Lee, Y. and Pawitan, Y. (2006). Multicomponent variance estimation for binary traits in family-based studies, *Genetic Epidemiology*, **30**, 37-47.

Paik, C.M., Lee, Y. and Ha, I. (2015). Frequentist inference on random effects based on summarizability. *Statistica Sinica*, **25**, 1107-1132.

Patefield, W.M. (2000). Conditional and exact tests in crossover trials. *Journal of Biopharmaceutical Statistics*, **10**, 109-129.

Patterson, H.D. and Thompson, R. (1971). Recovery of interblock information when block sizes are unequal. *Biometrika*, **58**, 545-554.

Pawitan, Y. (2001). *In all Likelihood: Statistical Modelling and Inference Using Likelihood*. Oxford: Clarendon Press.

Pawitan, Y. (2001) Estimating variance components in generalized linear mixed models using quasi-likelihood. *Journal of Statistical Computation and Simulation*, **69**, 1-17.

Pawitan, Y. and Lee, Y. (2016). Wallet game: probability, likelihood and extended likelihood. *American Statistician*, published online.

Pawitan, Y., Reilly, M., Nilsson, E., Cnattingius, S. and Lichtenstein, P. (2004). Estimation of genetic and environmental factors for binary traits using family data. *Statistics in Medicine*, **23**, 449-465.

Payne, R.W. and Tobias, R.D. (1992). General balance, combination of information and the analysis of covariance. *Scandinavian Journal of Statistics*, **19**, 3-23.

Perelman, E., Ploner, A., Calza, S., and Pawitan, Y. (2007). Detecting differential expression in microarray data: comparison of optimal procedures. *BMC Bioinformatics*, **8**:28.

Pettitt, A.N. (1986). Censored obevations, repeated measures and mixed effects models: An approach using the EM algorithm and normal errors. *Biometrika*, **73**, 635-643.

Phadke, M.S., Kackar, R.N., Speeney, D.V. and Grieco, M.J. (1983). Off-line quality control for integrated circuit fabrication using experimental design. *Bell System Technical Journal*, **62**, 1273-1309.

Pierce, D.A. and Schafer, D.W. (1986). Residuals in generalized linear models. *Journal of the American Statistical Association*, **81**, 977-986.

Pourahmadi, M. (2000). Maximum likelihood estimation of generalised linear models for multivariate normal covariance matrix. *Biometrika*, **87**, 425-435.

Press, S.J. and Scott, A.J. (1976). Missing variables in Bayesian regression, II. *Journal of the American Statistical Association*, **71**, 366-369.

Price, C.J., Kimmel, C.A., Tyle, R.W. and Marr, M.C. (1985). The developmental toxicity of ethylene glycol in rats and mice. *Toxicological Applications in Pharmacololgy*, **81**, 113-127.

Radchenko, P. and James, G. (2008) Variable inclusion and shrinkage algorithms, *Journal of the American Statistical Association*, **103**, 1304–1315.

Rao, C.R. (1973). *Linear Statistical Inference and Its Applications, 2nd Ed.* New York: Wiley.

Rasbash, J., Browne, W., Goldstein, H. and Yang, M. (2000). *A User's Guide to MLwin, 2nd Ed.* London: Institute of Education.

Reinsch, C. (1967). Smoothing by spline functions. *Numerische Mathematik*, **10**, 177–183.

Robbins, H. (1970) Statistical methods related to the law of the iterated logarithm, *Annals of Mathematical Statistics*, **41**, 1397-1409.

Robinson, G.K. (1991). That BLUP is a good thing: the estimation of random effects. *Statistical Science*, **6**, 15-51.

Robinson, M.E. and Crowder, M. (2000). Bayesian methods for a growth-curve degradation model with repeated measures. *Lifetime Data Analysis*, **6**, 357-374

Rosenwald, A., Wright, G., Chan, W.C., Connors, J.M., Campo, E., Fisher, R.I., Gascoyne, R.D., Muller-Hermelink, H.K., Smeland, E.B., Giltnane, J.M., *et al.* (2002). The use of molecular profiling to predict survival after chemotherapy for diffuse large-B-cell lymphoma. *New England Journal of Medicine*, **346**, 1937-1947.

Ridout, M., Demetrio, C. and Firth, D. (1999) Estimating intraclass correlation for binary data. *Biometrics*, **55**, 137-148.

Rubin, D.B. (1974). Characterizing the esimation of parameters in incomplete data problems, *Journal of the American Statistical Association*, **69**, 467-474.

Rubin, D.B. (1976). Inference and missing data. *Biometrika*, **63**, 581-592.

Rubin, D.B. and Wu, Y.N. (1997). Modeling schizophrenic behavior using general mixture components. *Biometrics*, **53**, 243-261.

Ruppert, D., Wand, M.P. and Carroll, R.J. (2003). *Semiparametric regression.* Cambridge: Cambridge University Press.

Sabatti, C., Service, S. and Freimer, N. (2003). False discovery rates in linkage and association linkage genome screens for complex disorders. *Genetics* **164**, 829-833.

Saleh, A., Lee, Y. and Seely, J. (1996). Recovery of inter-block information: extension in a two variance component model. *Communications in Statistics: Theory and Method*, **25**, 2189-2200.

Savage, L.J. (1976). On rereading R. A. Fisher.*Annals of Statistics*, **4**, 441-500.

Schall, R. (1991). Estimation in generalised linear models with random effects. *Biometrika*, **78**, 719-727.

Schefffe, H. (1956). A mixed model for the analysis of variance. *Annals of Mathematical Statistics*, **27**, 23-36.

Schmee, J. and Hahn, G.J. (1979). A simple method for regression analysis with censored data. *Technometrics*, **21**, 417-423.

Schumacher, M., Olschewski, M. and Schmoor, C. (1987). The impact of heterogeneity on the comparison of survival times. *Statistics in Medicine*, **6**, 773-784.

Seely, J. and El-Bassiouni, Y. (1983). Applying Wald's variance component test. *Annals of Statistics*, 11, 197-201.

Seely, J. Birkes, D. and Lee, Y. (1997). Characterizing sum of squares by their distributions. *American Statistician*, **51**, 55-58.

Self, S.G. and Liang, K.Y. (1987). Asymptotic properties of maximum likelihood estimators and likelihood ratio tests under nonstandard conditions. *Journal of the American Statistical Association*, **82**, 605-610.

Sham, P.C. (1998). *Statistics in Human Genetics*. London: Arnold.

Shepard, T.H., Mackler B. and Finch, C.A. (1980). Reproductive studies in the iron-deficient rat. *Teratology*, **22**, 329-334.

Shephard, N. (1996). Statistical aspects of ARCH and stochastic volatility. In: *Time Series Models in Econometrics, Ginance and Other Fields*. London: Chapman & Hall.

Shephard, N. and Pitt, M.R. (1997). Likelihood analysis of non-Gaussian measurement time series. *Biometrika*, **84**, 653-667.

Shoemaker, A.C., Tsui, K.L. and Leon, R. (1988). Discussion on signal to noise ratios, performance criteria and transformation. *Technometrics*, **30**, 19-21.

Shun, Z. (1997). Another look at the salamander mating data: a modified Laplace approximation approach. *Journal of the American Statistical Association*, **92**, 341-349.

Shun, Z. and McCullagh, P. (1995). Laplace approximation of high-dimensional integrals. *Journal of the Royal Statistical Society* B, **57**, 749-760.

Singh, D., Febbo, P.G., Ross, K., Jackson, D.G., Manola, J., Ladd, C., Tamayo, P., Renshaw, A.A. , D'Amico, A.V., Richie, J.P., Lander, E.S., Loda, M., Kantoff, P. W., Golub, T. R., and Sellers, W.R. (2002) Gene expression correlates of clinical prostate cancer behavior, *Cancer Cell*, **1**, 203-209.

Silvapulle, M.J. (1981). On the existence of maximum likelihood estimators for the binomial response models. *Journal of the Royal Statistical Society* B, **43**, 310-313.

Silverman, J. (1967). Variations in cognitive control and psychophysiological defense in schizophrenias. *Psychosomatic Medicine*, **29**, 225-251.

Silverman, B.W. (1986). *Density Estimation for Statistics and Data Analysis*. London: Chapman & Hall.

Smyth, G.K. (2004). Linear models and empirical Bayes methods for assessing differential expression in microarray experiments. *Statistical Applications in Genetics and Molecular Biology*, **3**, Article 3.

Spiegelhalter, D.J., Best, N.G., Carlin, B.P. and van der Linde, A. (2002). Bayesian measures of model complexity and fit. *Journal of the Royal Statistical Society B*, **64** , 583-640.

Stamey, T., Kabalin, J., McNeal, J., Johnstone, I., Freiha, F., Redwine, E. and Yang, N. (1989). Prostate-specific antigen in the diagnosis and treatment of adenocarcinoma of the prostate. ii: Radical prostatectomy treated patients, *Journal of Urology*, **16**, 1076-1083.

Steinberg, D.M. and Bursztyn, D. (1994). Confounded dispersion effects in robust design experiments with noise factors. *Journal of Quality Technology*, **26**, 12-20.

Stern, H.S. and Cressie, H. (2000). Posterior predictive model checks for disease mapping models. *Statistics in Medicine*, **29**, 2377-2397.

Stewart, W.E. and Sorenson, J.P. (1981). Bayesian estimation of common parameters from multiresponse data with missing observation. *Technometrics*, **23**, 131-146.

Stokes, M.E., Davis, C.S. and Koch, G.G. (1995). *Categorical Data Analysis Using the SAS System*. Cary, NC: SAS Institute.

Storey, J.D. (2007) The optimal discovery procedure: a new approach to simultaneous significance testing. *Journal of the Royal Statistical Society B*, **69**, 347-368.

Stram, D.O. and Lee, J.W. (1994). Variance components testing in the longitudinal mixed effects model. *Biometrics*, **50**, 1171-1177.

Sun, L., Zidek, J., Le, N. and Özkaynak, H. (2000). Interpolating Vancouver's daily ambient PM10 field. *Environmetrics*, **11**, 651-663.

Sun, W. and Cai, T. (2007). Oracle and adaptive compound decision rules for false discovery rate control. *Journal of the American Statistics Association.*, **102**, 901-912.

Taguchi, G. and Wu, Y. (1985). *Introduction to Off-line Quality Control.* Nagoya: Central Japan Quality Control Association.

Ten Have, T.R., Kunselman, A.R., Pulkstenis, E.P. and Landis, J.R. (1998). Mixed effects logistic regression models for longitudinal binary response data with informative dropout. *Biometrics*, **54**, 367-383.

Thall, P.F. and Vail, S.C. (1990). Some covariance models for longitudinal count data with overdispersion. *Biometrics*, **46**, 657-671.

Tibshirani, R.J. (1996) Regression shrinkage and selection via the LASSO, *Journal of the Royal Statistical Society B*, **58**, 267-288.

Tierney, L. and Kadane, J.B. (1986). Accurate approximations for posterior moments and marginal distributions. *Journal of the American Statistical Association*, **81**, 82-86.

Troxel, A.B., Harrington, D.P. and Lipsitz, S.R. (1998). Analysis of longitudinal measurements with non–ignorable non-monotone missing values. *Applied Statistics*, **47**, 425-438.

Tweedie, M.C.K. (1947). Functions of a statistical variate with given means, with special references to Laplacian distributions. *Proceedings of the Cambridge Philosophical Society*, **43**, 41-49.

Vaida, F. and Blanchard, S. (2005). Conditional akaike information for mixed effects models. *Biometrika*, **92**, 351-370.

Vaida, F. and Meng, X.L. (2004). Mixed linear models and the EM algorithm. In: *Applied Bayesian and Causal Inference from an Incomplete Data Perspective.* New York: Wiley.

Verbeke, G. and Molenberghs, G. (2003). The use of score tests for inference on variance components, *Biometrics*, **59**, 254-262.

Vu, H.T.V., Segal, M.R., Knuiman, M.W. and James, I.R. (2001). Asymptotic and small sample statistical properties of random frailty variance estimates for shared gamma frailty models. *Communications in Statistics: Simulation and Computation*, **30**, 581-595.

Vu, H.T.V. and Knuiman, M.W. (2002). Estimation in semiparametric marginal shared gamma frailty models. *Australia and New Zealand Journal of Statistics*, **44**, 489-501.

Wahba, G. (1979). How to smooth curves and surfaces with splines and cross-validation. *DTIC Document.*

Wahba, G. (1990). *Spline Models for Observational Data.* Philadelphia: SIAM.

Wakefield, J.C., Smith, A.F.M., Racine-Poon, A. and Gelfand, A.E. (1994). Bayesian analysis of linear and nonlinear population models using the Gibbs sampler. *Applied Statistics*, **43**, 201-221.

Wang, Z., Shi, J. and Lee, Y. (2016). Extended *t*-process regression models. In press.

Wedderburn, R.W.M. (1974). Quasi-likelihood functions, generalized linear models and the Gauss-Newton method. *Biometrika*, **61**, 439-447.

Whitehead, D. (1980). Fitting Cox's regression models to survival data using GLIM. *Applied Statistics*, **29**, 268-275.

Whittaker, E.T. (1923). On a new method of graduation. *Proceedings of the Edinburgh Mathematical Society*, **41**, 63-75.

Wilkinson, G.N. and Rogers, C.E. (1973). Symbolic description of factorial models for analysis of variance. *Applied Statistics*, **22**, 392-399.

Wolfinger, R.D. and Tobias, R.D. (1998). Joint estimation of location, dispersion, and random effects in robust design. *Technometrics*, **40**, 62-71.

Wolynetz, M.S. (1979). Maximum likelihood estimation in a linear model from confined and censored normal data (Algorithm AS139). *Applied Statistics*, **28**, 195-206.

Xie, M. and Singh, K. (2013). Confidence distribution, the frequentist distribution estimator of a parameter: a review. *International Statistical Review*, **81**, 3-39.

Xue, X. (1998). Multivariate survival data under bivariate frailty: An estimating equation approach. *Biometrics*, **54**, 1631-1637.

Yates, F. (1933). The analysis of replicated experiments when the field results are incomplete. *Empire Journal of Experimental Agriculture*, **1**, 129-142.

Yates, F. (1939). The recovery of inter-block information in varietal trials arranged in three dimensional lattices. *Annals of Eugenics*, **9**, 136-156.

Yau, K.K.W. (2001). Multivariate models for survival analysis with random effects. *Biometrics*, **57**, 96-102.

Yun, S. and Lee, Y. (2004). Comparison of hierarchical and marginal likelihood estimators for binary outcomes. *Computational Statistics and Data Analysis*, **45**, 639-650.

Yun, S. and Lee, Y. (2006). Robust estimation in mixed linear models with non-monotone missingness. *Statistics in Medicine*, **25**, 3877-3892.

Yun, S., Lee, Y. and Kenward, M.G. (2007). Using h-likelihood for missing observations. *Biometrika*, **94** , 905-919.

Yun, S., Sohn, S.Y. and Lee, Y. (2006). Modelling and estimating heavy-tailed non-homogeneous correlated queue pareto-inverse gamma HGLMs with covariates. *Journal of Applied Statistics*, **33**, 417-425.

Zeger, S.L. and Diggle, P.J. (1994). Semiparametric models for longitudinal data with application to CD4 cell numbers in HIV seroconverters. *Biometrics*, **50**, 689-699.

Zeger, S.L. and Liang, K.Y. (1986). An overview of methods for the analysis of longitudinal data. *Statistics in Medicine*, **11**, 1825-1839.

Zeger, S.L., Liang, K.Y. and Albert, P.S. (1988). Models for longitudinal data: a generalized estimating equation approach. *Biometrics*, **44**, 1049-1060.

Zou, H. (2006) The adaptive LASSO and its oracle properties, *Journal of the American Statistical Association*, **101**, 1418-1429.

Zou, H. and Hastie, T. (2005) Regularization and variable selection via the elastic net, *Journal of the Royal Statistical Society B*, **67**, 301-320.

Zyskind, G. (1967). On canonical forms, non-negative covariance matrices and best and simple least squares linear estimators in linear models. *Annals of Mathematical Statistics*, **38**, 1092-1109.

Data Index

Author Index

Subject Index